CRC SERIES IN MEDICINAL CHEMISTRY

Series Editor-in-Chief

Matthew Verderame
Professor of Medicinal Chemistry
Head, Department of Chemical and
Physical Sciences
Albany College of Pharmacy
Albany, New York

Handbook of Chemotherapeutic Agents (Volumes I and II)

Handbook of CNS Agents and Local Anesthetics

Handbook of Autonomic Drugs and Autacoids

Handbook of Cardiovascular and Anti-Inflammatory Agents

Handbook of Hormones, Vitamins, and Radiopaques

CRC Handbook of Autonomic Drugs and Autacoids

Editor

Matthew Verderame, Ph.D.
Professor of Medicinal Chemistry
Head, Department of Chemical and
Physical Sciences
Albany College of Pharmacy
Albany, New York

CRC Press, Inc.
Boca Raton, Florida

Library of Congress Cataloging-in-Publication Data
Main entry under title:

Handbook of autonomic drugs and autacoids.

Includes bibliographies and index.
1. Autonomic drugs—Handbooks, manuals, etc.
I. Verderame, Matthew. II. Title: Autacoids.
[DNLM: 1. Autonomic Agents. 2. Slow-Reacting
Substances. QV 120 H236]
RM323.H35 1986 615'.78 85-26990
ISBN 0-8493-3289-3

Direct all inquiries to CRC Press, Inc., 2000 Corporate Blvd., N.W., Boca Raton, Florida, 33431.

International Standard Book Number 0-8493-3289-3

Library of Congress Card Number 85-26990
Printed in the United States

DEDICATION

To my typist and companion and the mother of our four children — Fran.

PREFACE

Handbook of Autonomic Drugs and Autacoids is a companion to five other handbooks in this pharmaceutically related set published by CRC Press. The titles of the other five follow: *Handbook of Chemotherapeutic Agents* (Volumes I and II); *Handbook of CNS Agents and Local Anesthetics; Handbook of Cardiovascular and Anti-Inflammatory Agents;* and *Handbook of Hormones, Vitamins, and Radiopaques.*

The six handbooks represent a systematic collection of selective chemical and pharmacological reference data on drugs drawn from the major categories of therapeutically important agents. For the principal classes of drugs, or for individual drugs which do not belong to a particular class, the following topics are addressed: Mechanism of Action, Structure-Activity Relationships, Pharmacokinetics, Uses and Dosage, Toxicity (including Drug Interactions), and Physcial-Chemical and Other Data. Drug metabolism data are heavily emphasized in the pharmacokinetics sections. The term "toxicity" is used broadly to also include side effects. Generally included in the physical-chemical and other data sections is information on the drug such as the chemical name, formula, melting point, pKa value(s), solubility and stability data, physical description, salt forms, and any other property deemed worthy of inclusion.

Other topics generally considered include an introduction, historical aspects leading to or associated with the development of the class of drugs, possible causes of the condition or disease state for which the drugs are used, life cycles of organisms against which the chemotherapeutic agents are used, testing methods employed in the screening of potentially useful compounds, and review articles covering the fields. Chemical syntheses are not emphasized. However, this topic receives significant attention in a few chapters. Figures are used appropriately to illustrate a point or concept, especially in the mechanism of action sections.

In this series of reviews, emphasis is placed on the most useful drugs associated with each physiological class or disease entity. The monograph format is used extensively, not only to denote the most useful members, but also to provide a close examination of these agents. Comparative pharmacological or therapeutic data on the principal drugs within the classes provide additional, practical information.

Monographs of drugs with useful multiple functions are appropriately placed in more than one chapter and cross referenced. Though necessitating some repetition of data, these monographs generally complement each other and maintain a continuity of thought and application.

Newer and/or experimental drugs have also been included to round out the coverage and to present current knowledge and possible future trends.

This set of reference books should have appeal and utility for the practitioners and students of medicinal chemistry, pharmacology, pharmacy, and medicine. It should also be useful to researchers in the pharmaceutical industry and in other allied health professions.

A special debt of gratitude is extended to the many contributors who labored long and diligently to prepare their manuscripts for publication. Also, for their counsel and cooperation, the dedicated assistance of the Advisory Board is acknowledged with sincere appreciation.

The editorial assistance of the CRC staff and, in particular, the coordinating efforts of Ms. Amy Skallerup must not go unmentioned.

Finally, it is our hope that the reader will find these first edition handbooks interesting, useful, and valuable. Comments for improvement will be gratefully received.

Albany, New York

Matthew Verderame, Ph.D.
Editor

THE EDITOR

Matthew Verderame, Ph.D., is Professor of Medicinal Chemistry and Head of the Department of Chemical and Physical Sciences at the Albany College of Pharmacy, Union University. Dr. Verderame received his undergraduate education at the University of Connecticut, where he earned the Bachelor's degree in Chemistry in 1950. At the University of Wisconsin, he received an M.S. degree in Pharmacy in 1952 and the Ph.D. degree in 1955 in Pharmaceutical Chemistry. In 1954, he joined the faculty at the Albany College of Pharmacy as an Assistant Professor and moved through the ranks to his present position. Dr. Verderame's education was enhanced by his attendance of NSF-sponsored institutes at the University of North Carolina in 1962, Michigan State University in 1964, Duke University in 1969, and State University of New York in Albany in 1974. From 1955 through 1968, Dr. Verderame received consecutive year-long research grants from the Sterling-Winthrop Research Institute. In 1968 and again in 1973, he was cited for his contributions to teaching by receiving the Lederle Award. Courses which he regularly teaches include Medicinal Chemistry, Qualitative Organic Analysis, and Drug Analysis. The principal areas of synthetic-research of Dr. Verderame include the preparation of various derivatives of cysteine, methionine, and piperazine as possible anti-infective agents. He also synthesized organo tin compounds and ether derivatives of cyclic hydrocarbons as possible anti-infective agents. He has authored several publications. Dr. Verderame is a member of the American Chemical Society, Rho Chi, Sigma Xi, Phi Sigma, Rho Pi Phi, the American Association for the Advancement of Science, the New York Academy of Sciences, the American Association of University Professors, and Phi Lambda Upsilon.

CONTRIBUTORS

Raymond P. Ahlquist, Ph.D. (deceased)
Charbonnier Professor
Department of Pharmacology
Medical College of Georgia
Augusta, Georgia

Michael P. L. Caton, Ph.D.
Chemical Research Manager
May, Baker Ltd.
Dagenham, Essex, England

Lawrence W. Chakrin, Ph.D.
President
Sterling Research Group
Sterling-Winthrop Research Institute
Rensselaer, New York

James E. Gearien, Ph.D.
Professor of Medicinal Chemistry
College of Pharmacy
University of Illinois at Chicago
Chicago, Illinois

Michael N. Mimnaugh, Ph.D.
Assistant Professor of Chemistry
Department of Physical Sciences
Chicago State University
Chicago, Illinois

Donald C. Schlegel, Ph.D.
Senior Research Chemist
Department of Medicinal Chemistry
Sterling-Winthrop Research Institute
Rensselaer, New York

James T. Stewart, Ph.D.
Professor and Head
Department of Medicinal Chemistry
and Pharmacognosy
College of Pharmacy
University of Georgia
Athens, Georgia

Emmanuel B. Thompson, Ph.D.
Associate Professor of Pharmacology
Department of Pharmacodynamics
College of Pharmacy
University of Illinois at Chicago
Chicago, Illinois

TABLE OF CONTENTS

ADRENERGICS, ANTIASTHMATICS, AND ANOREXIGENICS

Michael N. Mimnaugh

INTRODUCTION

The Adrenergic Nervous System

The mammalian nerve anatomy can be divided into two major systems. The central nervous system (CNS) contains the brain, cerebellum, medulla, and spinal cord as well as those nerve fibers controlling sensory function and motor activity. The autonomic nervous system consists of the nerves responsible for controlling the heart, blood vessels, glands, viscera, smooth muscles, and all the so-called automatic or involuntary functions of the body.

The autonomic system in turn is composed of two subsystems: the parasympathetic system, which arises from the craniosacral region of the CNS, and the sympathetic system, which arises from the thoracolumbar region. Most organs receive innervation from both these autonomic systems, but the systems tend to function in an antagonistic fashion. Heart rate, for instance, will increase upon sympathetic stimulation, but will decrease upon parasympathetic stimulation. The overall activity, or tone, of an organ is therefore controlled by the balance between the degree of stimulation the organ receives from each of these systems. The primary difference between these two autonomic systems is the neurotransmitter, which carries the neuronal message from the nerve fiber to its target organ.

The Discovery of Epinephrine

The concept of autonomic neurohumoral transmission via a specific chemical substance developed as the result of the work of a number of investigators. In 1895, Oliver and Schaefer[1] prepared an extract from the suprarenal capsules which, when injected in animals, caused an increase in blood pressure and contraction of smooth muscle. Abel and Crawford,[2] in 1897, isolated a benzoyl derivative of the active component in the extract. They named the substance epinephrine. The pure base was isolated in 1901 by Von Furth,[3] who named it suprarenin, and Takamine,[4] who called it adrenalin. By 1905, Stolz[5] and Dakin[6] had independently synthesized the active compound as a racemic mixture and confirmed its structure as 2-(*N*-methylamino)-1-(3,4-dihydroxyphenyl)ethanol (Figure 1).

Elliott[7] observed in 1904 that the physiological effects of adrenalin were very similar to the physiological effects of sympathetic nerve stimulation. He therefore proposed that release of adrenalin by sympathetic nerves on depolarization may be responsible for mediating the action of sympathetic nerves in the body. Thus, adrenalin could be termed a neurotransmitter. In 1910, Barger and Dale[8] cast doubt on the role of adrenalin as a neurotransmitter by demonstrating that a number of different but chemically related amines were also capable of eliciting physiological activity similar to that of adrenalin. Barger and Dale thus termed all substances capable of producing these effects "sympathomimetics".

It was not until the 1940s, when von Euler[9] was able to isolate norepinephrine from bovine cardiac and neural tissues, that the identity of norepinephrine as the sympathetic neurotransmitter was confirmed.

Dale's recognition of the ability of chemical substances to mimic sympathetic activity strengthened and confirmed the concept of pharmacological receptors as mediators of physiological response.

The Concept of Adrenergic Receptors

The concept of pharmacological receptors was advanced years earlier by Ehrlich[10] in his attempt to explain the ability of chemotherapeutic agents to selectively attack parasites and

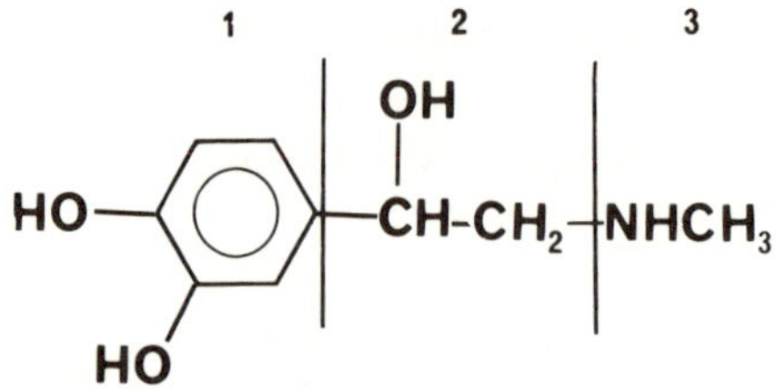

FIGURE 1. Epinephrine and its three structure-activity relationship divisions.

not their hosts. He envisioned chemicals as having two functional parts. One part was a selective group that governed distribution of the chemical in the body. The other part was a pharmacophore group that evoked the specific effect produced by the chemical when it interacted with certain specific component parts of the cell side chains, which I have characterized as "receptors". Dale[11] was the first to view the sympathetic neuromyal junction as the location of "the receptive mechanism for adrenaline".

Cannon and Rosenbleuth[12] attempted to apply the concept of neurotransmitters as mediators of sympathetic responses as a means for explaining the "motor" and "inhibitor" sympathomimetic activities of compounds studied by Barger and Dale. They postulated that there were two types of sympathetic neurotransmitters. One transmitter, sympathin E_1, was responsible for the excitatory responses of the sympathetic system, while another transmitter, sympathin I, was responsible for the inhibitory action of the system. A considerable amount of effort was expended in attempts to isolate and identify sympathins E and I.

No chemical substances having the properties attributed to sympathins E or I were ever isolated. Epinephrine (sympathin A) and norepinephrine (sympathin N), while having many sympathomimetic properties, could not satisfy the requirements for sympathin E or I inasmuch as both compounds were capable of demonstrating both excitatory as well as inhibitory responses, depending on the system studied.

Consequently, the idea began to evolve that the variations in activity of sympathetically innervated organs were not due to multiple neurotransmitters, but rather to the presence of multiple types of receptors.

The possibility of multiple receptors mediating the action of epinephrine was in fact suggested as early as 1906 in Dale's work[11] on the effect of ergot alkaloids on sympathetic activity. Dale found that ergot alkaloids could block many of the excitatory responses of epinephrine in smooth muscle (contraction), but not the excitatory responses in heart (increased rate and force). These observations suggested that multiple receptors may be capable of binding epinephrine but that they must be different in that they are not all blocked by the ergot alkaloids. Dale's work, while supporting the concept of multiple receptors, also demonstrated that a simple assignment of one type of receptor for mediating all excitatory responses of epinephrine and one receptor for mediating all inhibitory responses of epinephrine, was untenable. Ahlquist[13] sought to resolve this problem by suggesting that the union of a substance with the cell receptor is determined by the chemical structure of that substance, and that the nature of the response elicited by the union is determined by the organ. Ahlquist's postulate was based on his observation of the relative potencies of a series of six sympathomimetic amines, — L-epinephrine, D,L-epinephrine, α-methylepinephrine, norepinephrine, α-methylnorepinephrine, and isoproterenol — on a number of different organs. His results suggested that there were two distinct general types of receptors. One receptor, which he termed "α", showed an order of potency of L-epinephrine > D,L-epinephrine > norepinephrine > α-methylnorepinephrine > α-methylepinephrine > isoproterenol. This α-receptor mediated vasoconstriction in the viscera and skin, contraction of the nictitating membrane and uterus, relaxation of the intestine and ureter, and stimulation of the dilator pupillae. The other receptor, which Ahlquist termed "β", showed an order of potency of isoproterenol > L-epinephrine > α-methylepinephrine > D,L-epinephrine > α-methylnorepinephrine > norepinephrine. This β-receptor mediated vasodilation in skeletal muscle and the coronary vessels and viscera, relaxation of the uterus and bronchioles, and stimulation of the myocardium. Thus, the ultimate physiological effect of epinephrine on an organ was dependent on the kind of adrenergic receptor(s) it had, and whether or not those receptors controlled excitatory or inhibitory responses in that organ. In general, the α-receptor was found to control most of the excitatory functions and one inhibitory function (intestinal relaxation), while the β-receptor was associated with most of the inhibitory functions and one excitatory function (myocardial stimulation). Ahlquist's studies also supported the theory that there was only one neurohormone, although he incorrectly believed it to be epinephrine rather than norepinephrine.

Ahlquist's theory of α- and β-receptors was not immediately embraced. Lands noted that Ahlquist's order of potencies of the sympathomimetic amines often varied considerably with the species of animal and the experimental conditions used. Lands preferred to try and preserve the idea of separate receptors for the excitatory and inhibitory action of sympathomimetics. He therefore proposed that there were actually three types of adrenergic receptors: the Ac (excitatory), the Ar (inhibitory), and the Acr (undifferentiated).[14] The Acr receptor presumably represented a receptor which responded about equally to almost all sympathetic amines, and its presence in the heart and the intestine might explain the anomolous responses to those organs in Ahlquist's study.

Furchgott,[15] on the other hand, preferred to maintain Ahlquist's classification, but expanded it to include two additional types of receptors: the γ-receptor, which regulated glycolysis, and the Δ-receptor, which regulated inhibition of smooth muscle.

Over 30 years of pharmacological experimentation appear to support the original theory advanced by Ahlquist for the existence of the two fundamental receptors which he termed "α" and "β". Furthermore, many of the pharmacological and biochemical processes under the control of these receptors have now been described (Table 1). Some refinements in the pharmacological activity and structure-activity relationships of these receptors have led to recent proposals that subtypes of each receptor may exist. A detailed discussion of the evidence for these receptor subtypes will be presented later in this review.

Structural Organization of the Adrenergic Synapse

Thousands of compounds are known to have activity on the adrenergic nervous system. Many of these compounds, however, do not act by binding to adrenergic receptors. Instead, these compounds act by affecting the in vivo levels of norepinephrine available for interaction at adrenergic receptors. There are many ways in which a compound can affect the concentration of norepinephrine at its receptors. In order to understand these mechanisms, however, it is necessary to briefly review the structural and physiological organization of an adrenergic synapse.

Preganglionic sympathetic nerves originate in the anterolateral gray matter of the thoracolumbar segment of the spinal cord. These nerves terminate at the sympathetic ganglia where they impinge on the peripheral adrenergic neuron. The cell body of this neuron is generally found at the ganglia. Radiating out from the cell body are thin, unmyelinated axons which pass in bundles to the innervated tissue or organ. The axon branches into numerous terminal fibers as it approaches the target organ. Along these fibers, at irregular intervals, are nodes which contain high levels of vesicles and mitochondria. The greatest concentration of the adrenergic neurotransmitter, norepinephrine, can be found in these nodes.

Norepinephrine stored in the vesicles accumulates through biosynthesis from the amino acid, tyrosine (site 1, Figure 2). Upon depolarization of the neuron, the vesicle will migrate

Table 1
RESPONSES OF EFFECTOR ORGANS TO AUTONOMIC NERVE IMPULSES

Effector organs	Receptor type	Adrenergic impulses Responses[a]	Cholinergic impulses Responses[a]
Eye			
Radial muscle, iris	α	Contraction (mydriasis) + +	—
Sphincter muscle, iris		—	Contraction (miosis) + + +
Ciliary muscle	β	Relaxation for far vision +	Contraction for near vision + + +
Heart			
S-A node	β_1	Increase in heart rate + +	Decrease in heart rate; vagal arrest + + +
Atria	β_1	Increase in contractility and conduction velocity + +	Decrease in contractility, and (usually) increase in conduction velocity + +
A-V node	β_1	Increase in automaticity and conduction velocity + +	Decrease in conduction velocity; A-V block + + +
His-Purkinje system	β_1	Increase in automaticity and conduction velocity + + +	Little effect
Ventricles	β_1	Increase in contractility, conduction velocity, automaticity, and rate of iodioventricular pacemakers + + +	Slight decrease in contractility claimed by some
Arterioles			
Coronary	α,β_2	Constriction +; dilatation[b] + +	Dilatation ±
Skin and mucosa	α	Constriction + + +	Dilatation[c]
Skeletal muscle	α,β_2	Constriction + +; dilatation[b,d] + +	Dilatation[c] +
Cerebral	α	Constriction (slight)	Dilatation[c]
Pulmonary	α,β_2	Constriction +; dilatation[b]	Dilatation[c]
Abdominal viscera; renal	α,β_2	Constriction + + +; dilatation[d] +	—
Salivary glands	α	Constriction + + +	Dilatation + +
Veins (systemic)	α,β_2	Constriction + +; dilatation + +	—
Lung			
Bronchial muscle	β_2	Relaxation +	Contraction + +
Bronchial glands	?	Inhibition (?)	Stimulation + + +
Stomach			
Motility and tone	α_2,β_2	Decrease (usually)[f] +	Increase + + +
Sphincters	α	Contraction (usually) +	Relaxation (usually) +
Secretion		Inhibition (?)	Stimulation + + +
Intestine			
Motility and tone	α_2,β_2	Decrease[f] +	Increase + + +
Sphincters	α	Contraction (usually) +	Relaxation (usually) +
Secretion		Inhibition (?)	Stimulation + +
Gallbladder and ducts		Relaxation +	Contraction +
Kidney	β_2	Renin secretion + +	—
Urinary bladder			
Detrusor	β	Relaxation (usually) +	Contraction + + +
Trigone and sphincter	α	Contraction + +	Relaxation + +
Ureter			
Motility and tone	α	Increase (usually)	Increase (?)
Uterus	α,β_2	Pregnant; contraction (α); nonpregnant; relaxation (β)	Variable[g]
Sex organs, male	α	Ejaculation + + +	Erection + + +
Skin			
Pilomotor muscles	α	Contraction + +	—
Sweat glands	α	Localized secretion[h] +	Generalized secretion + + +
Spleen capsule	α,β_2	Contraction + + +; relaxation +	—
Adrenal medulla		—	Secretion of epinephrine and norepinephrine

Table 1 (continued)
RESPONSES OF EFFECTOR ORGANS TO AUTONOMIC NERVE IMPULSES

Effector organs	Receptor type	Adrenergic impulses Responses[a]	Cholinergic impulses Responses[a]
Liver	α,β_2	Glycogenolysis, gluconeogenesis[i] + + +	Glycogen synthesis +
Pancreas			
Acini	α	Decreased secretion +	Secretion + +
Islets (β-cells)	α	Decreased secretion + + +	—
	β_2	Increased secretion +	—
Fat cells	α,β_1	Lipolysis[i] + + +	—
Salivary glands	α	Potassium and water secretion +	Potassium and water secretion + + +
	β	Amylase secretion +	
Lacrimal glands		—	Secretion + + +
Lacrimal glands		—	Secretion + + +
Nasopharyngeal glands		—	Secretion + +
Pineal gland	β	Melatonin synthesis	—

[a] Responses are designated 1 + to 3 + to provide an approximate indication of the importance of adrenergic and cholinergic nerve activity in the control of the various organs and functions listed.
[b] Dilatation predominates *in situ* due to metabolic autoregulatory phenomena.
[c] Cholinergic vasodilatation at these sites is of questionable physiological significance.
[d] Over the usual concentration range of physiologically released, circulating epinephrine, β-receptor response (vasodilatation) predominates in blood vessels of skeletal muscle and liver; α-receptor response (vasoconstriction), in blood vessels of other abdominal viscera. The renal and mesenteric vessels also contain specific dopaminergic receptors, activation of which causes dilatation, but their physiological significance has not been established.
[e] Sympathetic cholinergic system causes vasodilatation in skeletal muscle, but this is not involved in most physiological responses.
[f] It has been proposed that adrenergic fibers terminate at inhibitory β-receptors on smooth muscle fibers, and at inhibitory α-receptors on parasympathetic cholinergic (excitatory) ganglion cells of Auerbach's plexus.
[g] Depends on stage of menstrual cycle, amount of circulating estrogen and progesterone, and other factors.
[h] Palms of hands and some other sites ("adrenergic sweating").
[i] There is significant variation among species in the type of receptor that mediates certain metabolic responses.

From Goodman, L. S. and Gilman, A., *The Pharmacological Basis of Therapeutics*, 6th ed., Macmillan, New York, 1981, 60. With permission.

to the cell membrane, fuse with it, and release its stores of norepinphrine into the synaptic cleft.

Norephinephrine which is released into the synapse can bind to postsynaptic α- or β-receptors to mediate adrenergic activity (site 2, Figure 2), or it can bind to presynaptic receptors to regulate its own release (site 3, Figure 2).

The action of norepinephrine can be terminated in several ways. The most efficient and important way is through its selective uptake into the presynaptic neuron and from there, into the storage granule (site 4, Figure 2). This uptake process is specific for the catecholamine and is energy requiring. Norepinephrine can also have its action terminated by being metabolized through the action of the enzymes monoamine oxidase and/or catechol-*O*-methyl transferase (site 5, Figure 2).

Numerous agents have been found which are able to modulate the various mechanisms involved in adrenergic function. Many of these agents and their mechanisms will be described in greater detail in other reviews in this text. For purposes of discussion, the rest of the material in this review will be divided into two major areas: presynaptic adrenergic function and postsynaptic adrenergic function.

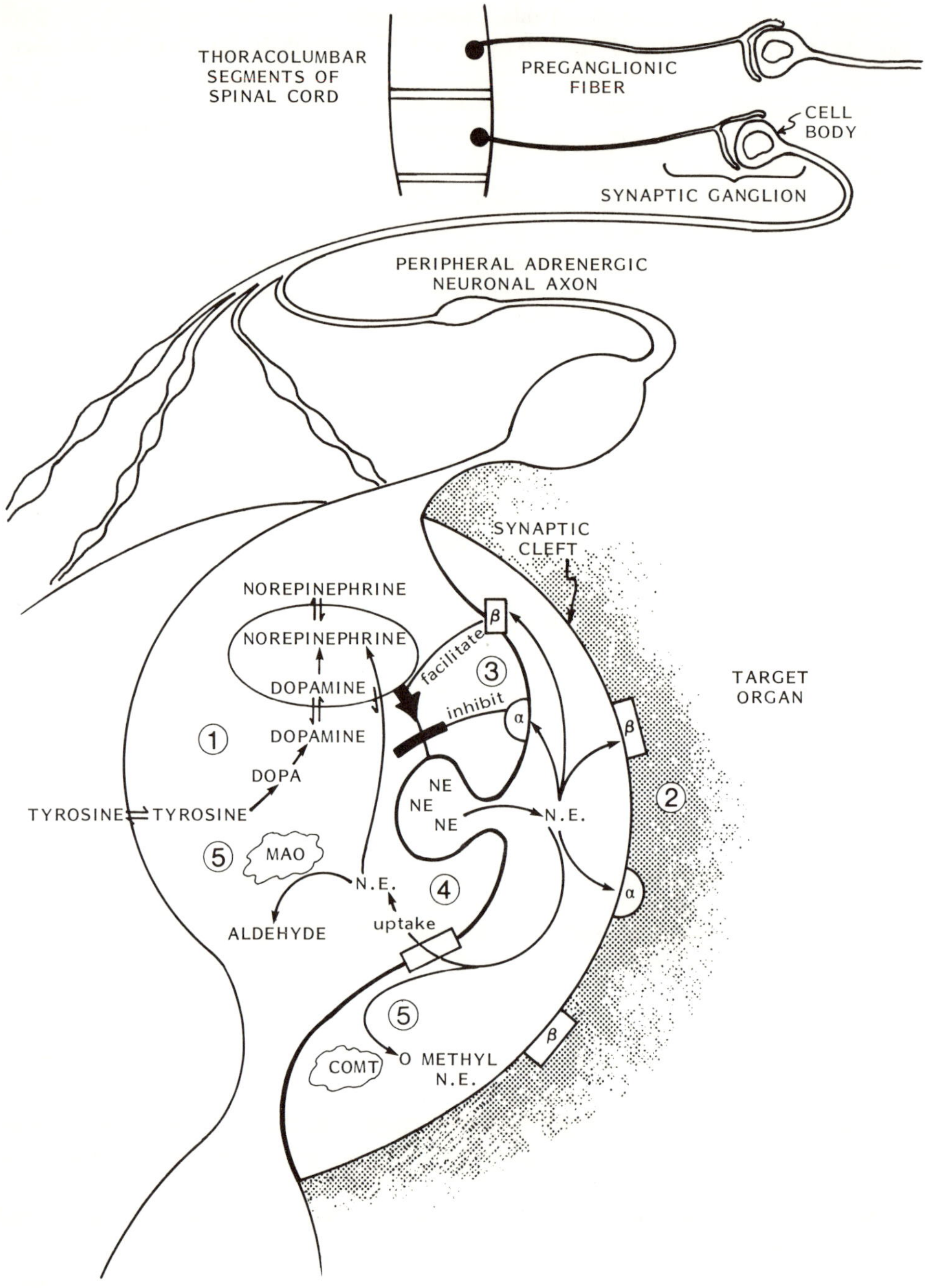

FIGURE 2. Structural organization of the adrenergic synapse. The circled numbers refer to the sites of action described in text.

POSTSYNAPTIC ADRENERGIC FUNCTION

As the concept of discrete pharmacological receptors for drug action gained wider acceptance, chemists began synthesizing numerous compounds in order to obtain information about the structural requirements a compound needed in order to bind to such receptors. The value of these studies was twofold: first, more selective and effective drugs could be prepared which acted on these receptors, and second, greater information about the chemistry and topography of the receptor was obtained which could be used to guide further research. Consequently, the development of therapeutically useful agents and our understanding of the mechanism and nature of adrenergic response are intimately related.

Even though Stolz[5] and Dakin[6] are given credit for determining the chemical structure of epinephrine, it was really Barger and Dale[8] in 1910 who took the first step in suggesting the basic structural requirements for sympathomimetic activity. The requirements they reported were: (1) sympathomimetics must be bases, (2) moreover, they must be primary or secondary amines, (3) the more closely a substance resembles adrenalin (epinephrine), the stronger the sympathomimetic activity, (4) the optimum carbon skeleton contains a benzene ring and a two-carbon atom side chain with an amino group attached, (5) the presence of hydroxyl groups at the 3 and 4 positions of the ring increases activity, (6) the presence of an alcoholic hydroxyl group on the carbon α to the ring increases activity, but only if the ring is catechol, and (7) that inhibitor and motor (agonist) activity vary to some extent independently when a hydrogen atom of the amino group is replaced by various alkyl radicals. This last statement was to foreshadow the eventual distinctions between α- and β-adrenergic receptors. It is perhaps ironic that Barger and Dale, who set forth the first set of structure-activity relationships for sympathomimetic activity, were themselves reluctant to accept the idea of the existence of "specific chemical receptors", partially because of the "somewhat irregular variations of activity in the series".

These irregularities in activity were eventually attributable to the existence of multiple adrenergic receptors, each having their own particular structure-activity requirements. In 1940, Konzett[16] discovered the unique and powerful bronchodilator activity of the *N*-isopropyl analog of norepinephrine (isoproterenol), and Ahlquist used the selectivity of isoproterenol action to support his idea of separate α- and β-adrenergic receptors.

Structure-Activity Relationships of β-Adrenergic Agonists

Despite the division of adrenergic receptors into the subtypes α and β, it became increasingly apparent that structure-activity relationships derived for adrenergic response did not fall into two well-defined classes. As early as 1949, Lands[17] noted that the structure-activity relationships for compounds acting at β-adrenergic receptors were not consistent. They studied the action of 15 compounds on the β-receptor-mediated systems, lipolysis, cardioacceleration, bronchodilation, and vascular relaxation. They found that identical activity sequences, (i.e., isoproterenol > norepinephrine > epinephrine) are found for lipolysis and cardioacceleration, while a different activity sequence (isoproterenol > epinephrine > norepinephrine) was found for bronchodilation and vascular relaxation. These observations led Lands et al.[18] to propose that there are two β-receptor subtypes which he classified as β_1, mediating lipolysis and cardioacceleration, and β_2, mediating bronchodilation and vasodilation.

This classification of β-receptor subtypes has continued to gain acceptance as the development of β_1 and β_2 selective agonists and antagonists has grown. Based on these observations, Arnold[19] has expanded the classification of β_1- and β_2-adrenergic responses to include those shown in Table 2.

In elucidating the structure-activity relationships of β-adrenergic agonists, it has been found that they belong to two major chemical classes: the phenethanolamines and the aryloxypropanolamines. This last class of compounds has only recently been determined as having agonist activity, and that activity appears to be more selective for the β_1-receptor subtype. The largest chemical class of β-agonists is the phenethanolamines.

The Phenethanolamine β-Agonists

As the name suggests, the parent structure for this class of compounds is the phenethanolamine nucleus. We shall consider structure-activity relationship trends for these compounds

Table 2
PHARMACOLOGICAL RESPONSES MEDIATED BY β_1- AND β_2-ADRENERGIC RECEPTORS

β_1-Receptor-Mediated Responses
Heart rate
Force of contraction
Relaxation of coronary arteries
Relaxation of intestinal smooth muscle
Calorigenesis of brown adipose tissue
Adipose tissue lipolysis
Salivary gland secretion
β_2-Receptor-Mediated Responses
Vascular smooth muscle relaxation
Tracheal and bronchial smooth muscle relaxation
Inhibition of release of anaphylactic mediators
Skeletal muscle contraction
Skeletal muscle glycogenolysis
Urinary bladder relaxation
Pancreatic insulin release
Gallbladder relaxation
Vas deferens relaxation
Diaphram contraction in rat

by dividing the parent structure into the three divisions shown in Figure 1 and considering each section individually.

Amine Substitution

As originally observed by Barger and Dale, a basic amine is essential for agonist activity. This is confirmed by the fact that carbon isosteres are nearly devoid of activity. The presence of a hydrogen-bonding proton on the amine also appears to be necessary for maximal activity. Tertiary and quaternary amines are considerably less active. The amine is almost completely in the ionized form at physiological pH.

Konzett[16] demonstrated that the size of the amino alkyl substituent is important for differentiating β-agonist from α-agonist activity. Increasing the size of the substituent favors β-receptor binding and decreases α-receptor binding. Aromatic rings, two to three atoms distant from the amine on the alkyl side chain, increase binding, possibly through some form of hydrophobic or pi bonding. Substitution of polar groups on this ring at the 3 and 4 positions is also favorable. Branching on the side chain is permissible if the alkyl substituents are small. The effect of these amine modifications on β-adrenergic activity is summarized in Table 3.

Aromatic Subsitution

A considerable number of analogs have been synthesized in which the catechol nucleus of norepinephrine has been modified. In almost all cases, elimination of one or both of the hydroxyls results in substantial or complete loss of activity. It does appear, however, that the catechol is more important for β-agonist activity than α-agonist activity.

If a single phenolic hydroxyl is present, its position on the ring may be important for selectivity. Thus, the *meta* hydroxyl in phenylephrine provides part of the α-receptor agonist selectivity of this compound, while the *para* hydroxyls in isoxuprine, nylidrin, salbutamol contribute to their activity as β-receptor selective agonists. None of these compounds, however, is as active as epinephrine at either the α- or β-receptor. Their therapeutic value lies in their selectivity.

Table 3
THE EFFECT OF SUBSTITUTION ON THE AMINE PORTION OF PHENETHANOLAMINES

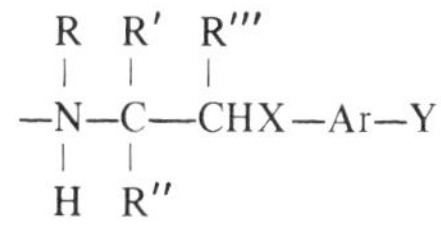

Substituent	Effect of substituent on β-agonist activity: Increase or maintain activity	Decrease or lose activity
R =	H	Alkyl
R′ =	H, CH_3, CH_2CH_3	Propyl, butyl
R″ =	H, CH_3	CH_2CH_3
X =	–O–	$-CH_2CH_2-$, propyl, butyl
R‴ =	H	CH_3, ethyl
Ar =	Phenyl, indole	Cyclohexyl, hexyl
Y =	pOH, CH_3, OCH_3, H	—

In general, the substitution of any group for a phenolic hydroxyl on an adrenergic agonist will convert that agonist to an antagonist. An exception to this generalization applies to those substituents classified as "phenol equivalent" groups.[20] Such groups can be substituted for the *meta* hydroxy group of a catechol β-agonist and retain significant β-agonist activity. These phenol equivalents encompass a wide range of size, functionality, and ionizability. Phenol equivalents include such substituents as sulfonamides, amides, amines, and hydroxymethyl groups. The unifying feature of the substituents appears to be the presence of a proton, one to three atoms from the aromatic ring, which is capable of hydrogen bonding. This generalization is by no means absolute, however, in that $-CO_2H$ and $-CONH_2$ groups are not active, while $-C{\equiv}N$ is active. The size of the phenolic equivalent also does not appear to be too important in that groups as small as NH_2 and as large as $-NHCH_2C_6H_4OCH_3$ are both active. It is interesting to note that these compounds have their greatest activity at β_2-receptors rather than as β_1-receptors. Furthermore, that action appears to be more significant at the bronchial β_2-receptors as opposed to the peripheral. Hence, salbutamol (Figure 3), which has a 3-position hydroxymethyl group, is used as a bronchodilator.

Another variation in the aromatic nucleus which appears to favor bronchial β_2-agonist activity is the 3,5-dihydroxyl-substitution pattern shown in terbutaline and metaproterenol (Figure 3). These two agents have also found use as bronchodilators in the treatment of asthma and allergies.

Many types of heterocyclic rings have been substituted for the benzene ring. With few exceptions, the substitutions have resulted in loss of agonist activity.

Compounds containing a phenol equivalent are also noteworthy in that such compounds retain their β-agonist activity whether or not there is also an OH on the ring. The ring position of the phenol equivalent, however, must be *meta*. Compounds with *para* substitution are inactive as agonists. This positional requirement also applies if there is an OH in the ring. Thus, 3-methane sulfonamido-4-hydroxy derivatives are agonists where 4-methane sulfonamido, 3-hydroxy derivatives are not.[21] One of the exceptions, however, is pirbuterol, in which a pyridine ring has been substituted for the benzene ring of salbutamol (Figure 3). This compound was found to have potency and β_2 bronchial selectivity equivalent to that of salbutamol.

Side Chain Substitution

There are five basic modifications that can be made on the side chain. These are (1) the

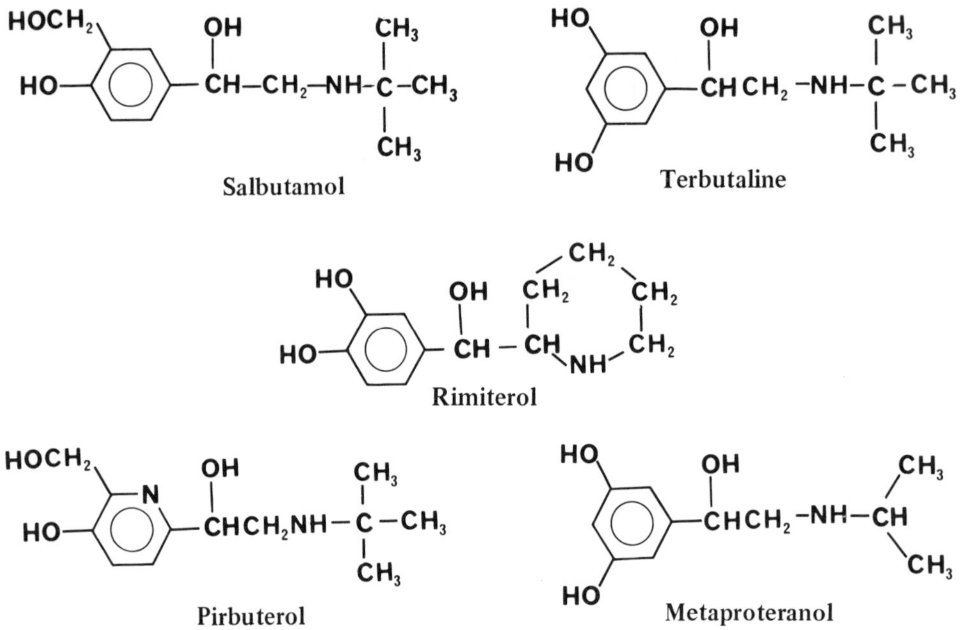

FIGURE 3. Aromatic substitution patterns for β_2-selective adrenergic agonists.

length of the carbon skeleton, (2) substitution at the α-carbon, (3) substitution at the β-carbon, (4) stereochemistry at the β-carbon, and (5) conformation about the carbon-carbon bond of the side chain.

Carbon Skeleton

Early studies showed that the two-carbon-atom distance between the aromatic ring and the nitrogen is very important for optimal binding to the receptor. Expanding or contracting the length of the chain invariably led to loss of agonist activity. Only recently has it been shown that an oxypropanolamine side chain may possess agonist activity. Nevertheless, the agonist activity of these compounds has been rationalized, in part, on their ability to assume a conformation which still allows the amine to be within a two-carbon-atom distance from the ring.

α-Substitution

Substitution on the α-carbon generally leads to decreased agonist activity. An exception is an ethyl group. Catecholamines with an α-ethyl substituent are able to maintain agonist activity. It has been shown that this ethyl substituent can also be tied back to the amine and still maintain agonist activity. Such a compound is rimiterol (Figure 3). It is both an agonist and selective for the β_2-receptor. The methyl substituent is also an interesting case. α-Methyl substitution allows for more efficient uptake of the compound from the synaptic cleft. Once taken up, the α-methyl compound displaces norepinephrine from its storage granule. This displaced norepinephrine then enters the synapse and contributes to the total adrenergic response observed.

In addition to potentiating the synaptic uptake of the compound, the α-methyl group also promotes better oral absorption of the compound, providing significant clinical advantage over i.v.-administered adrenergic agonists.

The ethyl side chain may also be tied back to the catechol ring and still maintain activity. However, the position which the side attaches to the ring is important. If the side chain is attached so as to give a 1,2,3,4-tetrasubstituted, ring-fused structure, the compound is an agonist. The 1,2,4,5-tetrasubstituted compound, however, is inactive.

Table 4
THE EFFECT OF SUBSTITUTION ON THE SIDE CHAIN PORTION OF PHENETHANOLAMINES

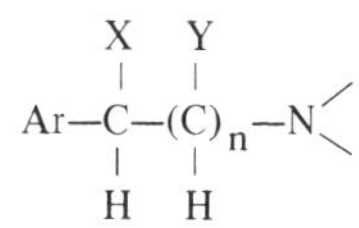

Direct β-Agonist Activity

	Increase or maintain activity	Decrease or lose activity
n =	1	0, 2, 3
X =	OH	H, NH_2, O
Y =	H, CH_2CH_3	CH_3, $-CH_2CH_2CH_3$, $-CH(CH_3)_2$

β-Substitution

The presence of a β-hydroxyl group is considered essential for full agonist activity. Conversion of the alcohol to a ketone causes loss of activity. Replacement of the hydroxyl in norepinephrine with a hydrogen yields the compound known as dopamine. While dopamine has β-adrenergic activity, it is less than that of norepinephrine. Dopamine also has α-agonist and serotonergic activity. Its pharmacological profile will be monographed elsewhere.

The effect of substitution on the side chain portion of the phenethanolamine nucleus is summarized in Table 4.

β-Carbon Stereochemistry

The stereochemistry at the asymmetric β-carbon is important in determining maximum binding to adrenergic receptors. The favored absolute configuration is R as defined by Patil et al.[22] Plane-polarized light passed through a solution of R-epinephrine will be rotated in a levorotatory (l) fashion.

The observation that one stereoisomer of epinephrine is substantially more active than the other isomer led Easson and Stedman[23] to postulate that adrenerigic agonists bind to their receptor using a three-point attachment. These attachment sites are an ionic bonding area for the amine, a hydrogen bonding area for the hydroxyl, and a hydrophobic or pi bonding area for the aromatic ring. Only the stereoisomer with the correct stereochemistry will allow for a full three-point attachment. The wrong stereoisomer will, at best, only accommodate a two-point attachment (Figure 4). If one assumes that the degree of binding to the receptor is a reflection of relative adrenergic activity, then the potency of R-epinephrine over S-epinephrine is apparent.

Side Chain Conformation

The carbon-carbon single bond of the side chain allows the side chain to assume a number of different conformational rotamers. Of the three gauche conformers, the conformer in which the aromatic group and the amine are *anti* or *trans* to each other appears to be the most energetically stable conformation in solution. Whether or not this is the conformation actually bound at the receptor is not known. It should also be noted that the *anti* conformation may gain some additional stability through the potential internal hydrogen bonding possible between the amine and the hydroxyl group (Figure 5). It is possible that the great lack of adrenergic agonist activity of tertiary amines may be due to the absence of this hydrogen bonding.

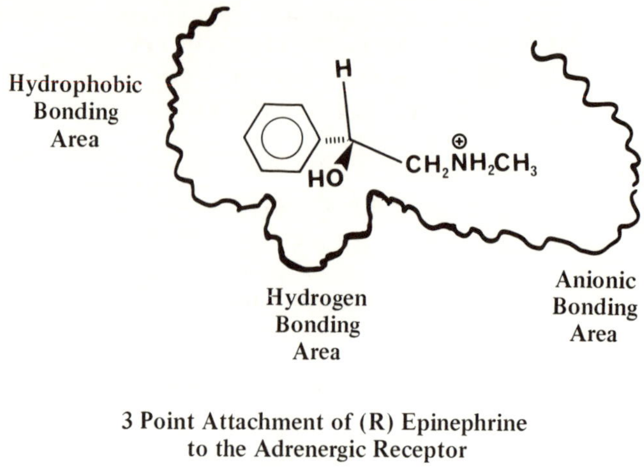

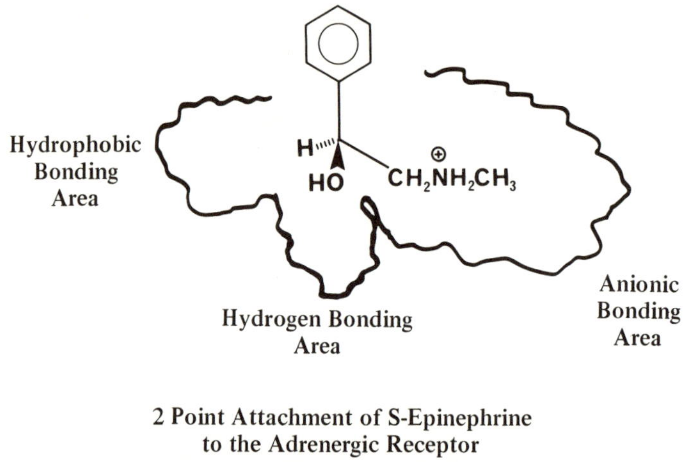

FIGURE 4. Easson-Stedman model of the adrenergic receptor showing stereospecificity of epinephrine binding.

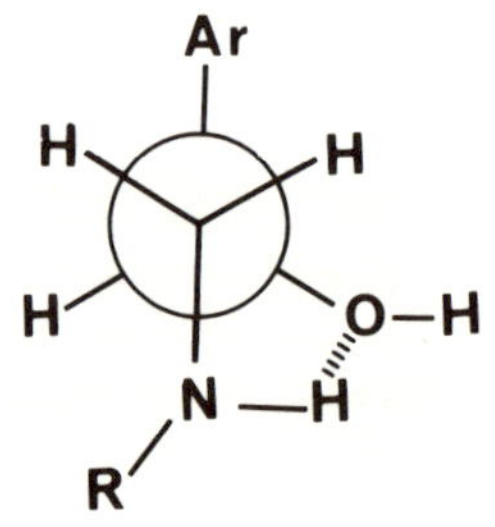

FIGURE 5. Internal H-bonding of anticonformation.

The Aryloxypropanolamines

Aryloxypropanolamines were considered for a long time to only be effective as β-adrenergic antagonists. Indeed, the most prescribed β-adrenergic antagonist, propranolol, is an aryloxypropanolamine. Within the last 10 years, however, several aryloxypropanolamines have been found which have β-agonist activity. Interestingly, these agonists also account for some of the first β_1 selective agonists discovered.

When a catechol ring was attached to the oxypropanolamine side chain, the β_1 selectivity

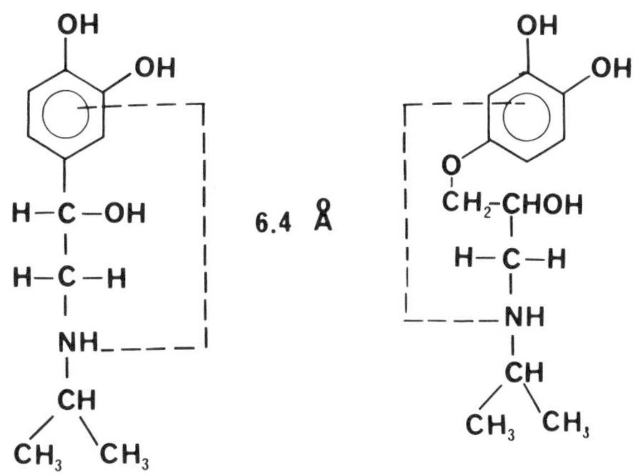

FIGURE 6. Side chain conformation of aryloxypropanolamines which mimics that of ethanolamines.

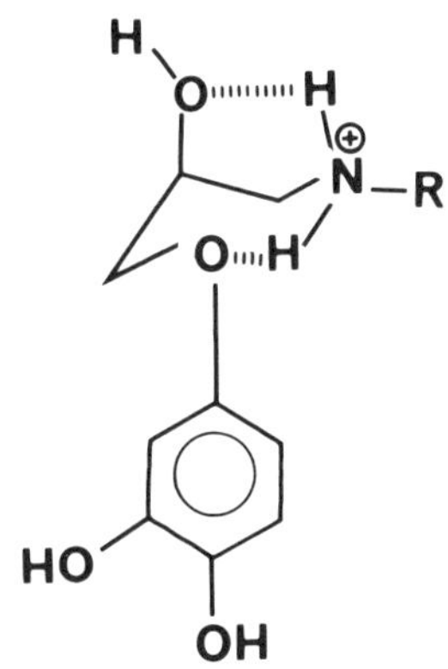

FIGURE 7. Possible internal hydrogen bonding in aryloxypropanolamines.

was lost, but the resulting compound was as good as or better than isoproterenol at both β_1- and β_2-receptors. In the preceding section on the structure-activity relationships of phenethanolamines, it was noted that the distance between the amine and the aromatic ring was crucial and thus, a two-carbon skeleton was all that could be accommodated and still maintain agonist activity. The agonist activity of the aryloxypropanolamines is thus surprising since four atoms separate the amine and aromatic ring. One rationale for their action might be the ability of the oxypropanolamine side chain to assume a conformation in which the amine and the aromatic ring are an effective distance of only two carbons from each other (Figure 6). Furthermore, this conformation might be additionally stabilized by internal hydrogen bonds. Jen and Kaiser[24] proposed that two internal hydrogen bonds are possible in a protonated aryloxypropanolamine (Figure 7). These hydrogen bonds include one between the hydroxyl and the amine, such as that proposed for the ethanolamines, as well as one between the amine and the etheric oxygen. This last hydrogen bond is presumably the one which brings the aromatic ring and the amine within the postulated two-carbon atomic distance required for receptor binding. The only difficulty with the proposed conformational model is that the spatial position of the hydroxyl group in the ethanolamine series is about 2 Å different from the spatial position of the corresponding hydroxyl in the aryloxypropanolamine series. Jen and Kaiser, however, suggest that this difference may not be crucial for adrenergic activity, and indeed may account for the fact that aryloxypropanolamine agonists tend to have greater selectivity for the β_1-receptor.

Aryloxypropanolamines are not exclusively active at β_1-receptors, however. As stated earlier, a catechol nucleus attached to the oxypropanolamine side chain imparts β_1 and β_2 activity to the compound. β_2 Selectivity can be obtained by the substitution of a "phenol equivalent" for the 3-position hydroxyl, just as was observed in the ethanolamine series. Furthermore, as was the case for the ethanolamines, reversing the position of the phenol equivalent to the 4 position severely decreases activity.

Structure-Activity Relationships of α-Adrenergic Agonists

Just as Lands and Brown noted that the structure-activity relationships for β-adrenergic agents were not consistent, so too have Starke, Endo, and Taube noted that the structure-activity relationships for α-adrenergic agents are not consistent. In 1975, Starke et al.[25] tested a number of α-adrenergic agonists on the postsynaptic α-receptors of rabbit pulmonary smooth muscle. They also tested these same compounds on the presynaptic α-receptors that mediate norepinephrine release (these receptors will be discussed at greater length in the presynaptic adrenergic function section of this review). Starke found that the order of potency for compounds acting on smooth muscle contraction was norepinephrine > oxymetazoline > naphazoline > phenylephrine > tramazoline > α-methyl norepinephrine > methoxamine. The potency order of these compounds on inhibition of norepinephrine release was epinephrine > oxymetazoline > tramazoline > α-methyl norepinephrine > norepinephrine > naphazoline > phenylephrine > methoxamine. Analysis of the data indicated that pre- and post-synaptic α-receptors may well be two different kinds of α-receptors. Based on the relative potencies of the compounds studied, phenylephrine and methoxamine were found to be most selective for the postsynaptic receptor, while oxymetazoline, tramazoline, and α-methyl norepinephrine were most selective for the presynaptic receptors. Norepinephrine, epinephrine, and naphazoline were relatively nonselective.

As more compounds were studied in more pharmacological systems, it became apparent that classifying these two different α-receptors simply by their anatomical location (i.e., pre- or postsynaptic) was not sufficient.

When the pressor responses of norepinephrine were studied in anesthetized cats, it was found that some of the α-agonist activity of norepinephrine could be blocked by prazosin (a postsynaptic selective α-blocker), but not all of the activity. Conversely, yohimbine (a presynaptic selective α-blocker was also able to block part of the agonist action of norepinephrine, but not all of it. This experiment, and others like it, suggested that α-receptors, previously classified as presynaptic, may also occur at postsynaptic sites. Consequently, the designation of α_1 and α_2 previously proposed by Langer[26] for the post- and presynaptic receptors, respectively, was adapted and expanded. In this classification, α_1 referred to those receptors selective for phenylephrine and methoxamine, and α_2 referred to those receptors selective for oxymetazoline and tramazoline, regardless of their anatomical location.[27]

In studying the structure-activity relationships of drugs acting at the α_1- and α_2-receptors, two chemical classes of compounds have been determined as having α-adrenergic activity. Those classes are the phenethanolamines and the imidazolines.

The Phenethanolamine α-Agonists

Many of the stucture-activity relationships for the phenethanolamine α-adrenergic agonists are the same as those for the β-adrenergic agonists. The two areas of major difference between the α- and β-agonists are in the nitrogen substitution and the aromatic substitution.

The α-receptors are quite sensitive to the size of nitrogen substituents. Unlike β-agonists, which can accommodate very large nitrogen substituents, α-agonist activity is severely decreased by anything larger than ethyl, and essentially abolished by anything larger than isopropyl.

The most important aromatic substitution requirement for α-adrenergics is the *meta* hydroxyl group. While the catechol nucleus appears more important for β-agonist activity, the *meta* hydroxyl appears to be sufficient for α-agonist activity.

α-Agonist activity can also be obtained when certain polar substituents are attached to the *meta* position. Methoxamine (*ortho, meta* dimethoxyphenyl-) and *meta* methanesulfonamido phenylethananolamine both show α selective agonist activity.

Finally, in terms of subreceptor selectivity, it appears that the ethanolamine α-agonists have more activity at the α_1-receptors than at the α_2-receptors.

The Imidazoline α-Agonists

The structure-activity relationships of imidazolines were first studied by Hartmann and Isler[28] in 1939, and it was from their work that the use of tolazoline as a hypotensive agent and naphazoline as a hypertensive agent developed.

The adrenergic agonist activity of the imidazoline class of compounds is surprising in view of the extensive structure-activity relationship work on adrenergic agents, which indicates that optimum adrenergic activity resides with 3,4-dihydroxyphenylethanolamine structures. Inspection of the imidazoline structure, however, reveals some similarity to other adrenergic agonists, and in fact a phenethylamine nucleus can be discerned.

The agonist activity of these compounds is also surprising in view of the lack of a β-hydroxyl group. Since lack of this functionality is often suggestive of indirect-acting agents, this activity was specifically tested for. Mujic and Van Rossum[29] found that cats, depleted of their norepinephrine stores by reserpine, still exhibited α-adrenergic responses when treated with naphazoline, and in fact, some potentiation of effect was observed. Use of this testing procedure also unmasked a partial agonist activity by the α-adrenergic blocker, and the structurally similar compound, tolazoline.

Imidazolines are known to have central, as well as peripheral, α-stimulant effects. Differentiation of these two activities appears to be highly dependent on the nature of the bridging atom between the two rings of the imidazoline structure. Stuctures in which the bridging atom is a nitrogen rather then a carbon have been found to have greater central activity. The differences in activity of these two families of compounds may be due to the fact that the 2-amino imidazolines exist in equilibrium with their tautomeric imino form, imparting a planarity and rigidity to the aminoguanidine center (Figure 8).

Clonidine, a potent centrally active α-adrenergic agonist, is an amino imidazoline which also possesses bulky chlorines at the 2 and 6 positions of the phenyl ring. Jackman and Jen[30] have suggested that steric hindrance between these chlorines and the coplanar amino imidazoline force the rings into a perpendicular conformation which is necessary for the central effects of the compound (Figure 8).

A number of structural modifications of imidazolines can be made and still retain α-agonist activity. All the useful modifications, however, involve substitution on the aromatic ring rather then on the imidazoline ring (Table 5).

Of the tabulated compounds, it is interesting to note that oxymetazoline, which has a phenolic OH at what would be the *meta* position of a phenethylamine agonist, is the most potent agonist of the series.

There is still a considerable amount of conjecture about whether imidazolines bind to the same receptor sites as the phenethanolamine α-adrenergic agonists. Ruffolo et al.[31] found that the contractile response of rat vas deferens could be desensitized to oxymetazoline upon repeated administration of the compound. After complete desensitization to oxymetazoline, Ruffolo found that he could still get a near normal contractile response upon addition of the phenethanoamines norepinephrine, phenylephrine, and methoxamine. Addition of the imidazolines, naphazoline and tetrahydrazoline to the desensitized preparation elicited no contractile response. The authors thus proposed that the phenethanolamines interacted with the classical Easson-Stedman[23] site while the imidazolines bound to a different site that may or may not overlap with Easson-Stedman site.

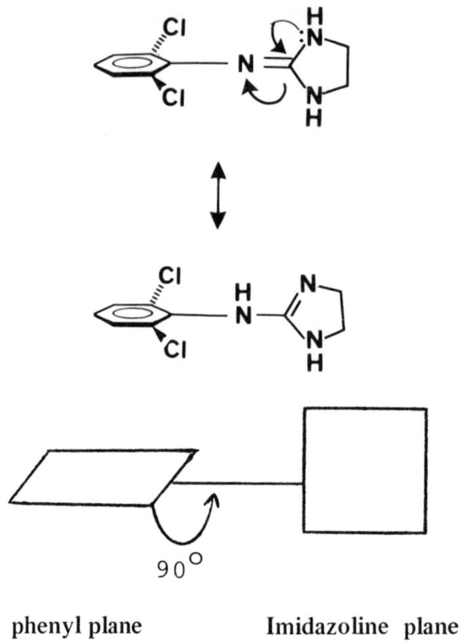

FIGURE 8. Tautomeric forms of clonidine, showing the perpendicular relationship of the two-ring systems.

Table 5
THE EFFECT OF AROMATIC SUBSTITUTION ON THE α-ADRENERGIC AGONIST ACTIVITY OF IMIDAZOLINES

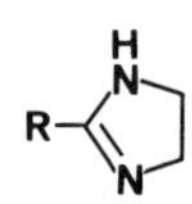

Name	R	Relative α-agonist activity (Norepi = 100)
Naphazoline	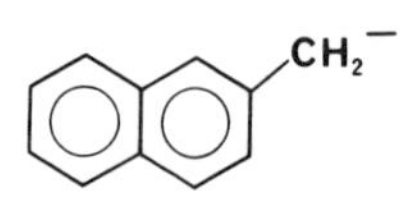	2
Tetrahydrazoline	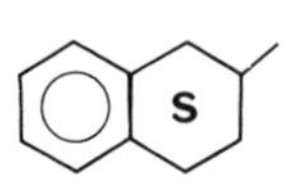	0.3
Oxymetazoline	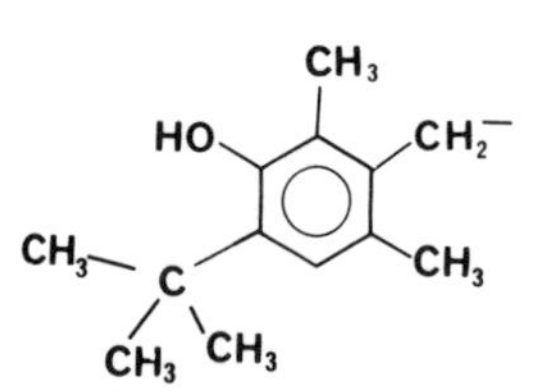	30

Finally, in view of the existence of the α-receptor subtypes α_1 and α_2, it is perhaps tempting to suggest that the phenethanolamine site is the α_1 site, while the phenethylamine (imidazoline) site is the α_2 site. In view of the lack of clear-cut α-receptor specificities for some of the imidazolines (i.e., naphazoline, oxymetazoline), it is perhaps too early to assume such structural specificities for these receptors.

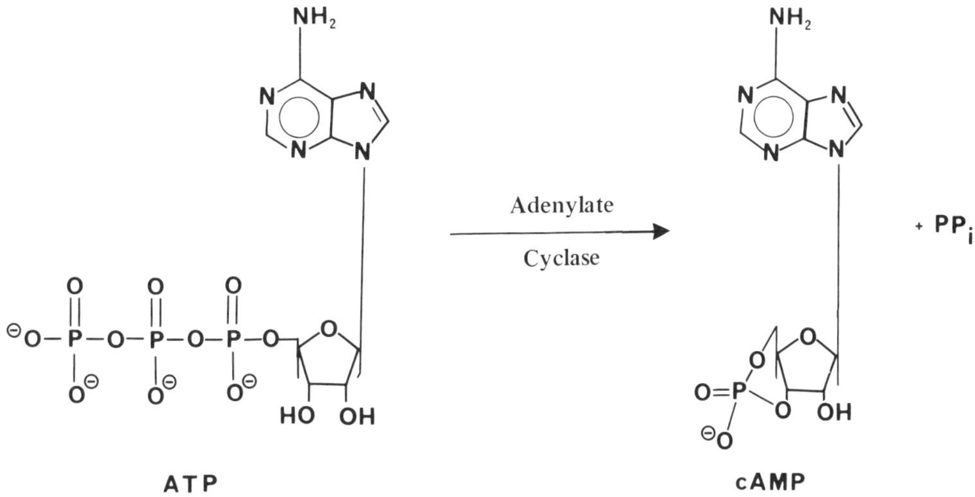

FIGURE 9. Biosynthesis of cAMP.

Mechanisms of β-Adrenergic Response

The Second Messenger Hypothesis

In 1957, Sutherland and Rall[32] discovered an "active factor" which could stimulate the phosphorylase system responsible for converting glycogen to glucose-1-phosphate. This active factor was identified as cyclic 3′,5′-adenosine monophosphate (cAMP). It is synthesized from ATP by the enzyme adenylate cyclase (AC) (Figure 9). AC was found to be bound to the inner wall of the cell membrane. What made the enzyme remarkable was the fact that its activity could be stimulated by a number of extracellular hormones such as norepinephrine, adrenocorticotropin hormone (ACTH), and glucagon. The fact that cAMP could be produced as a direct result of the action of a hormone on its receptor led Sutherland to postulate that cAMP is really a "second messenger" in the sequence of events between stimulus by hormone (first messenger) and physiologic response.[33] This concept became known as the "Second Messenger Hypothesis" and its validity has since been firmly established.

The importance of this hypothesis lies in the fact that it suggests that one can regulate some types of hormonal response simply by regulating intracellular cAMP levels. One of the hormones which regulates intracellular cAMP levels is norepinephrine through its action on β_1- and β_2-adrenergic receptors. A major question posed by this concept is how can binding of hormone to its receptor on the outer cell membrane influence the activity of an enzyme bound to the inner cell membrane? Answering this question, specifically with regard to the β-adrenergic receptor system, has been the goal of an intense amount of research.

It is known that cell membranes are composed mostly of lipids. These lipids are arranged in a bilayer such that the hydrophilic polar heads of the lipids face the intra- and extracellular aqueous fluid, while their hydrophobic tails are oriented toward each other (Figure 10). It is believed that the β-receptor protein is embedded in the extracellular surface of the lipid bilayer, with its hormone binding site facing the extracellular surface of the lipid bilayer. AC on the other hand, is embedded in the intracellular surface of the lipid bilayer with its catalytic site facing the intracellular fluid. Regulation of cAMP levels by hormones must, therefore, involve the interaction of the β-receptor protein and the enzyme AC through some process within the lipid bilayer of the membrane. Several models have been proposed to explain this receptor/enzyme interaction process.

Fundamental to the concept of the receptor enzyme interaction is the belief that proteins embedded in the cell membrane are capable of moving laterally and independently through

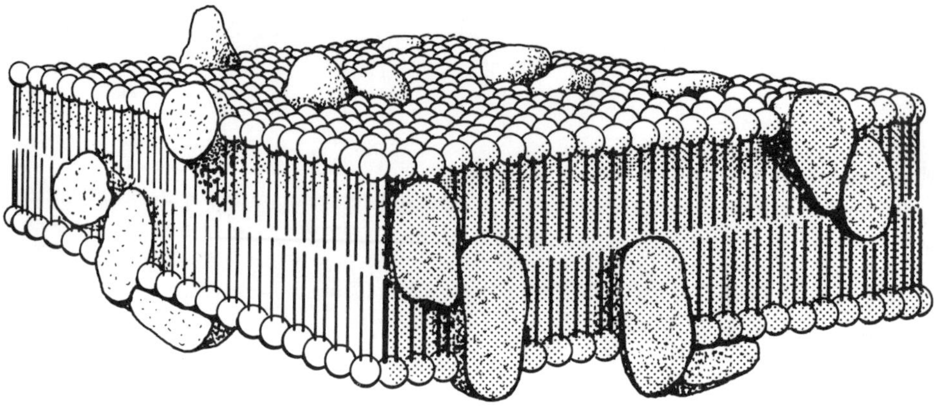

FIGURE 10. Lipid bilayer of cell membrane with embedded proteins.

the lipid bilayer. An interesting and pertinent example of how a protein can migrate through a lipid bilayer is afforded by cholera toxin, a protein exotoxin produced by the bacterium *Vibrio cholerae*. This toxin has two major components: a binding component and an active component. The binding component apparently is able to bind to gangliosides on the cell surface. Once this binding occurs, the active component can then be inserted into the lipid matrix of the membrane, leaving the binding component behind. The active component migrates through the membrane, attaches to a regulatory site on the AC, and activates the enzyme in an irreversible manner.[34] The resulting high levels of cAMP produced as a result of this activation cause severe fluid loss in the victim.

In addition to demonstrating how proteins can migrate through lipid bilayers, the action of cholera toxin on AC has also helped to explain some interesting regulatory properties of this enzyme which are directly applicable to the interaction of the enzyme and the β-adrenergic receptor.

In 1971, Rodbell and co-workers[35] discovered that guanosine 5′-triphosphate (GTP) was capable of stimulating AC by interacting at a regulatory site on the enzyme distinct from the catalytic site. The activation was very rapid and brief and was only observable if the GTP analog, GppNHp (Figure 10), was employed in the experiment rather then GTP itself.

The reason the activation was so brief was because as soon as GTP became bound to the regulatory site it was hydrolyzed there to GDP, thus terminating the activation. GppNHp, by virtue of the amide linkage between the β and γ phosphates, was resistant to hydrolysis at the regulatory site yet was still capable of stimulating cAMP formation. As a result of the prolonged stimulation by GppNHp, cAMP levels were allowed to accumulate to detectable levels.

It is now believed that the GTP regulatory site is actually part of a separate regulatory protein which is capable of interacting with AC. Furthermore, this nucleotide regulatory protein (NRP) appears to be the site of action by cholera toxin.

Moss and co-workers[36] have presented evidence that cholera toxin acts by catalyzing the formation of a covalent bond between the NRP and nicotinamide adenine dinucleotide (NAD) (Figure 11). As a result of this covalent modification, the NRP protein is able to bind GTP, but cannot hydrolyze it. Consequently, the NRP protein is irreversibly activated, and thus stimulation of AC also becomes irreversible.

The discovery of the NRP and its control by GTP opened the door to the exploration of a number of phenomena involving the β-adrenergic receptor and its control of AC.

The Phenomena of Desensitization

One of the most clinically significant phenomena involving the β-adrenergic receptor is

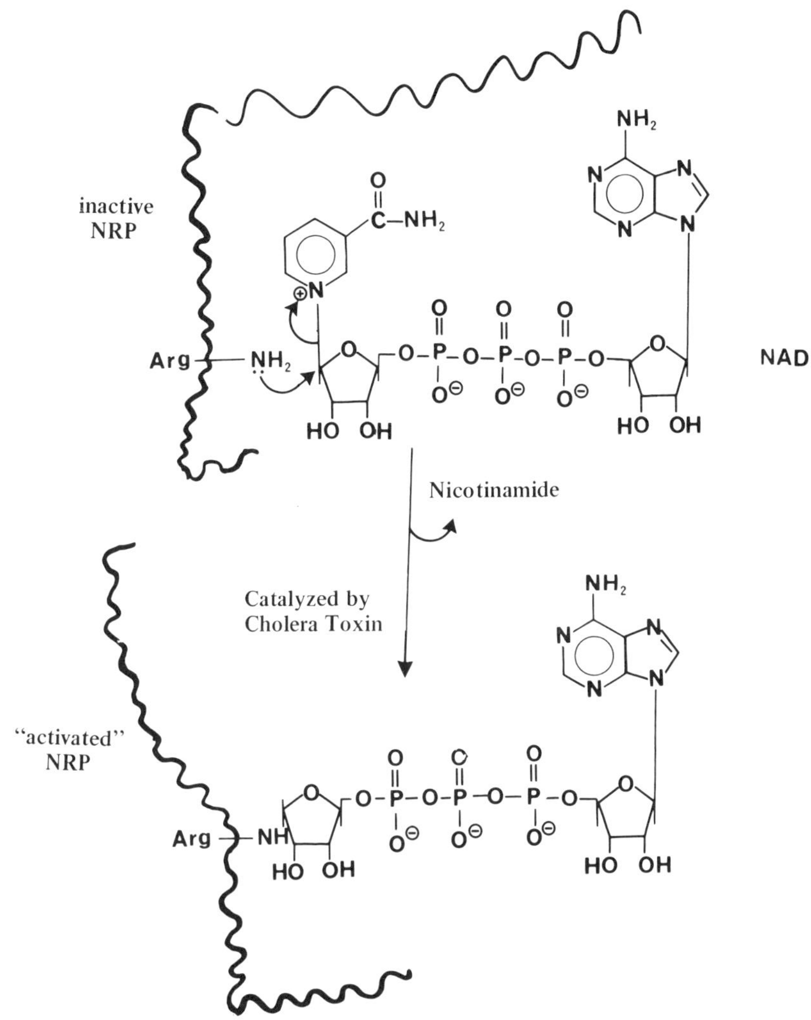

FIGURE 11. ADP ribosylation of the nucleotide regulatory protein.

desensitization. It has been observed that patients on prolonged β-agonist therapy have a tendency to increase the dosage of drug they use. This occurs because after prolonged administration, the β-receptors appear to become "desensitized" to the agonist, thus requiring larger doses of the compound to effect equivalent physiological response. The problem of desensitization is especially significant in the treatment of diseases such as asthma (see The β-Adrenergic Theory of Asthma, this review).

The treatment of asthma usually involves the selfadministration, either orally or by aerosol, of β-adrenergic agonists. These agonists mediate bronchodilation and promote easier breathing in the asthma sufferer. Asthmatic patients who use β-agonists for a prolonged time, gradually increase their selfadministered dosage in order to get a sufficient amount of bronchodilation and relief in breathing. As a result of this increased dosage, the agonist concentration builds to sufficient levels so as to enter the peripheral circulation where it acts on the β-receptors of the heart and vasculature to cause cardiovascular stress. Significant CNS effects and even death may occur as a result of overdosage (see monograph on isoproterenol toxicity).

Obviously, an understanding of the molecular mechanism of desensitization has clinical importance. Several lines of research have shed light on this problem. Mukherjee and

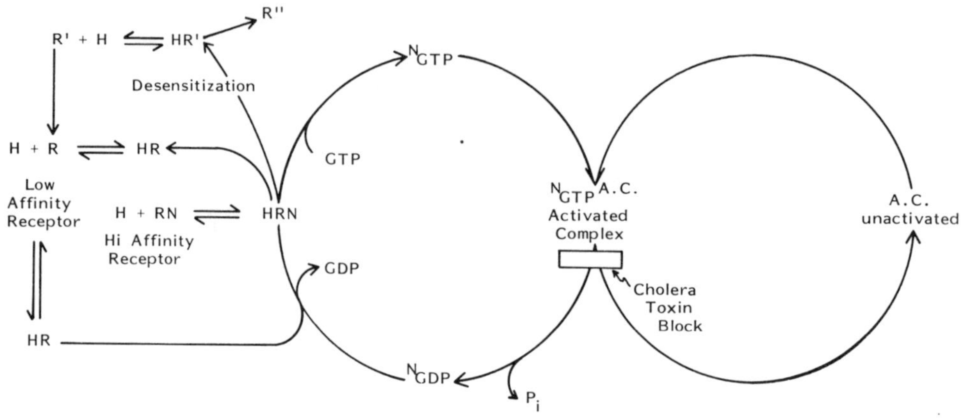

FIGURE 12. Proposed model for the mechanism of desensitization.

Lefkowitz,[37] using ligand binding techniques on cell membrane preparations (see the following chapter) have observed that during desensitization there is an apparent loss in the total number of β-adrenergic agonist binding sites on the cell membrane. However, this loss of binding sites can be completely reversed by the addition of GTP to the system. Furthermore, upon desensitization, there is an apparent shift in the binding constant (K_D) of the β-receptor for its ligands.

Based on these observations, and those of many other workers, Su et al.[38] have proposed the following model to explain the mechanism of desensitization (Figure 12).

Uncoupled β-receptors (R) are able to bind hormone with low affinity. This hormone-receptor complex can then couple with the nucleotide regulatory protein (N_{GDP}). As a result of this coupling, GDP bound to N can be ''cleared'' from N to form the ternary complex of hormone-receptor-nucleotide regulatory protein (HRN). Under normal cellular conditions this complex would avidly bind GTP and then quickly dissociate to the activated nucleotide regulatory protein-GTP complex (N_{GTP}) and the β-receptor-hormone complex (HR). The HR complex could then dissociate and/or reassociate to be recycled. The N_{GTP} complex can undergo association with adenylate cyclase to activate it for cAMP production. The length of the activation would be dependent on the rate at which the GTP is hydrolyzed to GDP by the regulatory protein. In the presence of cholera toxin, the life of the active complex would be indefinite due to the ability of the toxin to covalently modify the N protein so that it cannot hydrolyze GTP. In the absence of the toxin the hydrolysis of GTP would lead to formation of the N_{GDP} complex, which can be recycled with HR.

According to the authors, ''catecholamine induced desensitization causes a modification in the beta-adrenergic receptor per se or a modification in a portion of the nucleotide binding component that recognizes the beta adrenergic receptor''. In Figure 12, HRN would serve as the substrate for the formation of the uncoupled form of the receptor R′ during the desensitization reaction. Under normal conditions it is postulated that HRN never achieves significant levels in the cell because its rate of formation is the rate-limiting step in the catalytic cycle, and it is used as fast as it is formed. During desensitization, or under conditions of GTP depletion, it is postulated that HRN levels can build up. This buildup has two results. First, it allows for the formation of RN complex through dissociation of hormone. This RN complex is postulated as having a high affinity for hormone. The appearance of this high-affinity class of receptors may account for the observed shifts in binding constants for β-agonists in desensitized preparations. The second effect of the buildup of HRN is the formation of the uncoupled form of the receptor R′. It is postulated that R′ can be converted by an unknown mechanism to either R, which can be recycled, or R″, which is an irreversibly inactive nonligand binding form of the receptor. The ability of GTP to reverse the apparent loss of receptors due to the production of R″ would depend on its being added to the system in time to decrease HRN levels and shift the equilibrium for the $R \rightleftharpoons R' \rightarrow R''$ reaction in the direction of the useful R form. Once R″ is formed, the desensitization cannot be reversed by GTP.

It is apparent that the modulation of β-adrenergic receptor activity at the molecular level is an extremely complicated and sophisticated process which is only now starting to be unraveled. The very complexity of the model suggests the opportunity for more direct and selective control of the action of the β-adrenergic system.

Factors Affecting the Activity of Adenylate Cyclase

It is important to realize that while the β-adrenergic system utilizes adenylate cyclase and cAMP to relay its physiological instructions, and β-receptor is not the only regulator of adenylate cyclase activity. Indeed, there is a host of hormones and biological molecules whose activity is mediated via cAMP. These hormones include ACTH, glucagon, TSH, vasopressin, prostaglandin E (PGE), melanocyte stimulating hormone (MSH), and others. It appears possible that all of these hormones and/or their receptor proteins may compete for the use of a common pool of adenylate cyclase floating within the cell membrane. Thus, it is possible that the activity of one hormone may regulate the activity of another hormone simply by affecting the availability of the pool of adenylate cyclase.

Ions have also been shown to be effective in the regulation of adenylate cyclase activity. Magnesium is an absolute requirement for activity. In fact the actual substrate for the enzyme is a Mg-ATP complex, whereas free ATP is an inhibitor.

Divalent ions such as manganese can readily substitute for magnesium; cobalt is partially active; calcium, zinc, and copper are usually inhibitory.

One of the most striking examples of the effect of ions on adenylate cyclase activity is the effect of flouride ion. Flouride has been shown to be able to stimulate the enzyme 2-1 to 20-fold. The mechanism of this activation is not yet clear, but apparently does not involve the β-adrenergic receptor. The only other monovalent ion to have significant activity on the enzyme is lithium, which has been found to be inhibitory.

Mechanisms of α-Adrenergic Response

The molecular mechanisms of α-adrenergic activity are understood even less well than those of the β-adrenergic system. Several different hypotheses have been advanced.

It now appears that one of the primary actions of the α-agonist is the introduction of Ca^{2+} into the target cell. This Ca^{2+} may then regulate other cellular events. For this reason, it has been suggested that Ca^{2+} may function as a "second messenger" for the α-receptor in the same way that cAMP functions for the β-receptor.[39]

Changes in intracellular Ca^{2+} levels have been implicated in a number of physiological events controlled by α-receptors. The positive inotropic response in the heart caused by the α-agonist methoxamine has been explained as being due to an α-receptor-mediated influx of Ca^{2+} ions which then interact with contractile proteins. This effect was antagonized by slow channel blockers, as well as by the α-blocker phentolamine.[40]

Another intriguing action of Ca^{2+} is its effect, directly or indirectly, on intracellular cAMP levels. Ca^{2+} is known to stimulate cAMP phosphodiesterase through its interaction with calmodulin. This stimulation will lead to increased metabolism of cAMP and a decrease in intracellular cAMP levels.

Watanabe et al.,[41] however, suggest a more direct relationship between cAMP levels and α-adrenergic activity. They state that "catecholamines may increase or decrease cAMP levels in rat myocardium, depending on the intensity of stimulation of receptor types. Increases are mediated by beta-adrenergic receptors, whereas, decreases appear to be mediated by α-adrenergic receptors." It is not clear, however, whether or not the α-receptor-mediated decrease in cAMP is a result of direct receptor action on adenylate cyclase (as originally postulated by Robinson) or through some intermediary process such as inhibition of adenylate cyclase by calcium ion.

Finally, in another analogy with the β-receptor system, Amer[42] has reported that oxymetazoline, an imidazoline α-agonist, is capable of stimulating the formation of 3′,5′ cyclic guanosine monophosphate (cGMP). Formation of cGMP by oxymetazoline was inhibited by the α-blocker phentolamine. While decreased levels of cAMP were also observed in the tissue, the author suggests that the fall in cAMP levels might in fact be secondary to the rise in cGMP levels. This study might suggest a second messenger role for cGMP in the α-adrenergic system.

It is apparent that a clear-cut molecular mechanism for α-receptor action has not been found. A number of important cellular responses, such as increased Ca^{2+} and cGMP as well as decreased cAMP, have been reported for α-adrenergic agonists. It remains for future investigators to determine the interplay and importance of these events in mediating α-adrenergic activity.

CLINICAL APPLICATION OF POSTSYNAPTIC ADRENERGIC AGONISTS

Nasal Decongestants

Physiology of Nasal Congestion

The use of adrenergic agonists in the treatment of diseases of the respiratory system is an extremely important and profitable use of these agents. Approximately three fourths of the total clinical use of adrenergic agonists is in the treatment of respiratory problems. The use of adrenergic agents in the treatment of asthma, as it affects the respiratory system, will be discussed elsewhere in this review. This discussion will be concerned with the effects of adrenergic agents in treating congestion of the nasal passages and upper respiratory tracts.

Congestion is due to vasodilation and increased glandular secretion of the mucosa which lines the nasal and respiratory passages. The mucous membrane which lines the nasal cavities is of two types: the olfactory and the respiratory. The olfactory serves to relay olfactory stimuli via special nerve cells. The surface of the mucous membrane is kept moist by the secretions of serous glands. The respiratory mucous membrane lines the lower part of the nasal cavities. This membrane warms, moistens, and cleanses inspired air. This action is accomplished by heat exchange from veins in the submucosal connective tissue and mucous secreted by glands and goblet cells. Dust and particles trapped by mucous are moved backward by ciliary action to the pharynx and swallowed.

Other structures lining the passage are olfactory and ordinary sensation nerves, arterioles derived from the maxillary artery, and lymphatic vessels. Sympathetic nerves regulate the vasculature of the nasal and pharyngeal mucosa. Pure α-adrenergic stimulation causes vasoconstriction and decongestion, while β-adrenergic stimulation mediates vasodilation.

If an agent such as epinephrine is topically administered to the nasal passages, an initial decongestion and symptomatic relief is obtained. A second phase of vasodilation often occurs, however, due to β-adrenergic stimulation. For this reason, pure α-adrenergic agonists are the agents of choice in treatment of nasal congestion.

MONOGRAPHS

PHENYLEPHRINE

Mechanism of Action

Phenylephrine (Figure 13) is a direct-acting, α_1-selective, adrenergic agonist. Topical application of phenylephrine to the nasal mucosa results in local vasoconstriction, decreased nasal secretion, and increased sinus drainage.

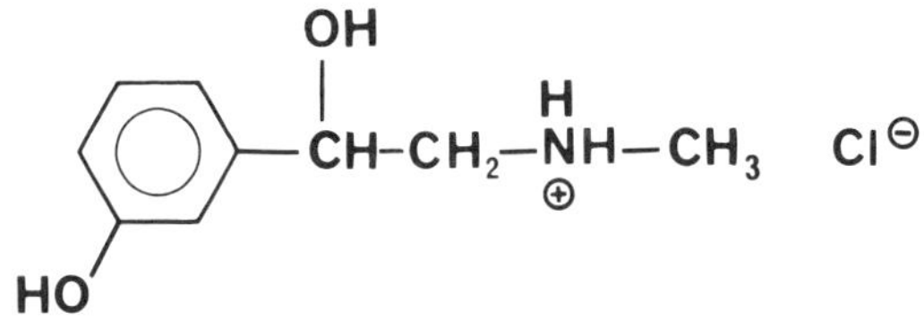

FIGURE 13. Phenylephrine hydrochloride.

Structure-Activity Relationships

Phenylephrine was originally prepared and studied by Barger and Dale[8] in 1910. As a direct-acting agonist it possesses those structural features required for a direct α-adrenergic agonist. Such features include a phenethylamine nucleus, a β-hydroxyl group, and an aromatic hydroxyl. Increasing the size of the amino substituents causes loss of α-agonist activity. The stereochemistry about the β-carbon must be levorotatory (R) for maximum activity. Loss of the β-hydroxyl group severely decreases but does not completely abolish activity. Perhaps most noteworthy is the presence of the *meta* hydroxy group on the aromatic ring. The *meta* rather than *para* position of this OH appears to be most responsible for the selective α-adrenergic activity of this compound.

Pharmacokinetics

Phenylephrine is rapidly and completely absorbed from the GI tract and undergoes significant first-pass metabolism. If applied topically, its effects occur almost immediately. Its duration of action is from 30 min to 4 hr. (Peak plasma levels are obtained 1 hr after oral application.) Sustained-release forms are available.

The majority of orally administered phenylephrine is eliminated as conjugates in the urine. The predominant conjugate found is the *O*-sulfate, with lesser amounts conjugated as the *O*-glucuronide.[44]

Because of the acid-base nature of these conjugated metabolites, renal excretion is strongly pH dependent. About 80% of an oral dose of phenylephrine could be eliminated in urine within 24 hr if the pH of the urine is maintained at 5.6 using ammonium chloride. Binding of phenylephrine to human serum albumin and plasma has not been studied; however, the structurally similar compound etilephrine was found to bind to HSA 8.5% and to human plasma 23%.

Other routes of metabolism of phenylephrine include microsomal *N*-demethylation and deamination by either plasma or mitochondrial monoamine oxidase. While these routes are not major ones in the overall body disposition of the drug, they can be locally important such as at effector organ synapses (see the toxicity section of this monograph).

Potency of the HCl salt form has also been reportedly lost in preparations containing aspirin.[45] Loss of potency was described as being due to acid-catalyzed acetylation of phenylephrine by aspirin. Such acetylation could occur either directly or through attack on an aspirin anhydride formed by an intramolecular transacetylation mechanism (Figure 14).

Uses and Doses

Phenylephrine is used primarily as a nasal and bronchial decongestant and is often combined with other agents such as antihistamines, analgesics, antitussives, and expectorants in cold preparations (Table 6).

Phenylephrine is also used in ophthalmic solutions as a vasoconstrictor and mydriatic. A 1% solution for injection is available for use as a vasopressor agent in shock, but the value of phenylephrine in this condition is questionable.

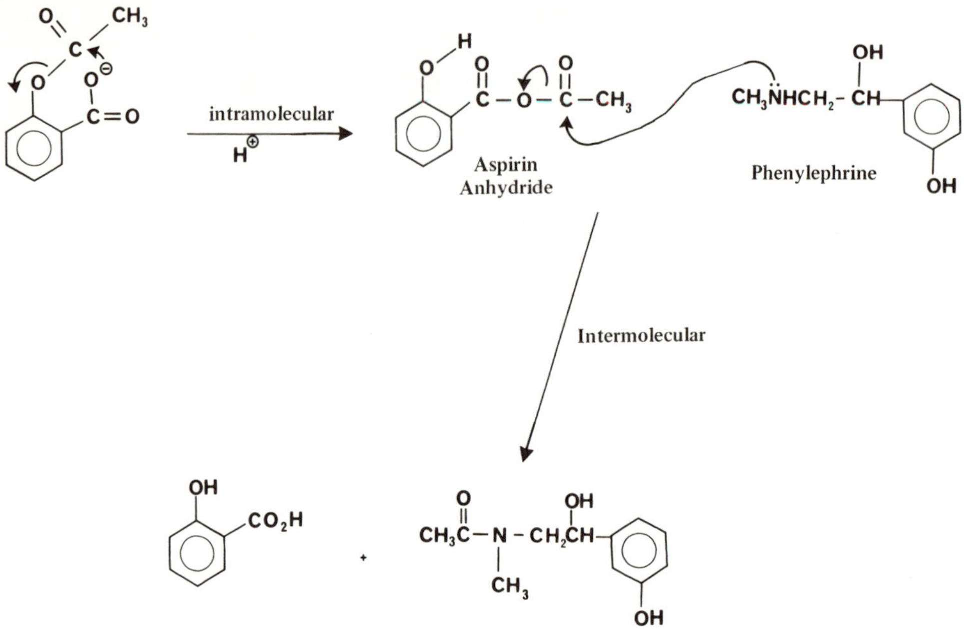

FIGURE 14. Acetylation of phenylephrine via intramolecularly generated aspirin anhydride.

Toxicity

The major toxicities associated with phenylephrine use arise when the compound is used in combination with other drugs which potentiate the α-adrenergic agonist effects of phenylephrine.

Phenylephrine should never be used with monoamine oxidase (MAO) inhibitors, such as the antihypertensive compounds pargyline and debrisoquine, or with the hydrazine-type antidepressant compounds. While deamination of phenylephrine by MAO is not a major route of metabolism (see the pharmacokinetics section of this monograph), the additive effect of increased norepinephrine levels due to MAO inhibition plus the agonist effect of phenylephrine at α-adrenergic receptors could precipitate a hypertensive crisis.[47]

Hypertensive crises can also be precipitated if ganglionic blockers such as hexamethonium or mecamylamine are used in conjuction with phenylephrine. These agents presumably block the feedback bradycardia which normally would occur due to the vasopressor effect of phenylephrine. Without this reflex slowing of heart rate, arterial pressure due to peripheral vasoconstriction can increase rapidly.

Caution should also be exercised in administering phenylephrine along with β-adrenergic blockers. Phenylephrine normally has little effect on bronchial air passages. Any α-adrenergic bronchoconstrictive effect by this agent usually is quickly compensated for via β-receptor-mediated bronchdilatory action. If this reflex action is blocked by β-receptor antagonists, respiratory crises may be precipitated.[48]

The oral LD^{50} in mice is less than 500 mg/kg.[49]

Physical-Chemical Data

Structure:	Figure 13
Empirical formula:	$C_9H_{13}NO_2 \cdot HCl$
Mol wt:	203.67
Melting point:	140 to 145°C

Table 6
PHARMACEUTICAL PREPARATIONS CONTAINING PHENYLEPHRINE

Proprietary name	Manufacturer	Other active components	Form of administration	Uses and action
Bronkometer/Bronchosol	Breon	Isoetharine saccharin menthol	Nebulizer/inhaler	1, Asthma
Demazin	Schering	A	Tablets and syrup	1, 2, Cold capsules
Donatussin	Laser	A, B	Syrup	1, 2, 3, Cold capsules
Duadacin	Hoechst-Roussel	A, D, C, E, F, G	Capsules	1, 2
Neotep Granucaps	Tutag	A	Capsules	1, 2
Sinovan Timed	Drug Industries	A, K	Capsules	1, 2, 5
Chlor-Trimeton	Schering	Pseudoephedrine	Tablets	1, 2
Citra	Boyle	Pheniramine, methapyrilene D, I, H, G, C	Capsules	1, 2
Citra	Boyle	Above composition plus hydrocodone	Capsules and syrup	1, 2, 3
Colrex	Rowell	H, F, A, C	Capsules and syrup	1, 2, 4, 5
Congesprin	Bristol-Meyers	G	Chewable tablets	1, 2
Coricidin	Schering	None	Nasal mist	
Coryban-D	Pfipharmecs	E, A, J	Capsules	1, 2
Dallergy	Laser	A, K	Capsules and syrup	1, 2, 5
Demazin	Schering	A	Capsules and syrup	1, 2
Dimetane	Robins	Brompheniramine, B, J		1, 2, 3
Duo-Medihaler	Riker	Isoproteranol	Aerosol	Asthma, 1
Dimetapp	Robins	Brompheniramine, J	Capsules	1, 2
Drinus Granules	Anfre-Grant	A, K	Capsules	1, 2
Entex	Baylor	J, B	Capsules and liquid	1, 2, 3
Extendryl	Fleming	A, K	Capsules, syrup, and tablets	1, 2
4-Way	Bristol-Meyers	Naphazoline, J, D	Nasal spray	1, 2
Guistrey Fortis	Bowman	A, B	Tablets	1, 2
Histabid	Meyer	A, J	Capsules	1, 2
Histalet	Reid-Provident	D, A, J	Syrup and capsules	1, 2
Histaspan	USV Pharmaceutical	A, K	Time-release capsules	1, 2
Marhist	Marlop	A, K	Time-release capsules	1, 2
NTZ	Winthrop	Thenyldiamine·HCL benzalkonium chloride	Nose drops	1

Table 6 (continued)
PHARMACEUTICAL PREPARATIONS CONTAINING PHENYLEPHRINE

Proprietary name	Manufacturer	Other active components	Form of administration	Uses and action
Naldecon	Bristol	Phenyltoloxamine, J, A	Tablets	1, 2
Narine	Dooner	A, K	Time-release capsules	1, 2
Nasahist	Keene	J, A	Capsules	1, 2
NeoSynephrine	Winthrop		1% Injection	Vasopressor
NeoSynephrine	Winthrop		Eye drops	Opthalmic vasocon strictor mydriatic
NeoSynephrine	Winthrop		Nasal spray	1
Oraminic	North American Pharmacol	J, A, Scopolamine	Time-release capsules	1, 2
Pediacof	Winthrop	H, A	Syrup	1, 2
Quelidrine	Abbott	Dextromethorphan, A, ephed-rine, ipecac	Syrup	1, 2, 3, 5
Respinol G	Misemer	J, B	Tablets	1, 3
Rhinex	Lemmon	G, A	Tablets	1, 3
Ryna-Tussadine	Mallinckrodt	Carbetapentane, A, ephedrine	Tablets	1, 2, 5
Sinarest	Pharmacraft	F, E, A	Tablets	1, 2, 4
Singlet	Dow	A, F	Tablets	1, 2, 4
S-T Forte	Scott-Tussin	Hydrocodone, J, pheniramine, B		1, 2, 3, 5
Trind	Mead Johnson	B, F		1, 3, 4
Tussanil	Misemer	J, D, A, Dihydrocodeinone	Syrup	1, 2, 3, 5
Tympagesic	Warren-Tead	Antipyrine, benzocaine	Ear drops	Analgesic decongestant
Rynatan	Malinckrodt	A, D	Tablets	1, 2

Note: A = Chlorpheniramine; B = guaifenesin; C = ascorbic acid; D = pyrilamine; E = caffeine; F = acetaminophen; G = salicylamide; H = codeine; I = phenactin; J = phenylpropanolamine; K = methoscopolamine; 1 = decongestant; 2 = antihistamine; 3 = expectorant; 4 = analgesic; 5 = antitussive.

Chemical name:	(−)-1-(*m*-hydroxyphenyl)-2-methylamino-ethanol hydrochloride
Generic name:	phenylephrine hydrochloride
Proprietary names:	Neosynephrine (Winthrop) for others see Table 8
Solubility (HCl salt):	freely soluble in water and alcohol, acid to litmus
Salt forms:	Hydrochloride, tannate
Physical description (HCl salt):	white crystals bitter taste
pKa:	base 8.77, acid 9.84
Specific rotation (HCl salt):	−42° to −47.5°
Stability:	prolonged exposure to air, heat or strong light results in oxidation and loss of potency

NAPHAZOLINE

Mechanism of Action

Naphazoline is a nonselective, direct-acting α-adrenergic agonist. Topical application of naphazoline to nasal passages results in constriction of arterioles and mucous decongestion.

Structure-Activity Relationships

The structure-activity relationships of imidazolines are reviewed in the structure-activity relationships section of α-adrenergic agonists.

Pharmacokinetics

Local vasoconstriction after topical application occurs within 10 min and may persist for 2 to 6 hr. Naphazoline is poorly absorbed through mucous membranes. Local application of the drug gives little or no systemic effects. As a result of its poor absorption, information on the human metabolism, distribution, and excretion of the compound is not available. However, it is likely that very little metabolism of the compound occurs. This assumption is based on the fact that Brodie et al.[50] have found that tolazoline, a structurally similar compound (phenyl instead of naphthyl), was found unchanged in the urine of dog and man. On the basis of comparison with tolazoline, it is likely that essentially all naphazoline is excreted via the urine.

Uses and Doses

Naphazoline is used primarily as a nasal decongestant in 0.05 and 0.10% solutions in spray dispersers (Table 7). Naphazoline is also used as an ophthalmic solution for relief of ocular congestion, itching, and minor irritation.[48] Ophthalmic solutions of 0.012 and 0.1% are available. Doses of 1 to 3 drops every 3 to 4 hr are normal. Treatment should not continue more than 3 to 4 days.

Toxicity

Due to poor absorption, toxicity is generally low and confined to such local effects as nasal irritation, burning, stinging, dryness, or ulceration of the mucosa. Systemic effects observed on prolonged usage include hypertension, bradycardia, sweating, and sedation. Rebound congestion may also occur on prolonged administration.[48]

The i.v. LD_{50} in mice is 170 mg/kg and in rabbits is 0.8 mg/kg.[49]

Physical-Chemical Data

Structure:	Figure 15

Table 7
PHARMACEUTICAL PREPARATIONS OF IMIDAZOLINE ADRENERGIC AGONISTS

Generic name	Proprietary name	Manufacturer	Other active components	Dosage and route of administration	Use
Naphazoline	4-Way Nasal Spray	Bristol-Myers	Phenylephrine, phenylpropanol, amine	0.05% Spray solution; nasal inhalation	Decongestant
Naphazoline	Privine	Ciba	None	0.05% Spray solution; nasal inhalation	Decongestant
Oxymetazoline	Afrin	Schering	Phenylmercuric acetate	0.05% Spray solution; nasal inhalation	Decongestant
Tetrahydrazoline	Tyzine	Pfizer	None	0.1% Spray solution; nasal inhalation	Decongestant
Tetrahydrazoline	Visine	Lederle	Sodium borate	0.05% Solution; eyedrops	Relieve eye irritation
Xylometazoline	Otrivin	Geigy	None	0.1% Spray solution; nasal inhalation	Decongestant

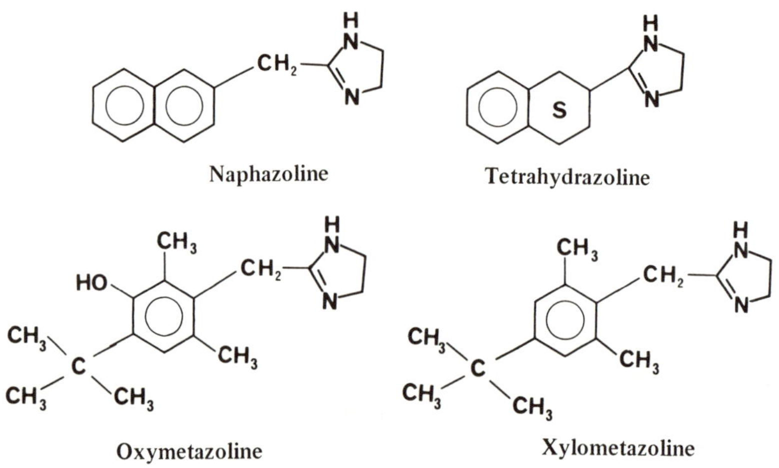

FIGURE 15. Imidazoline compounds used as nasal decongestants.

Empirical formula:	$C_{14}H_{14}N_2 \cdot HCl$
Mol wt:	246.74 (HCl salt)
Melting point:	253 to 258°C (HCl salt)
Chemical name:	2-(1-naphthylmethyl)-2-imidazoline monohydrochloride
Proprietary name:	naphazoline
Trade names:	Privine, for others see Table 7
Solubility:	freely soluble in water and alcohol (HCl salt), slightly soluble in chloroform, insoluble in ether
Physical description:	white crystalline power, bitter taste
Salt forms:	nitrate, hydrochloride
pKa:	10.14 (free base)
Stability:	store in airtight, light proof containers

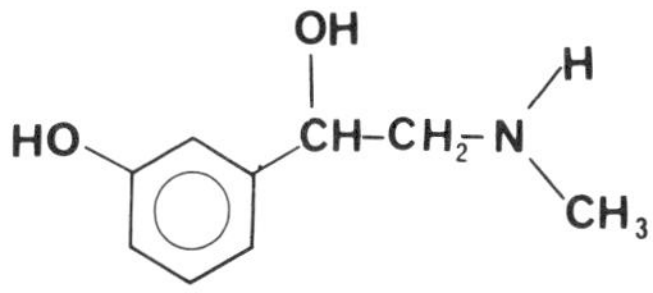

Phenylephrine

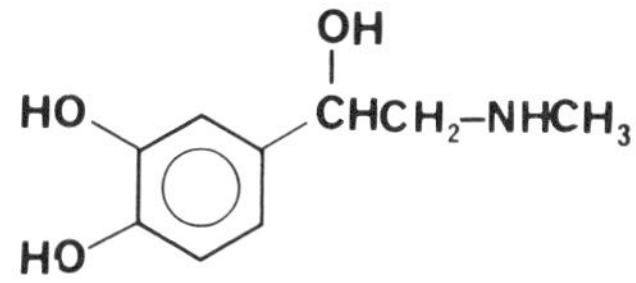

Epinephrine

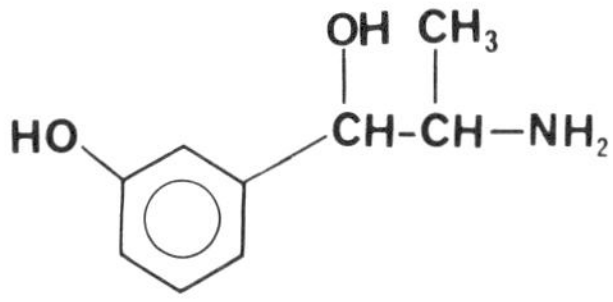

Metaraminol

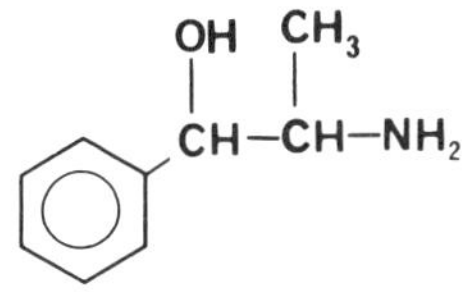

Phenylpropanolamine

FIGURE 16. Ethanolamine compounds used as nasal decongestants.

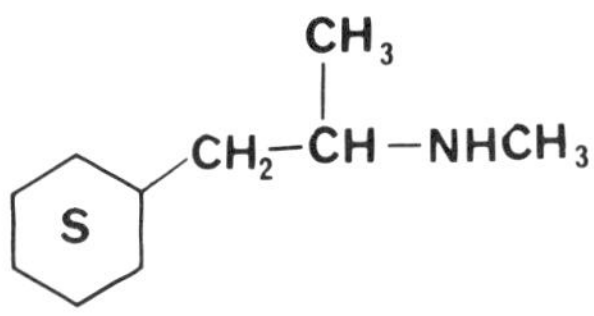

Propylhexadrine

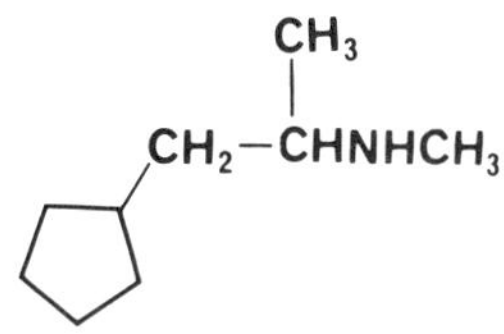

Cyclopentamine

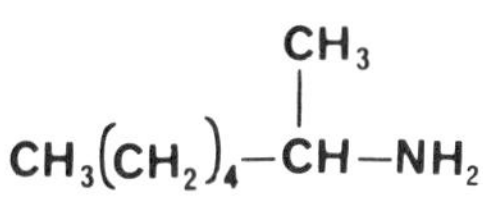

Tuaminoheptane

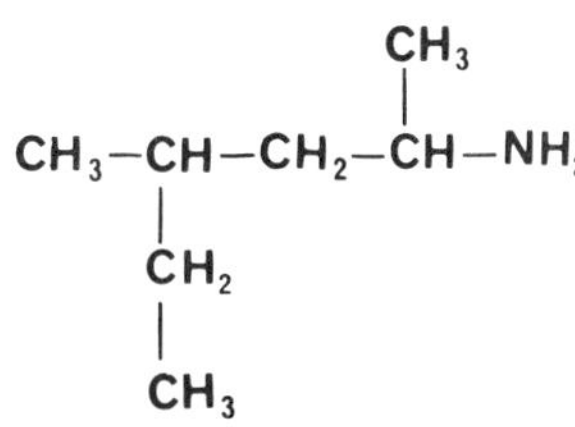

Methylhexaneamine

FIGURE 17. Aliphatic compounds used as nasal decongestants.

OTHER COMPOUNDS OF INTEREST

In addition to naphazoline, several other imidazolines are used as vasoconstricting nasal decongestants. These are oxymetazoline, tetrahydrazoline, and xylometazoline (Figure 15) (Table 7).

In the ethanolamine class of α-agonists there are several compounds other than phenylephrine which have been used as decongestants. These are epinephrine, metaraminol, and phenylpropanolamine (Figure 16). These compounds, however, are generally used for the treatment of other conditions (see appropriate monographs).

A number of ring-saturated and aliphatic derivatives have also been found to be vasoconstrictors on topical administration, and thus are used as decongestants. These compounds are propylhexadrine, tuaminohexidrine, cyclopentamine, and methylhexaneamine (Figure 17). Because they lack the structural characteristics of adrenergic agonists, it is believed that they derive their pharmacological activity through effecting the release of norepinephrine from its storage sites in the neuron. The released norepinephrine then mediates vasoconstriction and decongestion.

Agents Used in Cardiac Failure, Shock, Hypotension, and Peripheral Vascular Disease

Physiology of Cardiac Failure and Shock

Cardiac Failure

The term "cardiac failure" refers to a decreased efficiency of pumping by the heart. Such a condition usually arises from actual damage to the heart by some disease or mechanical process like coronary artery obstruction due to arteriosclerosis.

It is possible for only one half of the heart to fail. If the left half of the heart fails (cardiogenic shock) while the right half continues pumping normally, blood will accumulate in the pulmonary vessels. If the pulmonary pressure increases sufficiently, fluid will leak into the lung tissues, causing pulmonary edema.

If the right side of the heart fails, blood attempting to return to the heart will be dammed in the peripheral veins, causing an increase in venous pressure. As a result of this pressure, veins in the neck may become distended, and venous reservoirs such as the liver and spleen may become engorged with blood. Generalized systemic edema may also result due to high capillary pressure thoughout the circulatory system.

Complicating factors, such as decreased urine output due to decreased cardiac output and increased salt retention, cause an increase in body water and blood volume and increase the work load on the heart.

Shock

Circulatory shock results when cardiac output becomes so low that tissues almost everywhere in the body fail to receive adequate blood perfusion. There are two classes of circulatory shock: cardiac shock and low venous return shock.

Cardiac shock is caused by an extreme decrease in the effectiveness of the heart to pump blood as might occur during a heart attack. In this sense, it is an extreme case of cardiac failure.

Low venous return shock can be caused by excessive loss of blood volume (absolute hypovolemia) or by a reduction of peripheral resistance (relative hypovolemia).

Absolute hypovolemic shock occurs as a result of excessive hemorrhage and loss of blood plasma (hemorrhagic or traumatic shock) or excessive loss of water or body fluids through dehydration.

Relative hypovolemic shock may be due to several causes. Shock caused by sudden cessation of sympathetic impulses from the CNS results in general loss of vasomotor tone and diminished venous return, causing hypovolemia. Fainting is an example of this kind of shock. Allergic reactions can also cause hypovolemia and venous pooling shock. The release of histamine during such an anaphylactic, endotoxic, or septic shock results in peripheral vasodilation and decreased venous return. This type of hypovolemia can occur so rapidly and to such an extent that a person can die before therapy is started.

Shock progesses though several stages. In the early stages, the body is able to compensate for decreased venous return by peripheral vasoconstriction. This compensatory stage can occur as long as no more than 15 to 20% of the total blood volume is lost. Beyond this stage, one goes into progressive shock whereby the very factors that result from shock in turn continue to promote the shock. Such a condition occurs when myocardial oxygen perfusion falls due to lack of cardiac output. Because of the decrease in O_2 to the tissue, the heart becomes weaker, thus causing a greater fall in cardiac output, which continues and accelerates the problem.

Early stages of progressive shock can be treated by increasing the efficiency of pumping the heart. However, in the later stages of progressive shock the shock becomes irreversible. In such a situation, so much damage has been done to the brain, myocardium, and/or peripheral circulation, that no amount of therapy is effective.

MONOGRAPHS

EPINEPHRINE

Mechanism of Action

Epinephrine is a direct-acting adrenergic agonist having activity on both α- and β-receptors. Epinephrine has physiological activity on the myocardium, the peripheral vasculature, and the bronchial tree.

Systemically, epinephrine acts on the β_1-adrenergic receptors of the myocardium to cause a positive chronotropic and inotropic response. Cardiac output, oxygen consumption, and work are increased but at the expense of cardiac efficiency.

At low doses, epinephrine acts on the β_2-adrenergic receptors of the vasculature to cause vasodilation. At higher doses, epinephrine acts at the α_1-receptors of the vasculature to cause vasoconstriction. This vasoconstriction leads to increased blood pressure which may involve a compensatory reflex bradycardia which would counteract the direct stimulant action of epinephrine on the myocardial β_1-receptors.

Epinephrine has no direct effect on cerebral arterioles or cerebral blood flow. Increased blood pressure due to epinephrine administration may, however, cause increased cerebral blood flow and oxygen consumption.

Constriction of renal blood vessels leads to increased renal vascular resistance and a simultaneous decrease in urine flow with a subsequent decrease in sodium, potassium, and chloride excretion.

Metabolically, epinephrine causes increased glycogenolysis in the liver, reduces glucose uptake by the tissues, inhibits insulin release, increases lipolysis, and increases oxygen consumption by about 20 to 30%.

Epinephrine can be used in certain cases of cardiac arrest. Generally, when cardiac arrest occurs external cardiac compression or a mechanical defibrillator is first employed. If these techniques fail, then intracardiac injection of epinephrine is the next choice in reestablishing a heart beat, followed by thoroactomy and manual cardiac massage. Once some circulation has been restored, i.v. epinephrine may be used to restore intrinsic cardiac rhythm.

Epinephrine should not, however, be used in the treatment of acute attacks of left heart failure (see section on cardiogenic shock). This procedure would increase the demand by the heart for oxygen during a time when the heart is already partially anoxic due to the failure of the left heart function. Use of epinephrine in cardiac arrest following anesthetic accidents should also be avoided since general anesthetics predispose the myocardium to fibrillation upon injection of epinephrine.

Epinephrine is the drug of choice in treating anaphylactic shock. In addition to antagonizing the vasodilatory effects of released histamine on the peripheral vasculature, it also promotes bronchodilation via the β-adrenergic receptors of the bronchioles, allowing the patient to breath easier.

Epinephrine has also been used for the treatment of shock associated with trauma and hemorrhage on the assumption that the vasoconstrictive action on the peripheral vasculature will maintain a high blood pressure for perfusion of vital organs. The value of such therapy is not clearly established since the vasoconstriction caused by epinephrine may also impair circulation to vital organs. Fluid replacement is the therapy of choice in maintaining adequate blood volume and venous return in cases of traumatic or hemorrhagic shock.

Structure-Activity Relationships

The history and structure-activity relationships of epinephrine are discussed in the introductory section to this chapter.

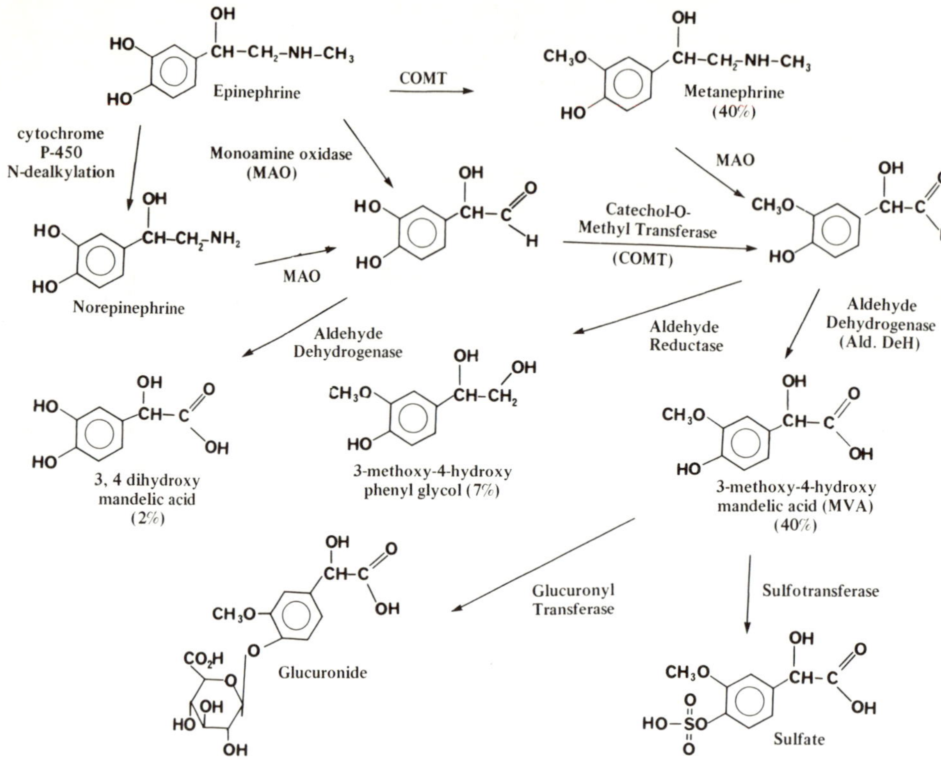

FIGURE 18. Metabolism of epinephrine and norepinephrine.

Pharmacokinetics

Epinephrine is well absorbed from numerous administration sites, including i.m. and s.c. injection, oral inhalation, and parenteral injection. Orally administered epinephrine, however, is rapidly metabolized in the GI tract and liver such that physiologically active blood levels cannot be achieved by this route of administration.

Epinephrine will not cross the blood-brain barrier, but will cross the placenta and can be excreted in breast milk.

Epinephrine has a rapid onset and short duration of action. Its half-life on inhalation is less than 1 min. The plasma half-life is about 2.5 min. Maximal activity upon s.c. injection occurs in about 10 to 20 min.

Epinephrine is metabolized at its site of action at the neuronal synapse, as well as by the liver and by circulating plasma enzymes. Synaptic metabolism of catecholamines has been discussed elsewhere. In addition to the synapse, the important metabolic enzymes monoamine oxidase and catechol-*O*-methyl transferase can also be found in the plasma and liver. The enzymes aldehyde dehydrogenase, aldehyde reductase, glucuronyl transferase, and sulfotransferase can be found in the liver and also are involved in catecholamine metabolism. Sulfotransferase is also found in the kidney. As a result of the action of these enzymes, parenteral administration of epinephrine results in its metabolism and excretion as the compounds shown in Figure 18.

Uses and Doses

The rational use of epinephrine in the treatment of cardiac failure and certain types of shock has been discussed in the mechanism of action section of this monograph. Epinephrine is also used in the treatment of bronchial asthma due to its bronchodilatory action through β-adrenergic stimulation.

The vasoconstrictive property of epinephrine is utilized by adding it to solutions of local anesthetics in order to decrease the rate of vascular absorption of the anesthetic. This allows better localization of the action of the anesthetic, prolonging its duration of action, and minimizing any possible systemic side effects of the anesthetic.

The vasoconstrictive properties of epinephrine can also be used to control superficial bleeding and have been used in this way to help keep surgical fields clear.

Epinephrine has been used for the control of hypotension during spinal anesthesia or after symphathectomy or ganglionic blockade. Hypotension in these cases occurs because of a lack of reflex vasoconstriction due to neuronal blockade. Epinephrine maintains vasoconstriction by its direct α-adrenergic agonist action, preventing hypotension.[48]

Dosage, routes of administration, and uses of epinephrine are shown in Table 8.

Toxicity

Systemic and CNS side effects of epinephrine include fear, anxiety, tenseness, restlessness, headache, tremor, dizziness, nervousness, sleeplessness, excitability, weakness, nausea, vomiting, sweating, pallor, and respiratory difficulties. Tolerance may develop when epinephrine is used in asthma therapy for prolonged periods of time. Rebound bronchospasm may occur when therapy is terminated.[48]

The minimum lethal dose given s.c. is 4 mg/g for adults. The lethal dose in rats is 30 mg/g orally.[49]

Epinephrine has a number of toxic interactions which must be considered when it is used in combination therapy.

Monoamine oxidase (MAO) inhibiting antidepressants should not be given with epinephrine. The potentiation of the effects of epinephrine, due to inhibition of metabolism by MAO, can lead to severe hypertensive crises. Tricyclic antidepressants should not be given with epinephrine due to their ability to inhibit norepinephrine uptake into synaptic neurons. The additive effect of increased synaptic norepinephrine levels along with administered epinephrine may result in cardiovascular crisis.[47]

The ability of epinephrine to stimulate glycogenolysis also contraindicates use with oral hypoglycemic and antidiabetic agents.

The use of epinephrine in the treatment of cardiac arrest has already been described. If cardiac arrest occurs during administration of general anesthetics, such as cyclopropane or the halogenated hydrocarbons, the use of epinephrine can be extremely hazardous due to the ability of these anesthetics to sensitize the myocardium to adrenergic stimulation.

Physical-Chemical Date

Structure:	Figure 1
Empirical formula:	$C_9H_{13}NO_3$
Mol wt:	183.21
Melting point:	211 to 212°C
Chemical name:	(−), 1-(3,4-dihydroxyphenyl)-2-methyl-aminoethanol
Generic name:	epinephrine
Proprietary names:	adrenalin, suprarenalin, nephridine, others (see Table 8)
Solubilities:	slightly soluble in water and alcohol, insoluble in ether and chloroform
Physical description:	white, crystalline, odorless powder
Salt forms:	hydrochloride, bitartrate
pK_a:	9.9 (acid), 8.5 (base)
Specific rotation:	−50°C (in 0.5 *N* HCl)
Stability:	will oxidize on exposure to air or light

Table 8
PHARMACEUTICAL PREPARATIONS OF MONOGRAPHED ADRENERGIC AGENTS USED IN CARDIAC FAILURE, SHOCK, HYPOTENSION, AND PERIPHERAL VASCULAR DISEASE

Generic name	Proprietary name	Manufacturer	Use	Dosage form, dose, and route of administration
Epinephrine	Adrenaline Hydrochloride	Parke-Davis	Cardiac stimulant and bronchodilator	1 mg/mℓ Solution, administer 0.2 to 1 mℓ i.m. or s.c. for brochospasm, i.v. for cardiac resuscitation
	AsthmaMeter	Rexall	Bronchodilator for treatment of allergic shock	Aerosol nebulizer, 0.2 mg delivered/dose administer 0.2 to 0.4 mg
	SusPhrine	Cooper	Bronchodilator for treatment of allergic shock	5 mg/mℓ Solution, 0.1 to 0.3 mℓ
Norepinephrine	Levophed Bitartrate	Winthrop	Antihypotensive peripheral vasoconstrictor	1 mg/mℓ Solution (weight based on free amine), 4 mℓ in 1000 mℓ; 5% dextrose administered by i.v. drip
Metaraminol	Aramine	Merck	Antihypotensive, peripheral vasoconstrictor for use in shock	10 mg/mℓ Solution, 2 to 10 mg i.m. or s.c.; 500 mℓ sodium chloride, i.v. infusion; 0.5 to 5 mg i.v. injection
Nylidrin	Arlidin	USV Pharmaceutical	Vasodilator for treatment of peripheral vascular disease	6- and 12-mg Tablets 3 to 12 mg 3 to 4 times a day
Isoxuprine	Vasodilan	Mead-Johnson	Vasodilator for treatment of peripheral vascular disease	10- and 20-mg Tablets 1 tablet 3 to 4 times a day; 5 mg/mℓ solution 1 to 2 mℓ 2 to 3 times daily

NOREPINEPHRINE

Mechanism of Action

Norepinephrine is a direct-acting adrenergic agonist which acts predominantly on α- and β_1-receptors. It has little or no effect on the receptors of the bronchioles or peripheral blood vessels.

The agonist action of norepinephrine on α-receptors causes peripheral vasoconstriction leading to increased diastolic and systolic blood pressure, decreased blood flow to vital organs, and reduced renal blood flow.

The action of norepinephrine on the β_1-receptors of the myocardium produces an increase in positive inotropic and chronotropic effects. The chronotropic response and increased cardiac output generally are not observed because of the reflex bradycardia which occurs as a result of increased blood pressure due to peripheral vasoconstriction by norepinephrine. Norepinephrine increases myocardial oxygen consumption, and work, and decreases cardiac efficiency.

In general, norepinephrine is less effective than epinephrine in stimulating the myocardium, stimulating metabolism, or precipitating CNS side effects.

Structure-Activity Relationships

The history and structure-activity relationships of norepinephrine and other direct-acting adrenergic agonists were discussed in earlier sections of this review.

Pharmacokinetics

Unlike epinephrine, norepinephrine is poorly absorbed on s.c. injection, and is destroyed by oxidation and metabolism if taken orally. If administered i.v., the onset of action is very quick but it suffers from significant first-pass metabolism, making its duration of action also very short (1 to 2 min). The drug crosses the placenta, but not the blood-brain barrier.

Norepinephrine is metabolized by the same enzymes as epinephrine (Figure 18). It therefore produces essentially the same metabolites, which are inactive as adrenergic agonists.

Uses and Doses

Norepinephrine, like epinephrine, is used to produce vasoconstriction and cardiac stimulation in the treatment of shock due to loss of plasma volume. Like epinephrine, its value in the therapy is questionable, and the replacement of fluid volume is the treatment of choice.[48]

Norepinephrine can also be used for treatment of hypotension due to spinal anesthesia, but other compounds are more long-lasting.

Norepinephrine has been used for the treatment of shock due to myocardial infarction, but its tendency to increase myocardial oxygen consumption and work tends to offset its usefulness.

Like epinephrine, norepinephrine can be used to constrict vascular beds for the purpose of prolonging the effects of local anesthetics. However, norepinephrine is less potent than epinephrine.

Dosage, routes of administration, and uses for norepinephrine are shown in Table 8.

Toxicity

Norepinephrine has almost all the same toxicities as epinephrine; therefore, like epinephrine, it should be used cautiously in combination therapy with MAO-inhibiting and neuron-depleting antihypertensive agents. These drugs, along with uptake-inhibiting antidepressant agents, may all potentiate the pressor effects of norepinephrine.[47]

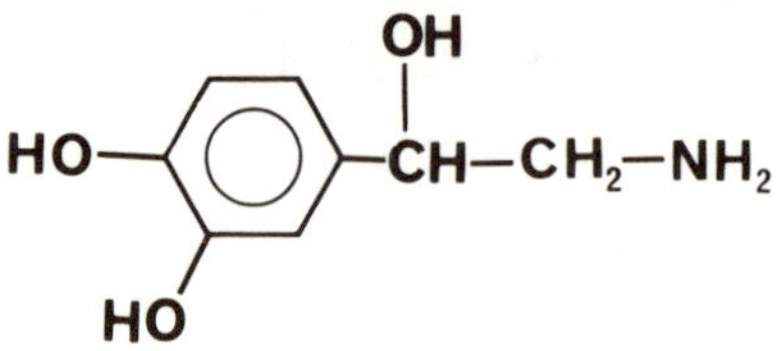

FIGURE 19. Norepinephrine.

While norepinephrine has some CNS side effects, such as headache, weakness, dizziness, and tremor, these side effects are generally less severe than those of epinephrine due to the fact that norepinephrine does not penetrate the CNS as well.

The estimated minimum lethal dose in children is 10 mg by i.m. or s.c. injection.[49]

Physical-Chemical Data

Structure:	Figure 19
Empirical formula:	$C_8H_{11}NO_3$
Mol wt:	169.2
Melting point:	217°C dec
Chemical name:	(−)-2-amino-1-(3,4-dihydroxyphenyl)ethanol
Generic name:	norepinephrine, 1-arterenol, levarterenol, noradrenaline
Proprietary names:	adrenor, aktamin, Levophed, noradrec
Solubility:	Slightly soluble in water, ethanol and ether
pK_a:	8.72 (acid) 9.72 (base)
Salt forms:	bitartrate, hydrochloride

METARAMINOL

Mechanism of Action

Metaraminol is a mixed-acting adrenergic agonist. It has a direct agonist effect predominantly on α-receptors without any significant direct β_1, action on the heart. It also has an indirect agonist effect due to its ability to displace norepinephrine from neuronal storage sites.

The major pharmacological responses to therapeutic doses of metaraminol are vasoconstriction and cardiac stimulation. Total peripheral resistance is increased, resulting in increased systolic and diastolic blood pressure. A corresponding decrease in blood flow to the vital organs and renal blood vessels is observed. On prolonged use, vasodilation and hypotension may occur.

The hypotensive effect appears to be due to the fact that metaraminol can also act as a false transmitter. The α-methyl group in the structure of metaraminol both increases its uptake as well as decreases its metabolism due to its ability to inhibit the catalytic action of MAOs. Thus, over prolonged periods of time, metaraminol will tend to accumulate in nerve terminal storage sites at the expense of the more readily metabolized norepinephrine. Upon depolarization of the nerve, metaraminol, along with norepinephrine, will be released. While metaraminol does have direct agonist activity, it is not as great as that of norepinephrine. Thus, the observed pharmacological response of adrenergic systems on prolonged metaraminol treatment would be one of decreased activity. When tested in rats, the relative

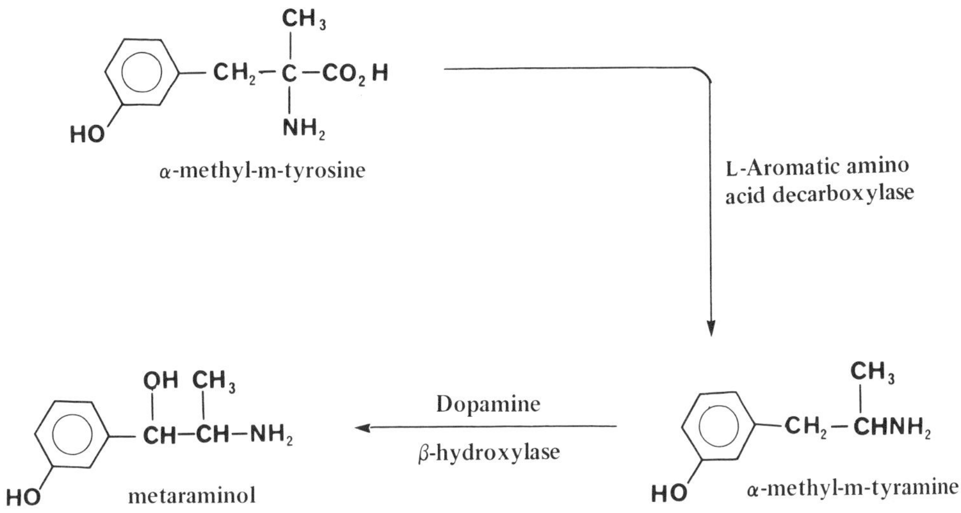

FIGURE 20. Biosynthesis of metaraminol from α-methyl-*m*-tyrosine.

pressor effects of norepinephrine vs. the false transmitters α-methylnorepinephrine and metaraminol, were 1 to 0.29 to 0.45, respectively.[51]

Interestingly, metaraminol can also be biosynthesized as a false transmitter upon administration of α-methyl-*meta*-tyrosine (Figure 20). This biosynthesis is analogous to that which generates the false transmitter α-methylnorepinephrine from α-methyl-*para*-tyrosine.

The observed gradual lowering of blood pressure due to norepinephrine depletion by metaraminol on prolonged therapy suggests its use as an antihypertensive agent. The drug cannot be used this way, however, due to the acute α-receptor-mediated vasoconstriction observed when the drug is first administered. Such a response could send a patient being treated by hypertension into a hypertensive crisis. Thus, the use of metaraminol in patients with high blood pressure is contraindicated. The total myocardial response to metaraminol is an unchanged or decreased cardiac output.

Oxygen demand by the heart increases as a result of an increased inotropic response to metaraminol. Although metaraminol enhances cardiac metabolism, the α-receptor-mediated vasoconstriction of coronary arteries tends to offset the beneficial effect of this increased metabolism. This results in an increased work load on the heart and decreased cardiac efficiency.

Among the metabolic effects mediated by metaraminol are increased glycogenolysis, inhibition of insulin release, increased lipolysis, increased oxygen consumption, and elevated body temperature.

Structure-Activity Relationships

In view of the numerous mechanisms of action utilized by metaraminol, there are numerous structural requirements necessary for its action. The direct α-adrenergic activity of the compound is dependent on the phenethanolamine parent structure. The *meta* phenolic rather than a *para* phenolic group defines the α-receptor specificity.

The fact that it is a phenol rather than a catechol, and the presence of the α-methyl substituent, give the compound increased lipid solubility relative to the prototype agonist norepinephrine. This allows for greater uptake into the granule and promotes the indirect agonist, norepinephrine displacing activity of the drug.

The methyl group α to the amine prevents metabolic destruction of the compound by MAO. This MAO inhibition, due to the methyl group, is necessary in order for the compound to be able to accumulate to sufficient levels such that it can function as a false transmitter.

The stereochemistry about the two asymmetric centers is important. Only the (−)erythro (1R, 2S) isomer of metaraminol shows appreciable agonist or norepinephrine-depleting activity.[52]

Pharmacokinetics

Metaraminol is absorbed well orally but it is not usually administered via this route due to the difficulty in controlling its vasopressor effects. When given by i.v. injection to raise blood pressure, the onset of action occurs within 1 to 2 min, peaks at 5 min, and has a duration of action of about 20 to 90 min. If given i.m. or s.c., the onset occurs within 5 to 10 min and peaks at 30 min.

Termination of the effects of metaraminol appears to be due largely to tissue uptake and urinary excretion rather than metabolism. Some metabolism can be locally important such as at effector organ synapses. As previously discussed, the methyl group α to the amine prevents the oxidation of metaraminol by MAO. Termination of the action of the drug at the synapse is accomplished primarily by uptake into the neuron and storage in the synaptic granule where it displaces norepinephrine.

Elimination occurs via the urine and such metabolites as are found are predominantly the *O*-sulfate and *O*-methyl conjugates.

Uses and Doses

Metaraminol is used primarily for inducing vasoconstriction in shock after appropriate fluid replacement. It has also been used for the treatment of hypotension in spinal anesthesia. It has been suggested that the longer duration of action of metaraminol, due to its decreased metabolism (see pharmacokinetics section of this monograph) is an advantage over norepinephrine in treatment of certain types of shock.

Dosage forms, dosage, and preparations of metaraminol are shown in Table 8.

Toxicity

Systemic effects of overadministration of metaraminol include hypertension, cerebral hemorrhage, pulmonary edema, cardiac arrest, and/or arrhythmias. Possible hypotension may occur on long-term therapy. This can be corrected by administration of norepinephrine to replace lost stores.[48]

Other signs of overdosage include pallor, palpitation, weakness, dizziness, tremor, and anxiety.

Metaraminol should not be used with MAO inhibitors. While metabolism of metaraminol itself is not affected by MAO inhibitors, the increase in neuronal norepinephrine levels as a result of decreased metabolism could potentiate the actions of metaraminol and precipitate a hypertensive crisis.

The norepinephrine-depleting effects of reserpine can make the action of metaraminol unpredictable, and it should therefore be avoided in combined drug therapy.

The cardiosensitizing property of cyclopropane and the halogenated hydrocarbon general anesthetics may predispose a patient to arrhythmias upon administration of metaraminol. Palpitations, hypertension, and/or arrhythmias may also occur if reflex bradycardia caused by metaraminol is blocked by atropine.

The toxic dose of metaraminol in children is approximately 60 mg i.m.[49]

Physical-Chemical Data

Structure:	Figure 21
Empirical formula:	$C_9H_{13}NO_2$
Mol wt:	167.2
Melting point:	174 to 178°C (tartrate)

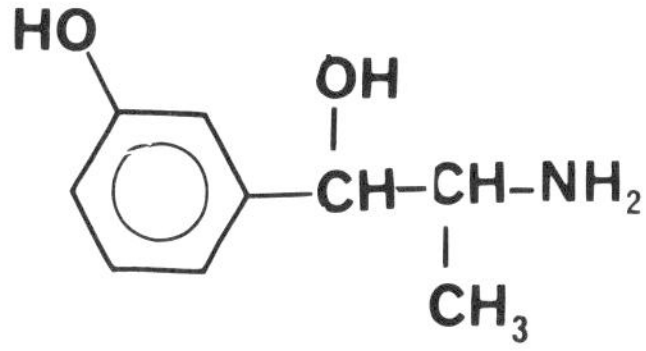

FIGURE 21. Metaraminol.

Chemical name:	(−)-2-amino-1-(3-hydroxyphenyl)propan-1-ol
Generic name:	metaraminol
Proprietary names:	Aramine, Pressorol
Solubilities:	Soluble in water and ethanol; insoluble in ether and chloroform
Physical description:	white, odorless, crystalline powder
Salt forms:	bitartrate
pK_a:	8.6 (base) bitartrate salt in 5% solution has a pH of 3.2 to 3.5
Stability:	will oxidize; store in dark airtight containers

NYLIDRIN/ISOXUPRINE

Mechanism of Action

Nylidrin and isoxuprine are direct-acting β_2-selective, adrenergic agonists. They cause vasodilation in skeletal muscle and relaxation of smooth muscle. As a result of these effects, nylidrin and isoxuprine are used for treatment of peripheral vascular disease and for management of premature or irregular labor (see dosage and use section of this monograph).

The metabolic effects of nylidrin and isoxuprine include increased lipolysis and increased glycogenolysis. Isoxuprine has also been reported by Falkay et al.[53] to induce the biosynthesis of PGE_2 and PGF_{2a}.

This observation is significant in that prostaglandins are known to stimulate adenylate cyclase and thus cause increases in intracellular cAMP concentration. Increases in cAMP also occur as a result of β-adrenergic receptor stimulation. It has been observed that the vasodilatory action of isoxuprine is not totally dependent on its β-agonist activity since vasodilation can occur with isoxuprine even in the presence of β-adrenergic blocking compounds. It may be that the vasodilatory action of isoxuprine is mediated by cAMP and that the cAMP levels in turn are controlled by the β-adrenergic receptor and/or prostaglandins formed as a result of isoxuprine.

The smooth muscle relaxation effect of isoxuprine is difficult to interpret since increased cAMP levels in the myometrium cause contraction rather than relaxation. The rebound contraction occasionally observed with obstetric use of isoxuprine, however, might be explained by the increased levels of cAMP mediated by the compound.

Structure-Activity Relationships

Nylidrin was developed as a result of the synthetic investigations of Kulz and Schneider[54] in 1950 on derivatives of epinephrine and ephedrine. It had long been recognized that pressor effects of epinephrine were only apparent at high concentrations of the compound. At low concentrations, epinephrine actually displayed a vasodilative action. Norepinephrine, on the other hand, only manifests vasoconstriction and reduced cardiac output, regardless of concentration. Since the only difference in structure between these two compounds is the *N*-methyl group, Kulz and Schneider undertook a study of the effect of *N*-substitution on the

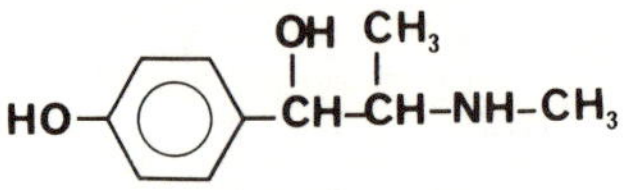

Suprifen
(p-hydroxy ephedrine)

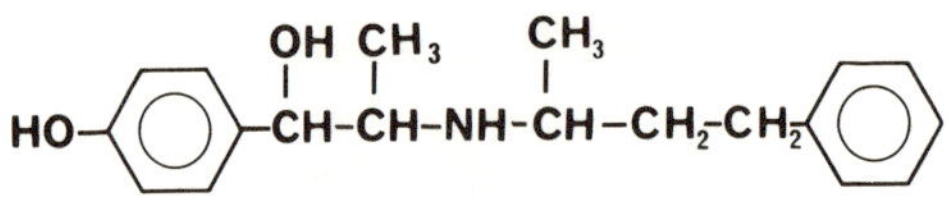

Suprifen-Psb
(Nylidrin)

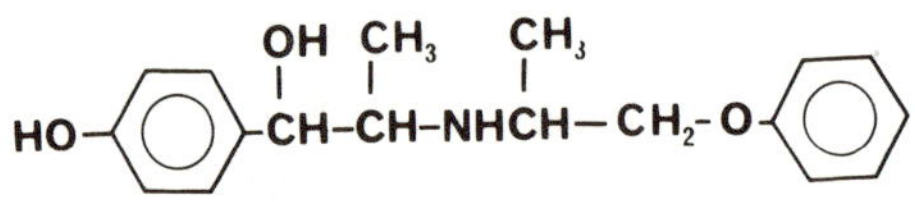

Isoxuprine

FIGURE 22. *N*-Substituted *para* hydroxy ephedrines.

vasodilative properties of epinephrine- and ephedrine-type compounds. These investigators selected suprifen (*p*-hydroxy ephedrine) (Figure 22) as their lead compound since it displayed the best vasodilative properties of the epinephrine and ephedrine compounds studied. By modifying the nitrogen substitution on suprifen, Kulz and Schneider showed that the vasodilatory effects of the compound could be potentiated by large alkyl groups of two or more carbon atoms. The presence of a benzene ring two to three atoms from the nitrogen further improved activity. The best derivative obtained was nylidin (Figure 22) (termed suprifen-Psb by the authors).

Wiemers[55] later confirmed the results of Kulz and Schneider and further showed that extending the arylalkyl side chain beyond three carbons decreased activity. Other adverse modifications included ring closure of the nitrogen side chain, oxidation or etherification of the benzylic hydroxyl, and ring position modification or substitution of the phenolic hydroxyl.

The application of the principle of isosteric replacement led to the development of isoxuprine. Isoxuprine differs from nylidrin in that an oxygen is substituted for a carbon in the nitrogen side chain (Figure 22).

Subsequent studies have shown that selectivity for the peripheral β_2-receptors is promoted by the presence of an α-methyl substituent. Futhermore, this methyl group inhibits oxidation and metabolism of the compound by the enzyme MAO. The fact that no *meta* hydroxyl group is present for methylation by catechol-*O*-methyl transferase (COMT), also causes a decrease in metabolism and a corresponding increase in duration of action. Enhanced oral absorption can also be attributed to the α-methyl group.

The optimal stereochemistry about the two asymmetric centers is believed to be 1R, 2S. This is similar to that observed for ephedrine and metaraminol.

As mentioned in the Mechanism of Action section, nylidrin also demonstrates smooth muscle relaxant properties, which are useful in the management of premature and irregular labor. Investigations on the structural requirements for this activity have been performed.[56]

The circulatory effects of catecholamines are more dependent on the *meta, para* dihydroxy aromatic substitution than are the smooth muscle relaxant effects. Satisfactory uterine-relaxing effects can be obtained with a single phenolic hydroxyl as occurs in nylidrin and isoxuprine. Loss of this hydroxyl or conversion to an ether causes loss of activity. Similarly, oxidation or etherification on the benzylic alcohol causes loss of activity.

Pharmacokinetics

Both nylidrin and isoxuprine are rapidly and almost completely absorbed from the GI tract. Onset of action occurs in 10 to 30 min. Peak plasma levels occur 1 to 3 hr after oral or i.m. administration. The plasma half-life is about $1^1/_2$ hr.

As discussed in the structure-activity relationships section, the α-methyl group in each structure inhibits metabolism by MAO. The lack of a *meta* hydroxyl group also prevents methylation by catechol-*O*-methyl transferase. As a result, these compounds are excreted either unchanged or as conjugates. The conjugates are most likely the sulfates of the phenolic hydroxyl. The primary route of excretion is via the urine with negligible fecal excretion.

Uses and Doses

Nylidrin and isoxuprine are used in the treatment of peripheral vascular diseases such as arteriosclerosis, thrombophlebitis, and diabetic vascular disease.

Both compounds have also been used in obstetric medicine for treatment of irregular uterine activity during labor, uterine hypertonus during the administration of oxytocin or prostaglandin, emergencies in delivery, operations during pregnancy, and long-term treatment for threatened premature birth and spontaneous abortion.[48]

See Table 8 for dosing and routes of administration.

Toxicity

The most important toxicities for these compounds derive from their systemic action. Such toxicity includes tachycardia, hypotension, and palpitations. As a result of these effects, the i.v. administration of these compounds is not recommended.

CNS toxicities include nervousness trembling, nausea, vomiting, and dizziness. GI irritation can be reduced by administering the compounds with meals, milk, or antacids.

Use of these compounds is contraindicated for patients with serious cardiac and circulatory disease as well as in glaucoma patients. Isoxuprine and nylidrin should not be used with halothane or other cardiac-sensitizing agents due to the danger of arrhythmia.

The estimated lethal dose of nylidrin in adults is 2 g, and in children, 200 mg.[49]

Physical-Chemical Data

Nylidrin

Structure:	Figure 22
Empirical formula:	$C_{19}H_{25}NO_2$
Molecular wt:	299.4
Melting point:	223 to 226°C
Chemical name:	1-(*p*-hydroxyphenyl)-2-(1-methyl-3-phenylpropylamino)propan-1-ol
Generic name:	Nylidrin
Proprietary names:	Arlidin, Perdilatal
Solubilities (HCl salt):	soluble in water and ethanol, insoluble in ether, slightly soluble in chloroform
Physical description:	white crystalline power
Salt forms:	hydrochloride
pH:	4.5 to 6.5 for a 1% solution of HCl salt
Stability:	subject to oxidation on prolonged exposure to light or air

Isoxuprine

Structure:	Figure 22

Empirical formula:	$C_{18}H_{23}NO_3$
Mol wt:	301.4
Melting point:	201 to 208°C HCl
Chemical name:	1-(*p*-hydroxyphenyl)-2-(1-methyl-2-phenoxy-ethylamino)propanl-o-l
Generic name:	isoxuprine
Proprietary names:	dilaunse, duvadilan, Vasodilan
Solubilities:	sparingly soluble in water, soluble in ethanol
Physical description:	white crystalline power
Salt forms:	hydrochloride
pK_a:	9.8 acid, 8.0 base
Stability:	will oxidize on exposure to heat, light, and/or air

OTHER COMPOUNDS OF INTEREST

A number of new compounds which are selective β_1-adrenergic agonists have been developed in recent years which may have application as myocardial stimulants in heart failure.

As discussed in the structure-activity relationships of the β-adrenergics section, β_1-selective agonists tend to be aryloxyproypanolamines. Several compounds of this class are currently being investigated for their potential as myocardial stimulants. Prenalterol (Figure 23) is an orally effective cardiostimulant that preferentially increases positive inotropic action over chronotropic action. It is currently undergoing clinical evaluation.[57] Two rigid analogs of the aryloxypropanolamines in which the side chain is tied back to the aromatic ring, have been reported to be β-adrenergic agonists. The position of the aromatic hydroxyl group appears to be important for determining receptor selectivity. If the hydroxyl group is *para* to the ether linkage the compound is selective for the vascular β_2-receptors, whereas if the hydroxyl is *meta* to the ether the compound acts at both myocardial β_1- as well as vascular β_2-receptors (Figure 23).[57]

Dobutamine has been used clinically to increase cardiac output in patients with cardiac decompensation due to heart disease or mechanical obstruction. It is a direct β_1-selective myocardial stimulant. The structure is noteworthy in that it is a direct agonist, yet lacks the β-hydroxyl group characteristic of direct stimulants. A derivative of dobutamine which has this hydroxyl group, but lacks one phenolic group, has recently been prepared (Figure 23).[58] This compound, butopamine, was found to have a pharmacological profile similar to dobutamine, but is orally effective.

ANTIASTHMATICS

Pharmacology of Bronchial Airways

The degree of bronchoconstriction or dilation in healthy and asthmatic subjects is dependent on the degree of balance between cholinergic and adrenergic stimulation of airway smooth muscle. Histochemical staining has shown dense cholinergic innervation of this smooth muscle. Cholinergic stimulation of the smooth muscle causes constriction, and it has been found that under normal conditions, airways are tonically constricted. Cutting the vagus nerve or administering cholinergic blockers such as atropine will cause bronchodilation by antagonizing this tonic constriction.

Histochemical fluorescence has also been used to visualize the presence of adrenergic fibers in these airways. The fact that β-adrenergic blockers are capable of causing bronchonstriction suggests that there is sympathetic bronchodilatory activity present in normal tissue. The source of this sympathetic activity, however, is not clear.

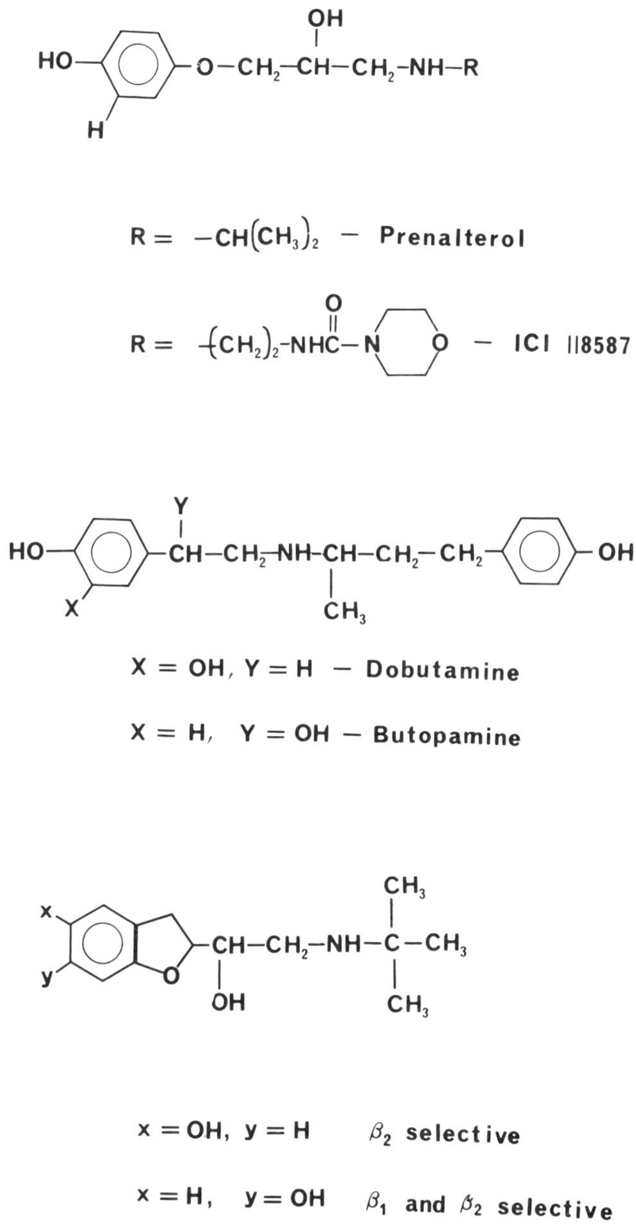

FIGURE 23. New cardiac stimulants.

Norepinephrine is the neurotransmitter mediating sympathetic stimulation of most organs. Release of this norepinephrine from nerve endings upon electrical stimulation of sympathetic nerves has been demonstrated in heart tissue. A corresponding release of norepinephrine upon electrical stimulation of thoracic symphathetic nerves, however, has not been demonstrated. Since catecholamines are not metabolized as rapidly as other neurotransmitters (such as acetylcholine), it has been suggested that sympathetic stimulation of airway smooth muscle might be mediated by catecholamines released from peripheral stores such as the adrenal medulla.

Aviado and Micozzi[43] have postulated that sympathetic nervous activation of the respiratory system may be mediated by epinephrine rather then norepinephrine. This suggestion is based on the fact that the pharmacological response profile of bronchial tissue to electrical stimulation more closely resembles that profile obtained by epinephrine administration rather than norepinephrine administration. Norepinephrine, in fact, fails to cause bronchodilation and instead causes bronchoconstriction. This bronchoconstriction is thought to be due to stimulation of α-adrenergic receptors since simultaneous administration of α-blockers along with norepinephrine prevents this bronchoconstriction. Thus, the role of norepinephrine as a neurotransmitter in bronchial tissues is questionable.

Bronchodilation appears to be a β-adrenergic receptor-mediated event. The significance and degree of α-receptor activity in bronchial smooth muscle is unclear. The presence of bronchial α-receptors has been invoked to explain the phenomenon of ''epinephrine reversal''. Repeated injections of large doses of epinephrine have been shown to cause bronchoconstriction rather than the bronchodilation normally observed with epinephrine. This phenomenon has been interpreted as being due to epinephrine stimulation of α-receptors with a simultaneous decrease in β-receptor responsiveness.

The cause of this decrease in β-receptor reponsiveness has not been determined, but appears to be rather complex in view of the fact that other β-agonists, such as isoproterenol, are still effective bronchodilators during epinephrine reversal. The role of α-receptors, if any, in the regulation of airway smooth muscle appears uncertain. It is known, however, that α-receptor activity is responsible for vasoconstriction of the blood vessels of the bronchial mucosa and for inhibiting secretions by bronchial glands. Thus, agents having α-adrenergic activity can be important in the treatment of disorders of the respiratory system.

The β-Adrenergic Theory of Asthma

Asthma is characterized by bronchoconstriction and wheezing accompanied by mucosal edema and mucous plugging. There appear to be a number of pharmacological mechanisms responsible for mediating these physiological effects. The two major mechanisms are immunological hypersensitivity and abnormal autonomic control.

The immunological basis of asthma has been extensively studied. In extrinsic asthma large amounts of tissue-sensitizing antibodies are present. These IgE-type antibodies are fixed primarily to mast cells in the lung. When an antibody-specific allergen binds to these cell-bound IgE antibodies, a perturbation of the mast cell membrane occurs which causes an influx of calcium into the mast cell and the release of histamine stored in granules within the cell. This released histamine then interacts with airway smooth muscle to cause bronchoconstriction. Histamine also enhances capillary permeability, contributing to mucosal edema. Other biological agents, such as SRS-A*, prostaglandins, and kinins, may also be involved in the physiological regulation of an asthmatic episode.

The treatment of asthma through the use of adrenergic agonists is of particular interest due to the recent development of the beta adrenergic theory of asthma. This theory, as advanced by Szentivanyi,[59] regards asthma, not as an immunological disease, but as a ''unique pattern of bronchial hyperactivity to a broad spectrum of immunological, psychological, infectious, chemical, and physical stimuli''. In this theory it is stated that all these diverse stimuli precipitate an asthmatic episode through a common biological process. This common process appears to involve the homeostatic balance between β- and α-adrenergic

* SRS-A has been shown to be a mixture consisting of leukotrienes which belong to a group of straight-chain fatty acids derived from arachidonic acid. Two are of particular interest, since they are potent smooth muscle vasoconstrictors and have been shown to be at least partly responsible for the bronchospasms of asthma. Recently,[58a] the structure of a SRS-A was elucidated and proved to be identical with one such leukotriene bronchoconstrictor. It is the product of the lipoxygenase metabolism of arachidonic acid and the glutathione pathway. Its chemical name is 5-hydroxy-6-cysteinylglycinyl-7,9,11,14-eicosatetraenoic acid. The first complete structure of an SRS-A had been previously reported by Hammarström et al.[58b] in 1979. The relationship between SRS-A and bronchospasm and the elucidation of the pathway of formation of SRS-A now make it possible for antagonists to be developed, which could mitigate or prevent asthmatic attacks by this novel mechanism.

response in the lung. Asthma, according to this theory, is therefore an abnormality in functioning of these two antagonistic systems.

The beta adrenergic theory evolved as an attempt to rationalize the observation that asthma, as opposed to other allergic conditions, is characterized not only by the release of the mediator substances such as histamine, SRS-A, prostaglandins, etc., but also by the existence of an enormously increased bronchial reactivity to these released mediators in the asthmatic.

In the bronchial tree it appears that the normal balance between α- and β-adrenergic response favors the β-receptors, which mediate bronchodilation. This bronchodilation serves to offset the tonic contraction of the smooth muscle due to vagal stimulation. β-Adrenergic blockade can result in bronchoconstriction, even without the provocation of mediator substances or *para*-sympathomimetics. If normal individuals are administered β-adrenergic blockers followed by any mediator substances, bronchial hypersensitivity to these mediators, similar to that seen in asthmatics is observed. It is noteworthy that this hypersensitivity occurs whether or not the vagus nerve is severed. The hyperreactivity of asthmatics to mediator substances may, therefore, be due to some physiological process which causes the asthmatic to have the biological equivalent of a continuous β-adrenergic blockade. It is believed that this biological β blockade is not the result of competitive β-receptor occupancy by some endogenous compound, but rather, the result of a decreased responsiveness of the receptor to adrenergic agonists and neurotransmitters.

In the discussion on the pharmacology of bronchial airways, it was noted that the phenomenon of epinephrine reversal or desensitization was mediated in part by a decrease in β-adrenergic receptor responsiveness to epinephrine. It is possible that this chemically induced decreased receptor responsiveness may serve as a model for the biologically decreased β-receptor responsiveness observed in asthmatics. Epinephrine fastness or desensitization cannot be accounted for by mechanical obstruction, since bronchodilator activity is still observed if other drugs such as theophylline are administered to the epinephrine fast patient. On the other hand, metabolic responses to β-adrenergic stimulation such as blood sugar levels, free fatty acids, and plasma cAMP levels are all depressed, suggesting an impairment of the β-receptor-adenylate cyclase system. Stimulation of adenylate cyclase by agents which by-pass the β-receptor show that the enzyme is still capable of normal response. It appears, therefore, that the decreased responsiveness of the β-adrenergic system to epinephrine stimulation is due to a functional impairment of the β-receptor protein itself. The nature of this functional impairment however is unclear. Several possible mechanisms might be involved.

One possible mechanism may involve an alteration in GTP regulation of receptor activity in a manner analogous to that described for desensitization (see previous section on desensitization for discussion). Alternatively, it is possible that the decreased sensitivity is due to a decrease in the absolute number of receptors available for binding of agonists. Perhaps one of the most intriguing explanations, however, may involve the recent discovery of autoantibodies to the β-receptor.

Venter and colleagues[60] have found antibodies to the β_2-adrenergic receptor in the serum of two patients with asthma. Binding of these antibodies to the β-receptors in the lung could cause inhibition of receptor-mediated relaxation of airway smooth muscle. Such inhibition of one of the mechanisms for antagonizing histamine-stimulated contraction of smooth muscle might explain the hyperreactivity of asthmatics to mediator substances.

It appears that all allergic responses result from the immunological release of smooth muscle-contracting substances. What distinguishes asthma from other allergic conditions is the hypersensitivity of bronchial smooth muscle to these released substances. This hypersensitivity may very well involve a functional alteration of the β-adrenergic receptors in the lungs of asthmatics. How this alteration occurs, and why it only occurs in asthmatics, are questions whose solutions may open the door to more effective agents for the treatment of asthma.

Asthma Therapy

Asthma can be treated in numerous ways employing agents that regulate either the mechanisms of the asthmatic response (mediator release inhibitors, antibody inhibitors) or the symptoms of the disease (bronchoconstriction, edema, congestion). The appropriate therapy is dependent of the severity of the asthmatic condition. It is therefore useful to be able to classify asthma by degree of severity and to apply appropriate therapeutic regimens for each classification. Such a classification has been suggested by Francis Lowell.[61]

Mild asthma — Mild asthma is characterized by brief episodes of wheezing which may be accompanied by coughing. Episodes are recurrent but at widely spaced intervals. Treatment may consist of oral sympathomimetics such as ephedrine or aerosol sympathomimetics such as epinephrine or isoproterenol. Theophylline may also be given orally to effect bronchodilation.

These agents are effective in the management of the symptoms of asthma such as bronchoconstriction and congestion. If the asthmatic episodes are persistent, then the administration of agents to prevent the onset of the asthmatic reaction may be warranted. Such therapy might include the prophylactic administration of cromolyn sodium.

Moderate asthma — Moderate asthma is characterized as wheezing, labored breathing, and congestion with or without cough and expectoration. The severity of the episode is such that it interferes with daily activities and possibly sleep. Therapy can include several different regimens. Bronchodilation can be achieved by oral administration of theophylline or the sympathomimetics ephedrine, epinephrine, or isoproterenol. The oral administration of expectorants such as potassium iodide may help to clear air passages of mucous. If the condition is sufficiently severe, then the prophylactic administration of cromolyn sodium, to prevent mediator release, or steroids to supress antibody formation, may be required.

Severe asthma — Severe asthma is characterized by extremely labored breathing, sufficient to incapacitate the patient. Cough and muscular pain in the chest and abdomen are frequent. Often the patient is unable to eat or sleep normally and suffers from exhaustion. In such situations quick and potent measures should be taken. Oral and aerosol administration of bronchodilators is often insufficient and should be replaced by s.c. or i.v. administration of epinephrine. Similarly, i.v. injection of aminophylline rather than oral theophylline is preferred. Expectoration through i.v. administration of sodium iodide may be alleviated by oxygen administered by nasal catheter. Oral steroids are also given in significant doses but must be tapered off as pulmonary function is restored.

The monographs in the following section will discuss only those agents whose mechanism of action involves the adrenergic system.

Adrenergic Agonists in Asthma Therapy

Epinephrine was first administered for the treatment of asthma in 1903. While originally given s.c., it was later formulated as an aerosol. The use of isoproterenol in asthma occurred in 1940. Its discovery not only provided a more effective agent for treatment of asthma, but also helped to substantiate Ahlquist's suggestion of the existence of the adrenergic receptor subtypes termed α and β.

Further refinement of structure-activity relationships in the β-adrenergic receptor system led to Land's proposal of the receptor subtypes β_1 and β_2. The search for selective agonists for the β_2-receptors in the lung led to the discovery of the selective agonists salbutanol and terbutaline, which have found wide use in asthma therapy.

MONOGRAPHS

ISOPROTERENOL

Mechanism of Action

Isoproterenol is a direct-acting nonselective β-adrenergic agonist. It has little or no effect on α-receptors. Through its action on β_2-adrenergic receptors, isoproterenol causes bronchial, GI, and uterine smooth muscle relaxation and also causes dilation of arteries and aterioles in skeletal muscle. The action of isoproterenol on β_1-adrenergic receptors includes an increase in chronotropic and intropic effects on the heart, increased cardiac output, increased myocardial oxygen consumption, and decreased cardiac efficiency.

As a result of both its β_1 and β_2 action, there is a net increase in systolic blood pressure, but a decrease in diastolic pressure, producing a mean blood pressure which is unchanged or slightly decreased. Metabolically, isoproterenol stimulates glycogenolysis and lipolysis and increases oxygen demand in most tissues,

The most clinically significant actions of isoproterenol are its ability to dilate bronchial smooth muscle and its ability to inhibit the release of bronchoconstricting mediator substances such as histamine and SRS-A. Both of these actions are useful in the treatment of asthma.

Structure-Activity Relationships

In the late 1930s, chemists at the firm of C. H. Boehringer Shon in Germany prepared a series of compounds which were structural modifications of epinephrine, phenylephrine, and ephedrine.[62]

It had previously been noted that enlarging the substituent on the amino group on ephedrine helped to differentiate the pressor effects of the compound from its smooth muscle relaxant effects. It was therefore hoped that similar structural modification of epinephrine or phenylephrine might help to decrease the pressor effects of the compounds while increasing their smooth muscle relaxant properties. Such a combination of properties would be of obvious benefit in the treatment of asthma. The compounds prepared by the Boehinger Shon chemists were evaluated for their bronchodilatory activity by pharmacologists Konezett and Rossler[63] at Vienna University in 1938. They found that the ethyl, *n*-propyl, and *n*-butyl derivatives of norepinephrine did indeed maintain bronchodilatory activity equal to that of epinephrine, but were totally devoid of the pressor effects of epinephrine. The isopropyl analog, isoproterenol, was ten times as potent as epinephrine in its bronchodilatory activity and also was devoid of pressor effects.

Due to the war, knowledge of the pharmacologial activity of the compound was limited until after 1945. Ahlquist's evaluation of isoproterenol as the most active, pure, β-adrenergic agonist discovered to date, and his use of the compound to verify his contention of the existence of separate α- and β-adrenergic receptors, brought the compound to the attention of scientists all over the world. As the prototype β-adrenergic agonist, a detailed discussion of its structure-activity relationships has already been presented in the appropriate section under β-adrenergic agonists.

Pharmacokinetics

Isoproterenol is readily absorbed and metabolized in the GI tract. Intravenous administration, however, is 100 times more potent than oral administration. This difference in activity appears to be due to significant metabolism of the compound when administered orally or by aerosol.

Most of the metabolism of the drug, upon oral administration, occurs in the intestinal wall and to a lesser extent in the liver. The major metabolite is the sulfate ester at the *para* position of the catechol.

Table 9
PHARMACEUTICAL PREPARATIONS OF SOME PHENETHANOLAMINES USED IN THE TREATMENT OF ASTHMA

Generic name	Proprietary name	Manufacturer	Other active components	Dosage, route of administration
Isoproteranol	Vapo-Iso	Fisons	—	0.5% Solution in nebulizer, 0.3 to 0.5 mℓ/dose
	Aerolone	Lilly	Cyclopentamine	0.25 g/100 mℓ Solution in nebulizer
	Duo-Medihaler	Riker	Phenylephrine	Nebulizer; 0.16 mg/dose
	Isuprel	Winthrop	—	1:5000 Solution, 0.01 to 0.02 mg i.v.
	Luf-Iso	Mallinckrodt	—	Nebulizer
	Norisodrine	Abbott	—	0.25% Solution in nebulizer
Isoetharine	Bronkometer Bronkosol	Breon	Phenylephrine	0.61% Solution in nebulizer
Metaproteranol	Alupent	Boehringer-Ingelheim	—	20 mg/Tablet, 1 tablet 3 to 4 times daily; 0.65 mg micronized powder in inhaler, 2 to 3 inhalations/dose
	Metaprel	Dorsey	—	0.65 mg Micronized powder/inhalation, 2 to 3 inhalations/dose
Terbutaline	Brethine	Geigy	—	5 mg/Tablet, 1 tablet every 6 hr
	Bricanyl	Astra	—	2.5 mg and 5 mg/tablet, 5 mg every 6 hr

Coadministration of salicylamide, which is also conjugated by sulfotransferases, has been shown to increase the bioavailability of orally administered isoproterenol.[64] This effect is due to the to the fact that there is a limited pool of sulfate available in the body for conjugation. If this pool is used up for salicylamide metabolism, then isoproterenol cannot be as readily conjugated to form its sulfate ester, and hence, will have a longer duration of action.

About 80% of an oral or aerosol administered dose of isoproterenol is eliminated via the urine within 48 hr of dosing. The remaining amount of compound is eliminated in the feces.

Intravenously administered isoproterenol undergoes significantly less metabolism than orally administered compound. This is probably due to the fact that it is not exposed to the aryl sulfotransferases found in the intestinal wall. Approximately 40% of an i.v.-administered dose is excreted unchanged in the urine. The major metabolite, accounting for about 30% of the i.v.-administered dose, is the 3-*O*-methyl, 4-sulfate ester.

After i.v. injection, plasma levels of isoproterenol were found to decline in a biphasic manner. The half-life of the rapid phase was about 5 min, while the half-life of the slow phase was $2^1/_2$ hr. After 10 min the compound was found to be concentrated in the kidney, adrenals, lungs, liver, and heart. Within 2 hr, however, drug concentrations declined in all tissues except for the liver and kidneys. Within 15 hr, approximately 75% of the i.v.-administered dose is eliminated in the urine.

Uses and Dosage

Isoproterenol is used primarily as a bronchodilator for the treatment of asthma and other causes of bronchospasm. It can be used in combination with other drugs such as phenylephrine (see Table 9).

Isoproterenol has also been used to produce cardiac stimulation and vasodilation in the treatment of shock. The use in shock is somewhat questionable since it increases oxygen demand and work by the myocardium.

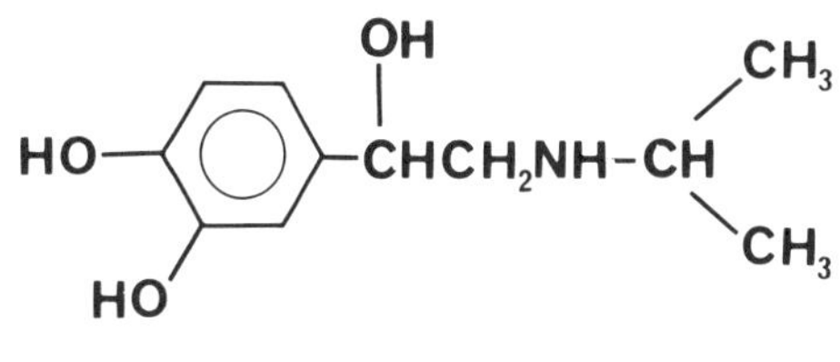

FIGURE 24. Isoproterenol.

Toxicity

In the 1960s it was noted that there was an increase in mortality rates for bronchial asthma sufferers in Australia, Great Britain, and Scandinavia.[66] The fatalities appeared to be limited to the excessive use of aerosol isoproterenol preparations. In some cases, isoproterenol actually caused bronchospasm rather than relaxation (locked-lung syndrome). The reasons this syndrome occurs is not known but several explanations have been advanced, including hyperreactivity of lung tissue to the aerosol propellents, bronchocongestion due to β-adrenergic-mediated vasodilation, and decreased numbers of β-receptors. Any or all of these factors might be involved. Caution should be observed in the use of aerosol isoproterenol in view of the potential for β-adrenergic agonists to cause desensitization (see previous discussion) and the tendency of asthmatics to increase the dosage of their medication in order to get relief.

Some centrally acting side effects of isoproterenol include nervousness, restlessness, insomnia, anxiety, tension, sweating, weakness, dizziness, tremor, and vomiting. Systemically, isoproterenol may cause disturbances of cardiac rhythm and rate, palpitations, and ventricular tachycardia.[48] Pulmonary edema, as a result of vasodilation and increased aortic blood pressure may occur.

The use of isoproterenol with β-adrenergic blockers such as propranolol is contraindicated since they compete for the same receptors. Isoproterenol should not be given with epinephrine since their effects on β-receptors could be additive leading to increased cardiotoxicity. Coadministration with the selective α-adrenergic agonist, phenylephrine, is permissible.

Antidepressant compounds such as imipramine tend to potentiate both the therapeutic as well as the toxic effects of isoproterenol. Similarly, MAO inhibitors may potentiate the toxic effects of isoproterenol. Use of isoproterenol with such agents should therefore be avoided.[47]

The ability of halogenated hydrocarbons to sensitize the myocardium may result in arrhythmias if coadministered with isoproterenol.

For children the minimum lethal dose is 100 mg when applied to the mucus membranes. In mice the LD_{50} for orally administered isoproterenol is 2 g/kg.[49]

Physical-Chemical Data

Structure:	Figure 24
Empirical formula:	$C_{11}H_{17}NO_3$
Mol wt:	211.3
Melting point:	155.5°C
Chemical name:	1-(3,4-dihydroxyphenyl)-2-isopropylamino ethanol
Generic name:	isoproterenol, isoprenaline
Proprietary names:	Aerotrol, aludrine, isuprel, asdrin, isupren, isovon, proternol
Solubilities:	1:3 water, 1:50 ethanol, insoluble in ether and chloroform
Salt forms:	hydrochloride and sulfate

Physical description:	both salt forms are white crystalline powers which will darken on exposure to air and light; the melting point of the HCl salt is 167 to 172°C and the melting point of the sulfate salt is 128°C with decomposition
pK_a:	8.57; the pH of a 1% solution of either salt form is about 5.0
Stability:	will decompose on exposure to air and light; solutions of the compound should be kept acidic to decrease rate of oxidation

EPHEDRINE/PSEUDOEPHEDRINE

Mechanisms of Action

Ephedrine and pseudoephedrine are mixed-acting agonists. As such they have direct stimulant activity on both α- and β-adrenergic receptors, as well as the ability to effect the release of norepinephrine from its storage sites in the neuron.

Ephedrine acts on the β_2-receptors of the bronchus to promote smooth muscle relaxation and bronchodilation. Bronchodilation is generally less strong but more long lasting than that afforded by administration of epinephrine.

α-Receptor-mediated constriction of blood vessels in the nasal mucosa promotes decongestion, but rebound congestion may occur.

Stimulation of the β_1-receptors of the heart causes an increase in the force of contraction and cardiac output. Heart rate may or may not be affected depending on vagal reflexes to peripheral resistance. Generally, ephedrine promotes β-receptor-mediated vasodilation, particularly in the skeletal muscle. α-Receptor vasoconstriction may occur in the skin arterioles, mucous membranes, viscera, and renal blood vessels.

Other actions of ephedrine include relaxation of GI and uterine smooth muscle, increased glycogenolysis in the liver, and CNS stimulation. These effects are less prominent with pseudoephedrine.

Structure-Activity Relationships

Ephedrine is the principal alkaloid of the plant species *ephedra*. The Chinese, who called the plant *Ma-huang*, used the plant for medicinal purposes for over 5000 years. Ephedrine was first isolated in pure form from the plant in 1887 by Nagai of Japan, and prepared synthetically in 1927.[67] Ephedrine was introduced into Western medicine after 1924 when Chen and Schmidt[68] reported on the sympathomimetic action of the compound.

As a mixed-acting adrenergic agonist, ephedrine has structure-activity relationships that incorporate aspects of both direct- and indirect-acting compounds. The functionalities contributing to the direct adrenergic activity of ephedrine include the phenethylamine nucleus, the β-hydroxyl group, and the primary amine. The lack of phenolic hydroxyl groups both increases the lipophilicity of the compound as well as contributes to the ability to act indirectly through displacement of norepinephrine from storage sites.

The introduction of an α-methyl group affects the mode of action of the compound. The methyl group generates a second asymmetric center in the molecule. Thus, four stereoisomers of the compound are possible (Figure 25). Ephedrine refers specifically to the erythro isomers, while pseudoephedrine refers to the threo isomers. Only one of the four isomers, (1R, 2S) ephedrine has a conformation which allows for a three-point receptor attachment necessary for optimum direct adrenergic activity as postulated by Easson and Stedman.[23] In the other isomers, at least one of the three binding functionalities is oriented away from the receptor surface, thus allowing only a two-point attachment (Figure 25). Consequently, a significant

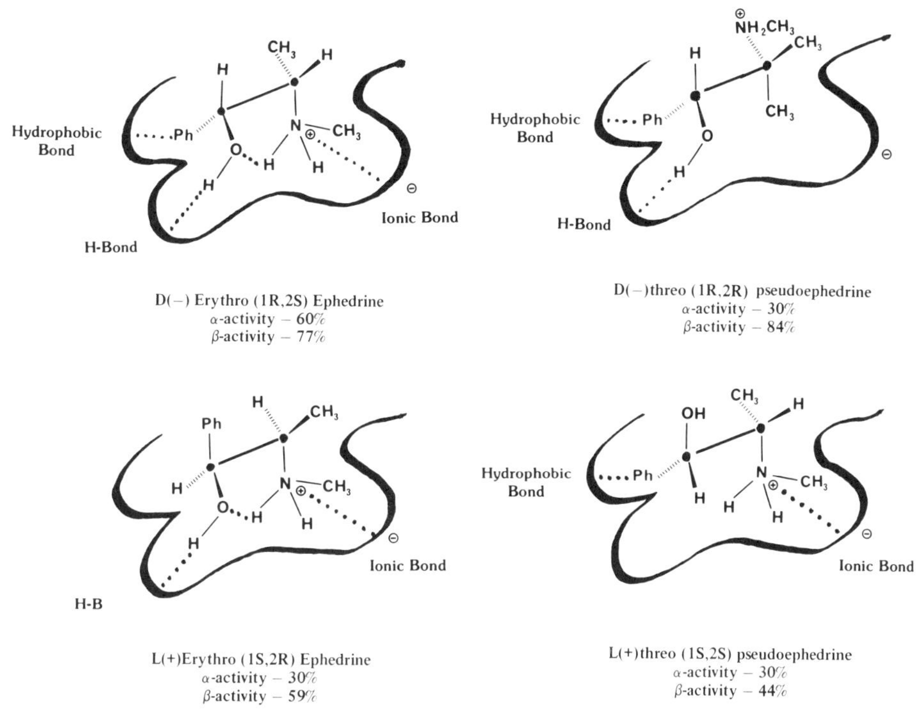

FIGURE 25. The four stereoisomers of ephedrine and their possible modes of interaction with adrenergic receptor sites.

amount of the α-adrenergic action of these three isomers is thought to arise from their indirect norepinephrine-releasing activity.

The stereochemistry about the β-carbon appears to be important for the receptor selectivity of the isomers.[69] In the 1R configuration, both the erythro and threo isomers have appreciable β-receptor agonist activity while α-receptor activity in the threo configuration is only about a third that of the erythro isomer. If the β-carbon is in the 1S configuration, the β-receptor activity is considerably less than in the 1R configuration regardless of whether or not the isomer is erythro or threo.

The reason for this apparent receptor selectivity is unclear. It may be due to the fact that the β-receptor is more "forgiving" of a two-point attachment provided that the two points are an OH and the aromatic ring. Alternatively, it may be that the β-receptor is more tolerant of steric interactions between the side chains of the molecule in each conformation than is the α-receptor. Finally, it may be that the α-methyl group affects the solution conformation of the 1R isomer population of the molecules such that more molecules are in conformations acceptable to the β-receptor than the α-receptor. It remains to be determined which of these, if any, are responsible for the greater β-receptor selectivity of the 1R isomers.

Finally, the α-methyl group also appears to help increase the lipophilicity of the compound while decreasing its metabolism by MAO. As a result of this, ephedrine and pseudoephedrine have appreciable more indirect- and CNS-stimulating properties than their non-methyl analogs.

Pharmacokinetics

Ephedrine was the first orally active sympathomimetic discovered. As such, it had tre-memdous value in the symptomatic treatment of certain physiological diseases such as asthma. The oral efficacy of ephedrine and pseudoephedrine arises from their lipophilicity. As discussed in the structure-activity relationships section, the lack of catechol hydroxyls and the presence of an α-methyl group contribute to this lipophilicity.

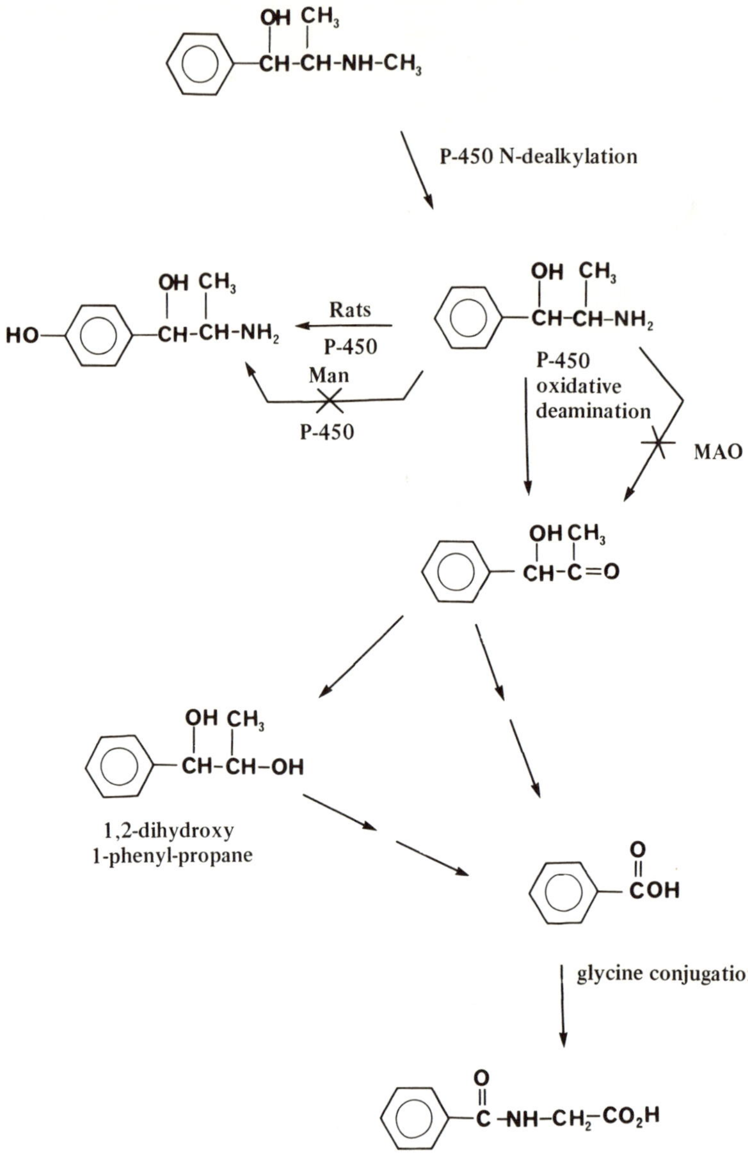

FIGURE 26. The major metabolites of ephedrine.

After oral ingestion of 60 mg of ephedrine or pseudoephedrine, physiological effects can be observed within 15 to 30 min. These effects may persist from 2 to 8 hr. Maximum plasma concentrations of ephedrine are about 200 mg/mℓ, occurring within 2 to 3 hr. The organs having the greatest uptake of the drug are the lungs, adrenals, spleen, and to a lesser extent, the heart.[70] Elimination from these organs and plasma is rapid, with a half-life of 1 hr. Ephedrine and pseudoephedrine may also cross the placenta and can be found in the milk.

The presence of the α-methyl group limits the metabolic pathways for ephedrine isomers, and thus they are excreted largely unchanged in the urine. In man, approximately 75% of the dose of the ephedrines is excreted unchanged. The major metabolites of ephedrine are the *N*-demethylated compound (5 to 20%) and the oxidatively deaminated ketone (Figure 26). Other in vivo metabolites include a dihydroxy compound from reduction of the ketone, and a glycine-conjugated benzoic acid derivative.

Urinary excretion of ephedrine and metabolites is very pH dependent.[71] Urinary recovery of unchanged ephedrines is much greater in acidic than in alkaline urine due to the lack of tubular reabsorption of the unchanged drugs. Under these conditions, the elimination half-life of unchanged ephedrine is 3 hr, with about 75 to 99% of an orally or i.v.-administered dose excreted within 24 hr. Conversely, urine alkalinized to pH 8 with sodium bicarbonate only eliminates 22 to 35% of an oral dose.

Uses and Dosage

Ephedrine is most effectively used as a bronchodilator in the treatment of mild bronchial asthma. Ephedrine has also been used parenterally for its cardiac stimulant and vasoconstrictive properties in the treatment of shock. The value in shock, however, is questionable.

Pseudoephedrine is used as a nasal decongestant and produces little if any of the rebound congestion observed with ephedrine. It is not as effective as ephedrine in producing bronchodilation.

Dosage and preparations of ephedrine and pseudoephedrine are outlined in Table 10.

Toxicity

Because of lipophilicity, ephedrine isomers can penetrate the CNS and cause nervousness, anxiety, weakness, irritability, insomnia, dizziness, and vertigo. Among its systematic side effects are respiratory difficulty, pallor, dryness of nose and throat, sweating, nausea, and vomiting.[48] As indicated earlier, rebound congestion may occur with ephedrine.

Because of its norepinephrine-releasing effects, ephedrine isomers should not be used with other sympathomimetic amines because of the possibility of additive agonist effects. Similarly, it should not be used in conjunction with MAO-inhibiting antidepressants as they may potentiate the indirect agonist action of ephedrine.[47]

Patients on norepinephrine-depleting antihypertensive agents such as reserpine and α-methyldopa may have decreased response to ephedrine therapy since part of the activity of ephedrine depends on the released norepinephrine stores.

Precipitates may occur if ephedrine sulfate is mixed in solution with the sodium salts of hydrocortisone, pentobarbitone, phenobarbitone, or thiopentone.

The minimum lethal dose in children is 200 mg. The minimum lethal dose in adults is 2 g.[49]

Physical-Chemical Data

Ephedrine

Structure:	Figure 27A
Empirical formula:	$C_{10}H_{15}NO$
Mol wt:	165.2
Melting point:	33 to 37.5°C
Chemical name:	(−)-2-methylamino-1-phenylpropan-1-ol
Generic name:	ephedrine
Salt forms:	sulfate, hydrochloride, also forms hydrates
Physical description:	(hemihydrate) colorless hexagonal crystals, m.p. 40 to 41°C; (sulfate salt) fine white crystals which darken on exposure to light m.p. 258°C (dec); (HCl salt) white crystalline power, m.p. 217 to 220°C
Proprietary names:	Mistrol, 1-sedrin (hemihydrate forms); glucofedrin (lactate); ectasule, slo-fedrin gyrocaps, isofedrol, ephedsol (sulfate salts)

Table 10
PHARMACEUTICAL PREPARATIONS OF EPHEDRINE AND PSEUDOEPHEDRINE USED IN THE TREATMENT OF ASTHMA

Generic name	Proprietary name	Manufacturer	Other active components	Dosage, route of administration
Ephedrine base	Amsec	Lilly	1, 2	25 mg Ephedrine/tablet 1 to 5 tablets/day
	Bronkolixir Bronkotabs	Breon	3, 4, 5	12 mg Ephedrine/tablet, 2 tablets 3 to 4 times daily; 12 mg ephedrine/5 mℓ, 2 teaspoons every 3 to 4 hr
	Quibron Plus	Mead-Johnson	4, 3	25 mg Ephedrine/1.5 mℓ of 15% alcohol elixer, 1 to 2 tablespoons 2 to 3 times/day; 25 mg ephedrine/capsule 1 to 2 capsules per day
Ephedrine-HCl	Calcidrine	Abbott	6, 7	4.2 mg Ephedrine-HCl/5 mℓ, 1 to 2 teaspoons every 2 to 4 hr
	Mudrane	Poythress	8, 1, 5, 3	16 mg Ephedrine/tablet, 1 tablet 3 to 4 times daily
	Quadrinal	Knoll	5, 4, 8	24 mg Ephedrine-HCl/tablet, 1 tablet 3 to 4 times daily
	Tedral	Warner/Chilcott	4, 5	24 and 48 mg Ephedrine-HCl/tablet, 1 tablet every 4 hr; 6 and 12 mg ephedrine-HCl/ 5 mℓ elixir; 1 to 2 teaspoons every 4 hr
	Verequad	Knoll	5, 4, 9	24 mg Ephedrine-HCl/tablet, 1 tablet 3 to 4 times daily
Ephedrine sulfate	Ectasule	Fleming	2	15 and 60 mg Ephedrine- sulfate/ capsule, 1 capsule every 12 hr
Ephedrine sulfate	Marax	Roerig	4, 11	25 mg Ephedrine-sulfate/tablet 1 tablet 2 to 4 times daily
	Slo-Fedrin	Dooner	2	30 and 60 mg Ephedrine-sulfate/ capsule, 1 capsule every 8 to 12 hr
Ephedrine tannate	Rynatuss	Mallinckrodt	10, 12, 13	10 mg Ephedrine tannate/tablet, 1 to 2 tablets every 12 hr
Pseudoephedrine-HCl	Actifed	Burroughs-Wellcome	6, 3	30 mg/5 mℓ Syrup, 10 mℓ 4 times daily; 60 mg/tablet, 1 tablet 3 times daily
	Bronchobid	Meyer	4	50 mg/Capsule, 1 capsule every 12 hr

Note: Only ephedrine and pseudoephedrine preparations specifically formulated for asthma are tabulated. These compounds are also used extensively in various cold remedies. The reader should consult the appropriate indexes for these preparations. 1 = Aminophylline; 2 = amobarbital; 3 = glycerol guaiacolate; 4 = theophylline; 5 = phenobarbital; 6 = codeine; 7 = CaI_2; 8 = KI; 9 = guaifenesin; 10 = chlorpheniramine; 11 = hydroxyzine; 12 = phenylephrine; 13 = carbetapentane; 14 = triprolidine.

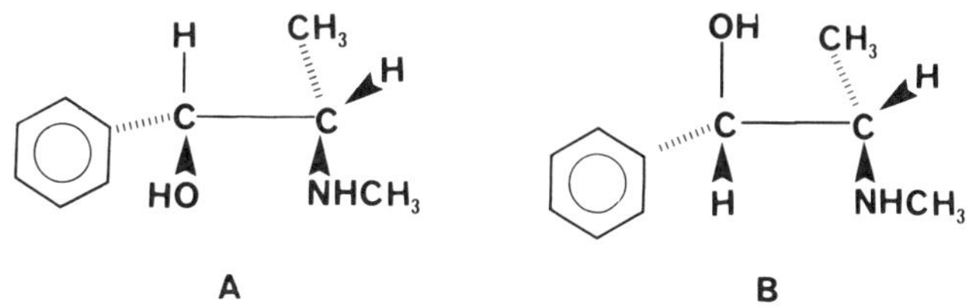

FIGURE 27. A = D(−) Erythro (IR,2S) ephedrine, B = L(+)threo (1S,2S) pseudoephedrine.

Solubility:	(free base and hemihydrate) 1 in 36 water, 1 in 1 in ethanol, soluble in ether and chloroform; salts are water and ethanol soluble and ether insoluble
pK_a:	9.63, solutions of sulfate have pH of 4.5 to 7
Stability:	ephedrine and its salts gradually decompose and darken on exposure to light and should be stored in light resistant containers; ephedrine base should be stored at or below 8°C

Pseudoephedrine

Structure:	Figure 27B
Empirical formula:	$C_{10}H_{15}NO$
Mol wt:	165.2
Melting point:	118 to 119°C
Chemical name:	(+)-2-Methylamino-1-phenylpropan-1-ol
Generic name:	pseudoephedrine
Salt forms:	hydrochloride
Proprietary name:	Sudafed (HCl salt)
Solubilities:	soluble in ethanol and ether; sparingly soluble in water; HCl salt is water and ethanol soluble but sparingly soluble in chloroform
Physical description:	HCl salt-white crystalline power; m.p. 183 to 186°C
pK_a:	9.73
Stability:	store in light-resistant containers

SALBUTAMOL

Mechanism of Action

Salbutamol is a direct-acting β_2-selective adrenergic agonist. As a β_2-agonist, salbutamol appears to have greater activity at the bronchial β_2-receptors than at the peripheral β_2-receptors. In contrast to isoproterenol, salbutamol does not decrease arterial oxygen tension.

Structure-Activity Relationships

Salbutamol belongs to a family of compounds referred to as saligenins. It was first synthesized in 1968 by Hartley et al.,[72] who reported on its bronchoselectivity. It belongs to the phenethanolamine class of direct-acting adrenergic agonists. In contrast to other members of the phenethanolamines, salbutamol contains a hydroxymethyl rather than a phenolic group on the *meta* position of the ring. As discussed in the structure-activity

Table 11
RELATIVE β_2-AGONIST ACTIVITY OF A SERIES OF SALIGENIN DERIVATIVES

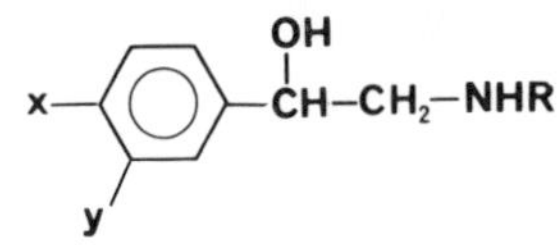

	x	y	R	Relative β_2-agonist activity
Isoproteranol	OH	OH	$-CH(CH_3)_2$	1
	OH	$-CH_2OH$	$-CH(CH_3)_2$	2.5
	$-CH_2OH$	OH	$-CH(CH_3)_2$	Inactive
Salbutamol	OH	$-CH_2OH$	$-C(CH_3)_3$	1.1
	$-CH_2OH$	OH	$-C(CH_3)_3$	Inactive

relationships of β-adrenergic agonists section, a hydroxymethyl group can be considered a "phenol equivalent" and as such contributes to the direct agonist effect of the compound. As shown in Table 11, the ring position of the hydroxymethyl group is important. Only the 3-hydroxymethyl phenol derivative was active, whereas the 4-hydroxymethyl phenol was inactive.[73] These results parallel the effects observed for the methanesulfonamido derivatives.

Pharmacokinetics

Salbutamol is administered by inhalation. Like isoproterenol, it is believed that significant amounts of the compound are swallowed during the inhalation. Depending on whether absorption occurs in the lungs or GI tract, peak plasma levels can be obtained within 10 min to 5 hr. The plasma half-life is 1.7 to 7 hr.

The onset of action occurs 5 to 15 min after oral inhalation, peaks at 30 min to 1 hr, and lasts 3 to 4 hr.

In man, 65 to 95% of an oral dose is excreted in the urine. About half of this dose is excreted as unchanged salbutamol. The major metabolite in man is the 4-*O*-sulfate ester, which is excreted in the urine. In rats, about 30% of the administered dose is excreted in the feces as the phenolic glucuronide. Because it lacks a *meta* hydroxyl group, salbutamol is not a substrate for catechol-*O*-methyl transferase.

Salbutamol has a longer duration of activity than isoproterenol, and because of its selectivity, suffers from fewer side effects.

Uses and Dosage

Salbutamol is used as a bronchodilator in the treatment of asthma and other bronchospasm due to its potent and selective action as a β_2-agonist in the bronchus. Salbutamol has also been found effective in the treatment of exercise induced bronchospasm.

Toxicity

Salbutamol has many of the same side effects as other sympathomimetics, although its β_2-receptor selectivity helps to reduce some of the cardiovascular side effects mediated by β_1-receptors. Like other sympathomimetics, tolerance to salbutamol can occur on prolonged use.

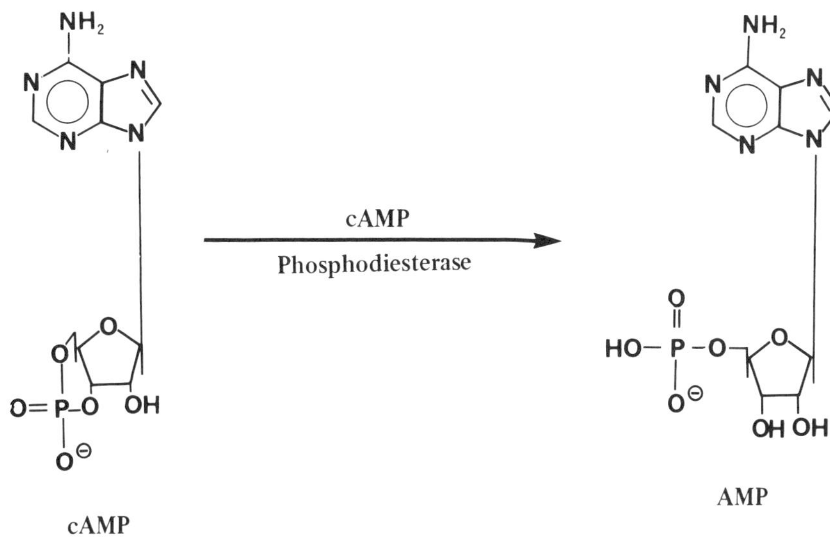

FIGURE 28. The metabolism of cAMP.

Salbutamol should not be coadministered with other sympathomimetics, MAO inhibitors, or tricyclic antidepressants due to the possibility of potentiating its effects.[47]

Physical-Chemical Data

Structure:	Figure 3
Empirical formula:	$C_{13}H_{21}NO_3$
Molecular wt:	239.3
Melting point:	156°C
Chemical name:	1-(3-hydroxymethyl-4-hydroxyphenyl)-2-tertbutylaminoethanol
Generic name:	salbutamol, albuterol
Proprietary names:	Proventil, Ventolin
Solubilities:	soluble in alcohol, sparingly soluble in water
Physical description:	white, tasteless, crystalline powder
pK_a:	9.3 (acid) 10.3 (base)
Stability:	store at 15 to 30°C

THEOPHYLLINE

Mechanism of Action

After the catecholamines, theophylline compounds are the most effective bronchodilator drugs. They are noteworthy in their mechanism of action in that their bronchodilatory activity arises, at least in part, through the inhibition of cAMP metabolism rather than through stimulation of its synthesis. β-Adrenergic agonists and neurotransmitter-releasing agents effect bronchodilation through increased cAMP production as a result of β-receptor stimulation. Under normal conditions, cAMP is metabolized by the enzyme cAMP phosphodiesterase terminating its physiological action (Figure 28). According to Sutherland's second messenger hypothesis (see mechanisms of β-adrenergic response section), it should be possible to mimic adrenergic activity by manipulating cAMP levels. Thus, the inhibition of cAMP phosphodiesterase should elevate intracellular levels of cAMP and produce the same physiological response as one would get from receptor stimulation. The bronchodilation observed on theophylline treatment appears to support this hypothesis.

It should be noted, however, that alternative mechanisms for the activity of theophylline have been proposed. Two of the major alternative theories involve the action of theophylline on Ca^{2+} movement and its effect on adenosine receptors.

Caffeine, a chemical analog of theophylline, is capable of causing nerve depolarization through modulation of Ca^{2+} release and uptake by the sarcoplasmic reticulum.[74] Such depolarization may explain the ability of theophylline to induce the secretion of catecholamines from the adrenal medulla. These released catecholamines, particularly epinephrine, could then be the actual agents of the bronchodilatory action observed on theophylline administration.

Adenosine receptors have been found in the plasma membranes of a variety of cells. Adenosine dilates blood vessels, slows the rate of discharge of cardiac pacemaker cells, and may inhibit the release of excitatory neurotransmitters in the CNS. Depending on the type of cell, stimulation of adenosine receptors can lead to either stimulation or inhibition of cAMP synthesis.[75] Theophylline has been found to be a universal antagonist of adenosine receptors. The modulation of intracellular cAMP levels, and thus bronchodilation by theophylline, may occur via adenosine receptor inhibition rather than inhibition of cAMP metabolism.

The major difficulty in definitively assigning a mechanism of action to the activity of theophylline lies in the fact that all the mechanisms suggested were observed only at high theophylline concentration. The effective blood levels of theophylline which give therapeutic results are considerably lower than those used experimentally, thus casting some doubt on the physiological significance of those results. Further study will be required in order to determine the most important mechanism for the action of theophylline. These studies, however, underline the importance of knowing the pharmacokinetic parameters of the activity of a drug in understanding its mechanism of action.

Pharmacologically, theophylline directly relaxes the smooth muscle of the respiratory tract, providing relief of bronchospasm, increased air flow rates, and increased vital capacity. However, its bronchodilator effect is minimal on uncontracted "normal" bronchial tissues. Theophylline also dilates pulmonary arterioles and increases pulmonary blood flow.

The CNS effects of theophylline include stimulation of the medullary respiratory centers and a lowering of the threshold of the respiratory centers to carbon dioxide, thus providing substantial increase in rates and depth of respiration in patients with respiratory impairment. This effect is not observed in normal patients.

At high doses, theophylline causes stimulation of both the chronotropic and inotropic responses of the heart. Heart rate is not generally affected. There is little or no increase in blood pressure due to the fact that increased cardiac output is generally balanced by peripheral vasodilation.

Unlike the sympathomimetics, tolerance to theophylline rarely occurs. Since they act by different mechanisms, sympathomimetics and theophylline may be given together to provide additive physiological effects.

Structure-Activity Relationships

Theophylline, and the closely related alkaloids, caffeine and theobromine, have been ingested by man for centuries. They are the principal alkaloids of a number of plants which have been used for making beverages as far back as prehistoric times. These plants include *Thea sinensis,* from which tea is brewed, *Theobroma cacao,* from which cocoa and chocolate are prepared, and *Coffea arabica* from which coffee is prepared. As a result of their widespread use, people have been aware of the pharmacological properties of these compounds since ancient times.

Structure-activity relationships for these compounds have been difficult to determine due to their diverse physiological effects and because the exact mechanism of action is unknown.

Recent work in the purification and study of cAMP phosphodiesterase, however, has generated structure-activity relationship trends for compounds inhibiting this enzyme. Since inhibition of cAMP phosphodiesterase has been suggested as a possible mechanism for the bronchodilatory action of theophylline, the structure-activity relationships for inhibition of this enzyme may be the equivalent to the structure-activity relationships of the bronchodilatory action of theophylline.

Cyclic AMP phosphodiesterase (cAMP-PDE) is actually a family of enzymes, or isoenzymes. Each isoenzyme differs in its V_{max}, K_m, and substrate specificities. Wells and coworkers[76] have done extensive work on the structural requirements for inhibitors of two isoenzyme forms of cAMP-PDE. One isoenzyme (PDEI) can be stimulated by Ca^{2+} and calmodulin, and is capable of hydrolyzing both cAMP and cGMP. The second isoenzyme (PDEII), is not stimulated by calmodulin and is relatively specific for cAMP. In pig coronary arteries, "total cAMP phosphodiesterase activity at 1-*M* substrate concentration appears to be equally divided between the calmodulin-sensitive form and the cAMP specific form." It is not clear at present which, if either, form is most important in the intracellular regulation of cAMP levels.

Based on the activity of a series of substituted xanthines, Wells has proposed the following structure-activity relationships for inhibition of PDEI and II[77]: (1) Nonpolar substituents larger than methyl in the 3 position of the xanthine nucleus enhance inhibition of both PDEI and II. (2) A substituent other than hydrogen is required at position 1 for inhibition of PDEI and II. Large substituents at position 1 favor inhibition of PDEII over PDEI. Neither PDEI or II can tolerate polar groups at position 1. (3) PDEI can be inhibited by xanthines with bulky substituents at position 8, whereas PDEII cannot. Electron withdrawing groups at position 8 decrease inhibitory potency against both PDEI and II. Opening the imidazole or pyrimidine ring of the xanthine nucleus decreases inhibition of both PDEI and II.

Using these results, a topographical model was described[78] which proposes a lypophilic pocket in PDEI and II for 3-postion substituents. A sterically restricted area was suggested for position-1 substituents in PDEI, whereas a sterically restricted area for position-8 substituents was cited for PDEII.

As discussed previously, it is not known whether or not PDE inhibition is the mechanism most responsible for the beneficial bronchodilatory action of theophylline. Therefore, modification of theophylline according to PDE structure-activity trends may not necessarily lead to more effective agents for the treatment of asthma. In fact, a more important parameter for the effectiveness of theophylline compounds appears to involve pharmacokinetic properties.

Pharmacokinetics

The effectiveness of theophylline as a therapeutic agent is directly related to its bioavailability. While it is readily absorbed after oral, rectal, or parenteral administration, the effect may be erratic due to poor water solubility. Because of this, theophylline is often administered as a water-soluble salt. Bases which are often employed as salts of theophylline include ethylenediamine, ephedrine, choline, *N*-methylglucamine, ethanolamine, and glycine.

In the absence of food, oral theophylline will reach maximal blood plasma levels within 2 hr of administration. Distribution to all body components occurs. The volume of distribution is between 400 and 550 mℓ/kg. Approximately 50% of theophylline is bound to plasma proteins. For maintenance therapy, minimal blood levels of theophylline should be maintained between 0.5 and 1.0 mg/100 mℓ.

Theophylline is eliminated primarily through metabolism in the liver. Only 10% of the administered dose is excreted unchanged in the urine. The half-life for elimination is about 3.5 hr for children and 8 to 9 hr for adults. Metabolism can be accelerated by metabolic

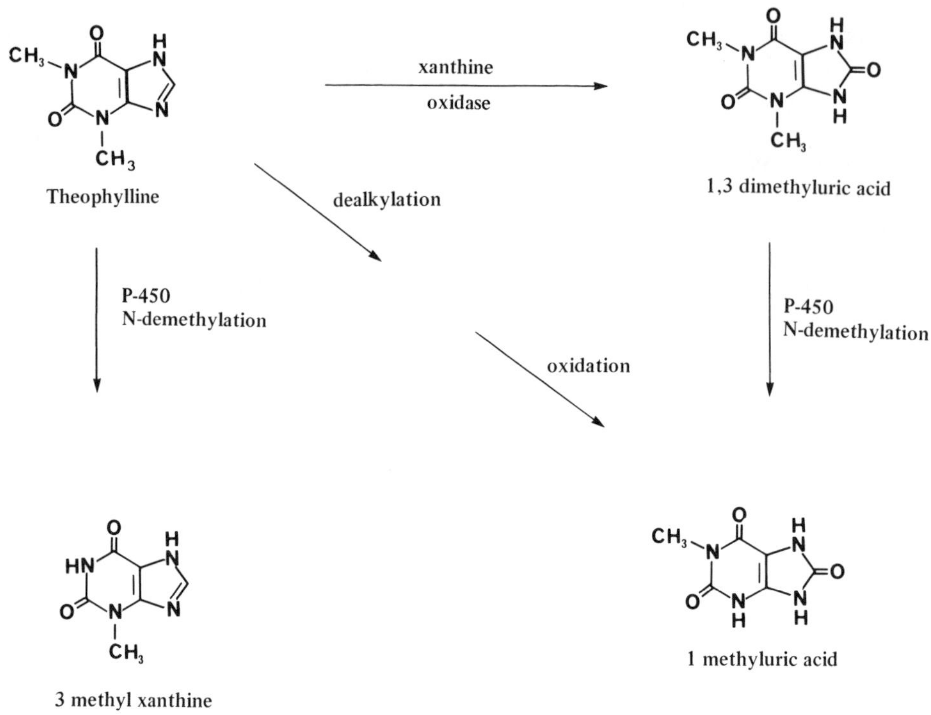

FIGURE 29. The major metabolites of theophylline.

enzyme inducer such as smoke and polycyclic aromatic hydrocarbons. The principal urinary metabolites of theopylline are 1,3-dimethyluric acid, 1-methyluric acid, and 3-methylxanthine (Figure 29). The 3-methyl xanthine metabolite is also active as a bronchodilator, and may contribute to the activity of theophylline.[79] Theophylline, however, is not completely demethylated to uric acid and therefore is not contraindicated in gout.

Uses and Dosage

Theophylline is used primarily as a bronchodilator in the treatment of bronchial asthma. It is available in numerous salt forms which can be administered orally, i.v., or rectally (Table 12).

Toxicity

The major toxic effects of theophylline involve the CNS, the GI system, and the cardiovascular systems. Like other xanthines, theophylline is a strong CNS stimulant. The toxic effects include headache, irritability, nervousness, insomnia, dizziness, and possibly convulsions. The cardiovascular toxicities include increased cerebral vascular resistance, decreased cerebral blood flow, heart palpitations, increased pulse rate, and sinus tachycardia.[48]

The GI toxicity of theophylline arises from irritation of the gastric mucosa due to stimulation of gastric acid secretion causing nausea, vomiting, pain, and cramps. Theophylline also has a diuretic action which may lead to dehydration.

Acute toxicity may occur if plasma levels of theophylline exceed 20μg/mℓ. Children are more susceptible than adults. Adults with renal, hepatic, or cardiovascular complications are susceptible. Toxicity is observed as extreme restlessness leading to convulsions and/or seizures due to CNS stimulation. Toxic cardiovascular effects include acute hypotension, ventricular fibrillation, and cardiac arrest.

Since theophylline is often administered in combination with sympathomimetics for treat-

Table 12
PHARMACEUTICAL PREPARATIONS OF THEOPHYLLINE USED FOR BRONCHODILATION

Generic name	Proprietary name	Manufacturer	Other active components	Dosage, route of administration
Theophylline	Aerolate	Fleming	—	130 mg and 260 mg Theophylline/capsules, 1 capsule every 12 hr, timed release
	Bronchobid	Meyer	1	260 mg Theophylline/capsule, 1 capsule every 12 hr
	Bronkolixir, Bronkotabs	Breon	2, 3, 4, 2, 3, 4	15 mg Theophylline/5 mℓ, 2 teaspoons every 4 hr; 50 mg theophylline/tablet, 2 tablets every 3 to 4 hr
	Elixophyllin	Cooper	—	80 mg Theophylline/15 mℓ, 75 mℓ every 6 hr; 200 mg theophylline/capsule, 1 capsule 3 times daily
	Marax	Roerig	2, 5	130 mg Theophylline/tablet, 1 tablet 2 to 4 times daily; 32.5 mg theophylline/5 mℓ syrup, 1 teaspoon 3 to 4 times daily
	Quibron	Mead-Johnson	3	150 mg Theophylline/15 mℓ, 1 to 2 tablespoons 2 to 3 times daily; 150 mg theophylline/capsule, 1 to 2 capsules 2 to 3 times daily
	Slo-phyllin	Dooner	—	125 mg and 25 mg Theophylline/capsule, 1 capsule every 8 to 12 hr; 100 mg and 200 mg theophylline/tablet, 1 tablet every 6 hr
	Tedral	Warner/Chilcott	2, 4	130 and 180 mg Theophylline/tablet, 1 to 2 tablets every 2 hr
Theophylline	Theobid	Meyer	—	260 mg Theophylline/capsule, 1 capsule every 12 hr
	Theolair	Riker	—	125 mg Theophylline/tablet, 1 to 2 tablets every 6 hr
	Theonar	Key	6	100 mg and 200 mg Theophylline/tablet, 1 to 2 tablets every 8 to 12 hr

Table 12 (continued)
PHARMACEUTICAL PREPARATIONS OF THEOPHYLLINE USED FOR BRONCHODILATION

Generic name	Proprietary name	Manufacturer	Other active components	Dosage, route of administration
	Theo-organidin	Wallace	7	120 mg Theophylline/15 mℓ, 1 to 2 tablespoons, 3 times daily
Aminophylline (theophylline ethylene diamine salt	Amesec	Lilly	2, 8	130 mg Aminophylline/capsule, 1 capsule every 8 hr
	Aminodur	Cooper	—	300 mg Aminophylline/tablet, 1 to 2 tablets every 8 to 12 hr
	Lixaminol	Ferndale	—	250 mg Aminophylline/15 mℓ, 1 tablespoon every 8 hr
	Mudrane	Polythress	9, 4, 2, 3	130 mg Aminophylline/tablet, 1 tablet 3 to 4 times daily
	Somophylline	Fisons	—	300 mg Aminophylline/5 mℓ, 5 mℓ rectally, 1 to 3 times daily
Dyphylline (dihydroxypropyl theophylline)	Airet	Baylor	—	400 mg Dyphylline/tablet, 1 tablet every 8 hr
	Dilor	Savage	—	200 and 400 mg Dyphylline/tablet, 1 to 4 tablets daily; 160 mg dyphylline/15 mℓ, 1 to 5 tablespoons 3 to 4 times daily; 250 mg dyphylline/mℓ injectable, 1 to 2 mℓ i.m.
Dyphylline (dihydroxypropyl theophylline)	Emfaseem	Saron	3	200 mg Dyphylline/capsule, 1 capsule every 6 hr; 100 mg dyphylline/15 mℓ, 1 tablespoon every 6 hr
	Lufyllin	Mallinckrodt	—	200 mg Dyphylline/tablet, 1 tablet every 6 hr; 100 mg dyphylline/15 mℓ, 2 tablespoons every 6 hr; 250 mg dyphylline mℓ injectable, 1 to 2 mℓ i.m.

	Neothylline	Lemmon	—	200 and 400 mℓ Dyphylline/tablet, 1 tablet 3 to 4 times daily; 160 mg dyphylline/15 mℓ, 1 to 2 tablespoons every 6 hr; 500 mg dyphylline/2 mℓ injection, 2 mℓ i.m.
Theophylline calcium salicylate	Quadrinal	Knoll	2, 4, 9	130 mg/tablet, 60 mg/teaspoon, 1 tablet or 2 teaspoons every 6 hr
	Theokin	Knoll	9	448.5 mg Theophylline calcium salicylate per tablet or tablespoon, 1 tablet or tablespoon 2 to 3 times daily
	Verequad	Knoll	4, 2, 3	130 mg Theophylline calcium salicylate per tablet or 10 mℓ, 1 tablet or 2 teaspoons, 2 to 3 times daily
Theophylline monoethanolamine	Fleet, Theophylline	Fleet	—	625 Theophylline monoethanol amine/37 mℓ, 30 mℓ rectally, every 8 hr
Theophylline sodium glycinate	Asbron	Dorsey	—	300 mg Theophylline sodium glycinate/tablet or tablespoons, 1 to 2 tablets or tablespoons 2 to 3 times daily
	Synophylate	Central	—	330 mg Theophylline sodium glycinate/tablet or tablespoon, 1 to 2 tablets or tablespoons every 8 to 12 hr

Note: 1 = Pseudoephedrine; 2 = ephedrine; 3 = glycerylguaiacolate; 4 = phenobarbital; 5 = hydroxyzine; 6 = noscapine; 7 = organidin (iodine); 8 = amobarbital; 9 = KI.

Table 13
PHENETHANOLAMINE BRONCHODILATORS

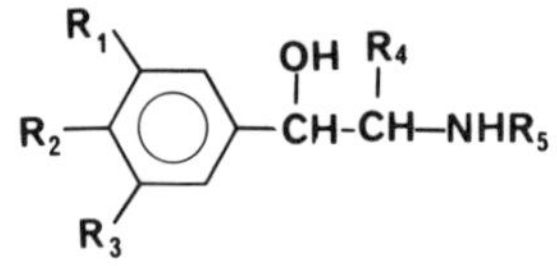

Name	R_1	R_2	R_3	R_4	R_5
Isoproteranol	OH	OH	H	H	Isopropyl
Isoetharine	OH	OH	H	Ethyl	Isopropyl
Metaproteranol	OH	H	OH	H	Isopropyl
Terbutaline	OH	H	OH	H	*t*-Butyl
Fenolterol	OH	H	OH	H	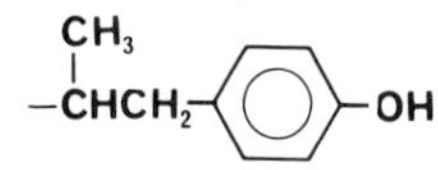
Colterol	OH	OH	H	H	*t*-Butyl
Bitolterol	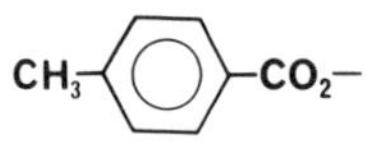	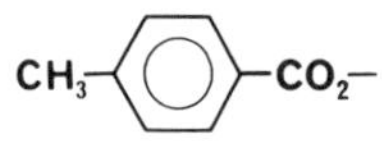	H	H	*t*-Butyl

ment of asthma, it is important to be aware of the additive or synergistic toxicity with these agents. This is particularly true with regard to the stimulant effect on the myocardium. This effect also enhances the sensitivity and toxic potential of cardiac glycosides.

Erythromycin and clindamycin may increase theophylline plasma levels by decreasing hepatic clearance.

The LD_{50} for oral aminophylline in mice is 540 mg/kg.[49]

Physical-Chemical Data

Structure:	Figure 29
Empirical formula:	$C_7H_8N_4O_2$
Molecular wt:	180.2
Melting point:	269 to 274°C
Chemical name:	1,3-dimethylxanthine
Solubilities:	1 in 120 of water, 1 in 80 of ethanol, 1 in 86 chloroform
pK_a:	8.6 (acid) 3.5 (base)
Preparations:	see Table 12

OTHER COMPOUNDS OF INTEREST

In addition to isoproterenol, a number of other phenolic derivatives of the ethanolamine family have been developed for use in asthma therapy (Table 13). Isoetharine, metaproterenol, and terbutaline have already been used clinically. A newer compound, fenolterol, has recently been developed which is even more effective than metaproterenol as a bronchodilator. Another interesting, newly developed compound is bitolterol. Bitolterol is the *para* methylphenyl ester of another phenolic β-agonist, colterol. In vivo, the pharmacologically inactive bitoletrol is hydrolyzed by plasma esterases to the active compound colterol.[80]

A number of rigid analogs of bronchodilating ethanolamines have been synthesized in an attempt to elucidate the stereochemical requirements of the β_2-receptor. A series of tetrahydronaphthalenes, which mimic β-agonists with their side chains tied back to the ring (Figure 30), were shown to be as much as 120 times more active than isoproterenol on the

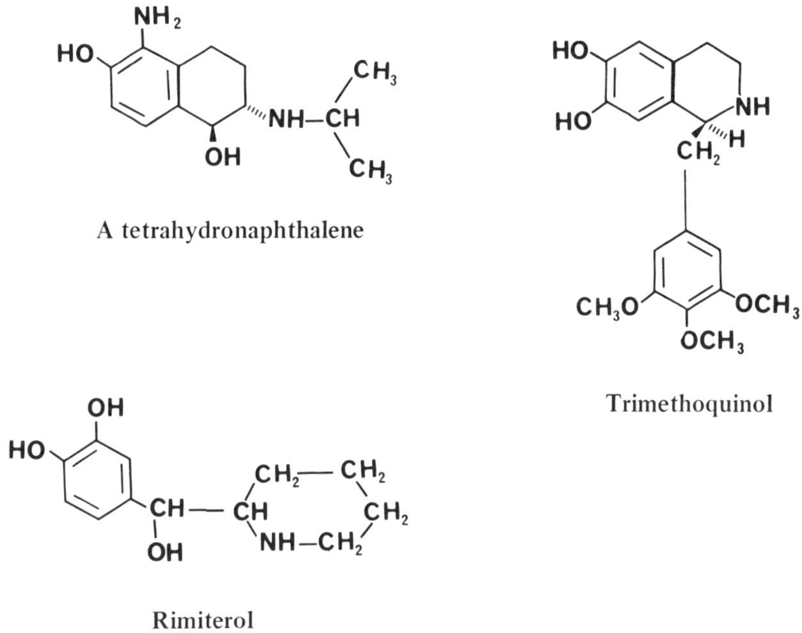

FIGURE 30. Rigid analogs of phenethanolamine bronchodilators.

β_2-receptor.[81] Similarly, trimethoquinol (Figure 30), a rigid side chain derivative of the phenethylamine, dopamine, was also shown to have agonist activity. Rimiterol, which has the amino group tied up in a cyclic side chain rather than tied back to the ring, also has β-agonist bronchodilatory activity.

TESTING METHODS FOR POSTSYNAPTIC ADRENERGIC AGENTS

Proper evaluation of the pharmacological activity of a compound is necessary for understanding the mechanism of action. As our understanding of the adrenergic system has improved, it has become apparent that the pharmacological action of a given compound can be the result of many processes other than straightforward receptor binding. Even receptor binding processes have become complicated in light of the existence of multiple receptor subtypes. This section is presented in order to describe the various preferred methods for evaluation of adrenergic agonists. An excellent review of this material has been written by Patil and Ruffolo.[82]

In Vivo Tests

1. Spinal cat preparation — used for measuring cardiovascular pressor effects of adrenergics. It can measure three parameters: carotid blood pressure, nictitating membrane, and heart rate.
2. Denervated cat nictitating membrane — prepared from spinal cat preparation.
3. Guinea pig preparation — useful for measuring bronchodilatory activity of adrenergic agents.

Organ-Perfusion Tests

In this in vitro method, a major blood vessel leading to a specific organ is catheterized and drugs are directly administered through this blood vessel. Direct effects on a test organ can then be measured by standard procedures. The advantages of this procedure are that many pharmacokinetic parameters (i.e., metabolism and whole body distribution) are minimized. Blood vessels infusing such organs as the heart, spleen, hind limb, aorta, or kidney are often employed. The classical heart-lung preparation of Liebman[83] is an example of this type of test. This method is particularly useful for analyzing the effects of adrenergic drugs on the sinus node and pacemaker.

Isolated Tissue Tests

In isolated tissue tests, tissues or organs are obtained from freshly killed animals and suspended in a tissue bath containing an oxygenated physiological salt solution maintained at a constant temperature (30 to 37°C). Drug effects on the preparation are measured using a physiograph, transducer, and recorder.

1. Spirally cut blood vessels: α-adrenergic vasoconstriction can be measured by dissecting a rabbit aorta and cutting it in a spiral fashion to produce a strip.
2. The mydriatic effect of adrenergic agonists can be observed by measuring dilation of the iris of isolated guinea pig eye.
3. Both α- and β-adrenergic effects can be observed in a smooth muscle strip from the urinary bladder.
4. The relaxant effect of β-adrenergics on bronchial smooth muscle can be measured in the guinea pig tracheal chain preparation of Castillo and deBeer[84] or the spirally cut tracheal preparation of Akcasu.[85]
5. Lipolytic activity can be determined by the measurement of free fatty acids or glycerol in isolated cells from rat epidymal fat pads.

PRESYNAPTIC ADRENERGIC FUNCTION

All the adrenergic agents discussed thus far exert their activity by direct action on α- or β-adrenergic receptors (with the exception of theophylline).

In the previous discussion on the structural organization of the adrenergic neuron, however, it was noted that many adrenergic agents exert their activity by regulating the in vivo levels of norepinephrine in the synapse rather than by acting directly on adrenergic receptors themselves.

The following discussion will provide an overview of typical agents which act on adrenergic systems by mechanisms other than direct postsynaptic receptor interaction.

Many of these agents and their analogs are extremely important therapeutic agents. Their detailed structure-activity relationships and mechanisms of action, however, will be monographed elsewhere in this text. For this information, the reader is particularly directed to those chapters covering antihypertensive agents, MAO inhibitors, and antidepressants.

Agents that Affect Neurotransmitter Storage

Neuronal Uptake Inhibitors

The primary mechanism for the termination of the activity of norepinephrine in the synapse is through uptake of the compound into the neuron (site 4, Figure 2).

This uptake process appears to be moderately specific for phenethylamine compounds, but will accept many structural analogs. However, the greater the structural deviation from norepinephrine the less likely the compound will be a substrate for the uptake process.

The uptake process is saturable, stereoselective, and energy dependent. The energy may be supplied by a sodium-potassium ATPase pump, since the process is dependent on sodium and can be inhibited by ouabain (a sodium pump inhibitor).

The ability to modulate the uptake process is of obvious importance in regulating adrenergic activity. The longer the agonist remains in the synapse, the longer it will be able to interact with its receptor, and hence the longer the duration of action.

Similarly, the ability of a compound to block the uptake process may potentiate the effect of an adrenergic agent. This effect is the basis of the contraindications for many compounds which might be coadministered with adrenergic agents.

Numerous compounds of varying chemical and pharmacological classes have been identified as inhibitors of the uptake process. Among these are cocaine, tricyclic antidepressants, some alkylating α-adrenergic blockers, and other miscellaneous compounds. Phenethylamine compounds which have the greatest affinity for the uptake process generally have: (1) phenolic hydroxyl groups, (2) small *N*-alkyl substituents, (3) α-methyl groups, and (4) do not have a hydroxyl group.[86]

In addition to the uptake of phenethylamines from the neuronal synapse, there also appears to be a second extraneuronal uptake mechanism. This process, which occurs in cardiac muscle, smooth muscle, and glands, has a lower affinity for norepinephrine, is not stereoselective, and is not inhibited by cocaine or desipramine. The physiological significance of this process is not fully known; however, it is unlikely that it is important in controlling norepinephrine levels. Since epinephrine appears to be the preferred substrate for the process, it is possible that it may play a role in the rapid removal of circulating epinephrine.

Biosynthesis Inhibitors

Agents which can inhibit the synthesis of norepinephrine (site 1, Figure 2) reduce the levels of neurotransmitter which can be released on depolarization. Such an effect thus mimics adrenergic blockade.

A number of compounds have been discovered which are capable of inhibiting the enzymes of the biosynthetic pathway to norepinephrine and epinephrine. These inhibitors are shown in Figure 31. Several of these agents are used medicinally. α-Methyldopa has been used in the treatment of Parkinsonism. Disulfiram is useful in alcoholism, although its value is due to nonspecific inhibition of oxidative enzymes rather than inhibition of norepinephrine biosynthesis.

The biosynthetic pathway for norepinephrine synthesis is also important because of the role it plays in mediating the activity of adrenergic agents whose mechanism of action involves false transmitters. These agents will be discussed in greater detail later.

Metabolic Inhibitors

The two enzymes which are of greatest importance in the metabolism of norepinephrine and epinephrine are monoamine oxidase (MAO) and catechol-*O*-methyl transferase (COMT). The biotransformations these enzymes perform on norepinephrine and epinephrine are shown in Figure 18.

Both enzymes are found widely distributed in the body. Both enzymes are also relatively nonspecific about the substrates they will accommodate.

At the sympathetic neuronal junction, most COMT will be found in the synaptic cleft, while most MAO is located in the axoplasm. Although these enzymes are responsible for the metabolism of neurotransmitter, it should be recalled that the primary route for termination of neurotransmitter activity is not metabolism but rather uptake. If, however, the uptake mechanisms are inhibited, the metabolic fate of neurotransmitter becomes more important. Furthermore, depending on which uptake mechanism is affected, neuronal uptake or storage granule uptake, one can get selective metabolism by either COMT in the synapse or MAO in the axoplasm.

A number of inhibitors of both MAO and COMT have been identified. For the most part, inhibitors of COMT are not of therapeutic interest. On the other hand, inhibitors of MAO have found widespread utility in the treatment of problems such as hypertension.

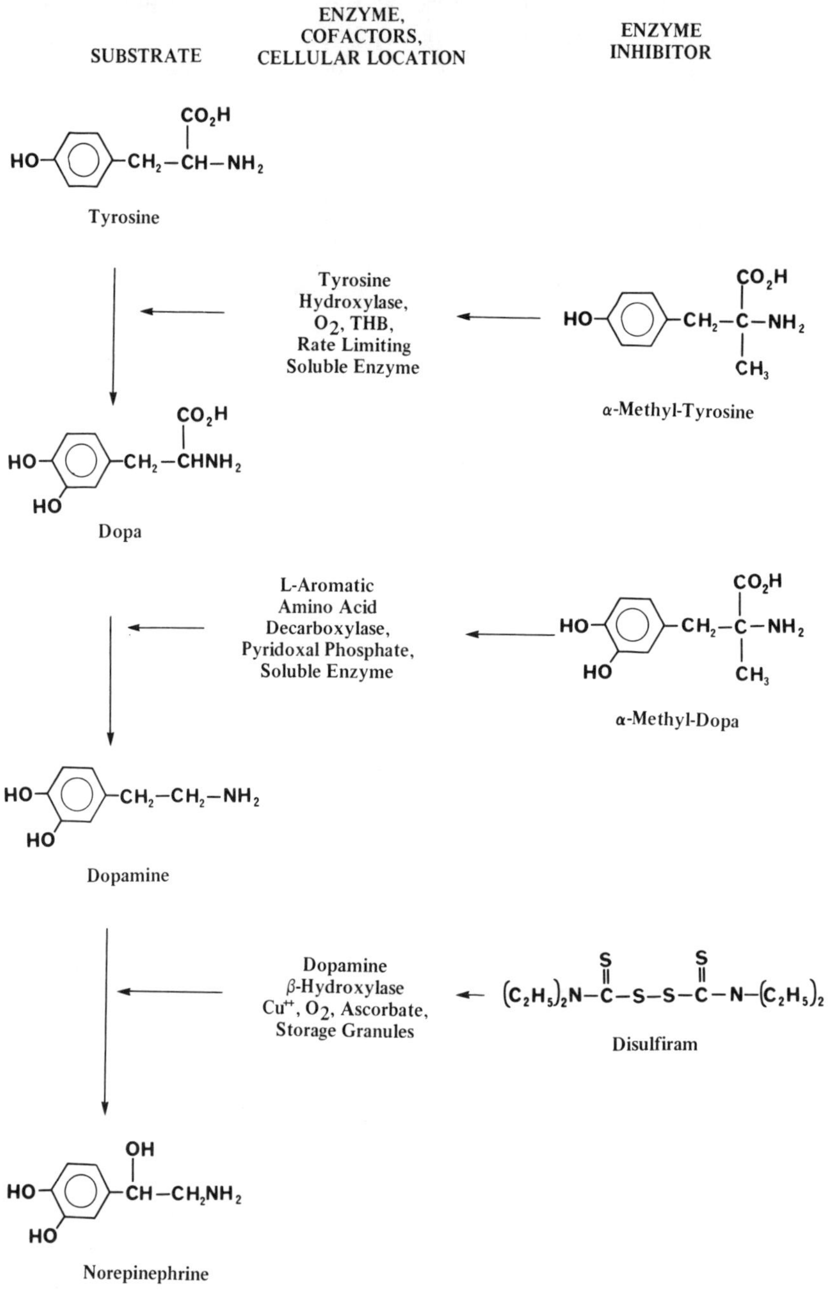

FIGURE 31. Enzymes and enzyme inhibitors of the norepinephrine biosynthetic pathway.

Agents that Affect Release of Neurotransmitters

Catecholamines biosynthesized in sympathetic nerve terminals are stored in organelles known as granules or vesicles. The granules perform several functions. They contain the enzyme dopamine β-hydroxylase and are the site for the last biosynthetic step in the biosynthesis of norepinephrine. Once synthesized, norepinephrine is held within the granule by equilibrium binding to an ATP-protein complex. The granule possesses an active transport system which selectively concentrates norepinephrine against a 200-fold gradient. The process is energy dependent and requires ATP and magnesium ion. The process can also be blocked by very low concentrations of reserpine. This uptake process is much more structurally specific than the neuronal uptake process. Upon depolarization of the nerve, these granules migrate to the cell membrane, fuse with it, and empty their contents, norepinephrine and ATP, into the synaptic cleft. Thus, regulation of adrenergic function can be manipulated by affecting processes involved in granular storage or depolarization and release of norepinephrine.

Adrenergic Neuron Blocking Agents

Depolarization of sympathetic neurons can be blocked at several points. Nicotinic receptors in the synaptic ganglion can be inhibited by ganglionic blockers, such as hexamethonium, or stimulated by nicotine. Stimulation of the postganglionic, peripheral adrenergic neuronal axon, can be suppressed by bretylium, guanethidine, or bethanidine. The amidine compounds, guanethidine and bethanidine, also cause depletion of norepinephrine stores in the granule.

There is now considerable evidence that neuronal blocking agents depend on uptake into the nerve in order to exert their activity. Also, this uptake mechanism may very well be the same one which takes up norepinephrine from the synapse. Therefore, inhibition of the uptake mechanism can antagonize the activity of the neuron blocking agent.

Indirect- and Mixed-Acting Amines

As early as the 1920s, it was observed that not all sympathomimetic amines behaved the same pharmacologically. Tainter and Chang[87] in 1927 noted that while cocaine could potentiate the sympathomimetic effects of adrenaline, it reduced the sympathomimetic effects of tyramine. Fleckenstein and Burn[88] further observed that the sympathomimetic effects of tyramine required a functionally responsive neuron. Chronic denervation of the neuron led to abolition of the effects of tyramine but not adrenaline.

In the 1950s, Burn and Rand[89] used reserpine to deplete granule stores of norepinephrine in the neuron and demonstrated that, in the absence of these stores, tyramine had no sympathomimetic activity. This suggested that the activity of tyramine was mediated through the displacement of norepinephrine from the neuron rather than from direct action in adrenergic receptors. Tyramine was thus labeled an indirect agonist.

Studies have since shown that the indirect release of norepinephrine by tyramine is considerably different from the release obtained by depolarization. While depolarization involves calcium influx, vesicle migration, and membrane fusion, the tyramine-mediated release appears to involve a type of equilibrium displacement of norepinephrine from the granule and eventually the neuron. It should be noted that because of this equilibrium process, norepinephrine spends an appreciable amount of time in the axoplasm, where it can be metabolized by MAO.

Numerous compounds have been shown to owe either part or all of their activity to this indirect mechanism. Compounds with both direct and indirect actions are termed "mixed-acting agonists". Structural types for these compounds encompass a continuum of structural variations. A given compound may have varying amounts of direct and indirect activity depending on the organ tested, species used, or other variables.

Some structural generalizations can be made, but are not absolute or clear-cut. These generalizations are based on the observations that for phenethylamines:

1. Catechols tend to be more direct acting and less indirect.
2. A β- (benzylic) alcohol favors direct over indirect activity.
3. The less polar the compound (i.e., few OHs) the more likely the compound will act indirectly and be orally effective.
4. An α-methyl group increases lipophilicity, decreases metabolism, and promotes indirect action.

Table 14
THE EFFECT OF SUBSTITUTION ON THE MODE OF ACTION OF SYMPATHOMIMETIC AMINES

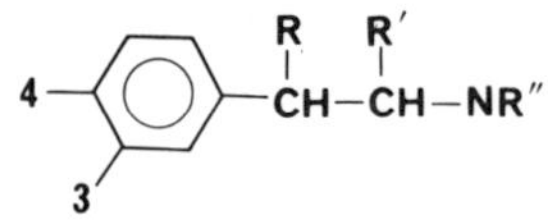

4	3	R	R′	R″	Type of action on heart rate
OH	OH	OH	H	H	D
OH	OH	OH	H	$CH(CH_3)_2$	D
H	OH	OH	H	H	D,M
OH	H	OH	H	H	M
H	OH	OH	CH_3	H	M
H	H	OH	CH_3	CH_3	M,I
OH	H	H	H	H	I
H	H	H	H	H	I
H	H	H	CH_3	H	I

Note: D = Direct; M = mixed; I = indirect.

Mixed-acting agonists tend to have functionalities intermediate to direct and indirect. Ephedrine, a mixed-acting agonist, for example, has an α-methyl group and lacks catechol hydroxyls, thus promoting indirect action. However, ephedrine also has a β-hydroxyl group which promotes direct activity. A representative list of direct, mixed, and indirect agonists is shown in Table 14.

Research during the past 20 years has indicated that even the indirect agonist explanation for the activity of tyramine was not sufficient to entirely explain all of its effects. In fact, several mechanisms are likely to be involved. These mechanisms are part of the theory of false transmitters.

The Theory of False Transmitters

The concept of false transmitters developed as a rationale for the mechanism of action for the antihypertensive agent, α-methyldopa (αMD). αMD was known to be an inhibitor of DOPA decarboxylase. After a thorough pharmacological investigation, however, Carlsson and Lindquist[90] suggested that the antihypertensive activity of αMD was due not so much to an inhibition of norepinephrine synthesis, but rather to the fact that αMD could also serve as an alternate substrate for the enzyme, and thus be converted to α-methyldopamine (αMDA). αMDA could then be taken up by storage granules where it would serve as an alternate substrate for dopamine-hydroxylase and be converted to α-methyl norepinephrine (αMNE) (Figure 32). Production and storage of αMNE in neuronal granules thus led to depletion of norepinephrine and antagonism of its vasopressor action.

In 1963, Day and Rand[91] modified this theory to include the concept that stored αMNE could be released on depolarization of the neuron, and that the released αMNE behaved as a weak agonist at the adrenergic receptor. In this way then, αMNE, formed from αMD, was acting as a "false transmitter" of adrenergic response.

Kopin[92] later defined the criteria which must be met in order to classify a compound as a false transmitter:

1. The false transmitter must be present in the adrenergic terminal fibers at the same site as norepinephrine (i.e., in the granules).

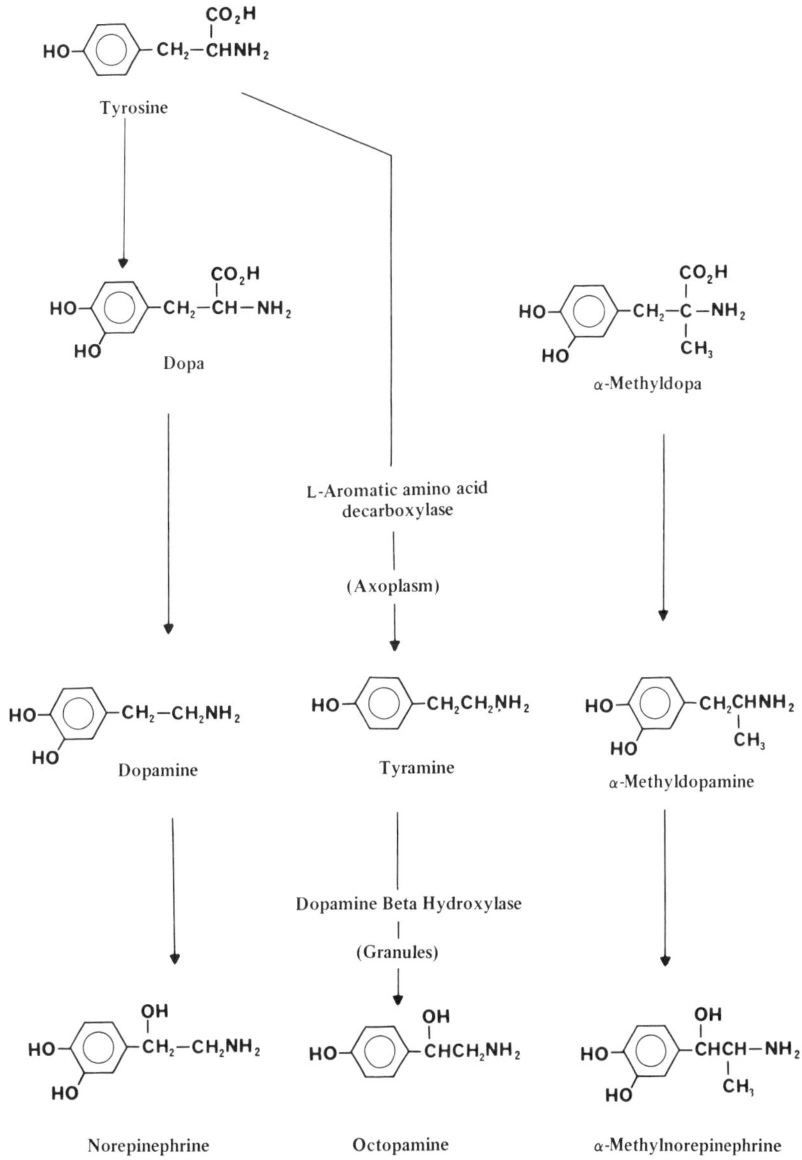

FIGURE 32. The biosynthetic pathways for false transmitters.

2. The false transmitter must be released by adrenergic nerve stimulation and be depleted by drugs which deplete norepinephrine.
3. In contrast to the requirements for identification of a physiological neurotransmitter, designation of a false transmitter does not imply that is as active as norpinephrine. In fact, most false transmitters are less potent at adrenergic receptors than norepinephrine.

The mechanism of action of the indirect agonist tyramine appears to include the production of false transmitters. As tyramine is taken up and concentrated in the granule, at the expense of norepinephrine, it is exposed to the enzyme dopamine-hydroxylase. The tyramine then acts as an alternate substrate for the enzyme and is converted to octopamine which, by virtue of its two hydroxyls, is tightly bound within the granule in preference to tyramine. On depolarization, octopamine, rather than norepinephrine, is released into the synaptic cleft. Octopamine, however, is a weaker adrenergic agonist than norepinephrine and provides a weaker adrenergic response than norepinephrine. An apparent antagonism of adrenergic response is thus observed. Octopamine formed from tyramine thus serves as a false transmitter.

The concept of false transmitters has been used to explain the mechanism of action of a number of sympathetic amines, including metaraminol (see metaraminol monograph). It has also been used to explain some of the toxicity and tolerance observed with the use of amphetamine (see amphetamine monograph).

One last component of the mechanism of action of tyramine appears to be the ability to inhibit uptake of catecholamines from the synapse. This initially potentiates the action of released norepinephrine, but in prolonged therapy leads to depletion of norepinephrine reserves due to metabolism by synaptic COMT.

Presynaptic Adrenergic Receptors

In 1956, Brown and Gillespie[93] reported that the α-adrenergic receptor blockers, phenoxybenzamine and dibenamine, produced increases in the amounts of norepinephrine appearing in the venous effluent from cat perfused spleen following stimulation of the sympathetic nerves to that organ. Further experiments indicated that the ability to increase stimulation induced efflux of norepinephrine appeared to be a general property of α-adrenergic antagonists. However, the effect was only noticed at high concentrations of antagonists. β-Adrenergic antagonists did not cause norepinephrine efflux.

In 1971, Kirpekar and Puig[94] proposed that the mechanism for this efflux involved the blockade of prejunctional (presynaptic) α-receptors, which normally act to inhibit the release of norepinephrine from the neuron. This concept was supported by the observation that the α-blocker-mediated increase in stimulation-induced efflux of norepinephrine was also accompanied by an increased efflux of dopamine-hydroxylase, which is normally present in synaptic granules along with norepinephrine. Classical inhibitors of neuronal uptake mechanisms, such as cocaine, do not increase norepinephrine efflux as strongly as α-blockers, nor do they stimulate efflux of dopamine-hydroxylase.

Thus, it appears that adrenergic neurons possess a mechanism for selfregulation of neurotransmitter release. According to this model, when synaptic levels of norepinephrine become high enough, they can stimulate α-receptors on the presynaptic terminal, and in this way inhibit further release of norepinephrine from neuronal granule stores (site 3, Figure 2).

The discovery of these prejunctional receptors suggests new avenues of research toward therapeutic agents. The use of prejunctional α-receptor blockers as therapeutic agents will be discussed in greater detail in the following chapter.

In the previous discussion on the structure-activity relationships of α-adrenergic receptors, it was noted that the presynaptic α-receptors possess structure-activity relationship trends different from those of postsynaptic α-receptors. Phenylephrine and methoxamine were the most selective agonists for the postsynaptic receptor termed α_1. Oxymetazoline, tramazoline, and α-methyl norepinephrine were more selective agonists for the presynaptic receptor termed α_2. α_2-Receptors could also be found at postsynaptic sites, thus clouding the generalization that all postsynaptic α-receptors are the same (i.e., α_1).

Perhaps the most noteworthy α_2-receptor agonist is clonidine (Figure 8). Clonidine has been shown to be effective in the treatment of hypertension. However, while part of its mechanism of action appears to be inhibition of the release of peripheral norepinephrine stores, the major antihypertensive activity is mediated through its action in the CNS.

It has also been recently reported that presynaptic β-receptors may exist.[95] In an analogous but opposite manner to the presynaptic α-receptors, these β-receptors may play a role in facilitating the release of norepinephrine stores from the neuron (site 3, Figure 2). Isoproterenol has been shown to increase the stimulation-induced overflow of norepinephrine, while propranolol blocks it. The receptor mediating the effect appears to have structure-activity relationship characteristics closer to β_2-receptors than β_1-receptors. The exact role or purpose of these presynaptic β-receptors is not clear. Given a choice, norepinephrine in the synaptic cleft appears to bind more readily to presynaptic α-receptors rather than presynaptic β-receptors. Thus, norepinephrine inhibits rather than promotes further release of neurotransmitter.

CLINICAL APPLICATIONS OF PRESYNAPTIC ADRENERGIC AGENTS

Anorexigenics

The Physiology of Feeding

Food intake is regulated by the sensations of hunger and satiety. Both these sensations appear to be under the control of specific neural centers for the regulation of food intake. The lateral hypothalamus (LH) controls the hunger center. Electrical stimulation of the LH nucleus causes an animal to eat voraciously. Conversely, lesioning the LH results in the complete loss of appetite, which can lead to starvation. The ventromedial hypothalamus controls the satiety center. Electrical stimulation of this nucleus results in inhibition of feeding, and lesioning of the area results in increased feeding. If both the lateral and the ventromedial hypothalamus are lesioned, complete loss of appetite results.

The sensation of hunger or the physiological demand for food thus arises from the net balance of signals produced by the hunger center of the lateral hypothalamus and the satiety center of the ventromedial hypothalamus.

Lesions of several other areas of the brain, such as the amygdala, septum, and midbrain, have also been shown to experimentally produce alterations in feeding behavior. The exact role of these structures in the feeding regulatory system, however, is unclear.

Food intake regulation can be classified as either short term or long term. Long-term regulation refers to the amount of food intake in relation to the amount of nutritive stores in the body. Subjects who have been either starved or overfed for prolonged periods of time will experience correspondingly intensified feelings of hunger or satiety due to the depletion or overabundance of nutritive supplies.

The precise mechanism by which the nutritive stores affect hunger is not known. However, any likely mechanism probably involves some method for physiologically monitoring blood levels of glucose, amino acids, and/or free fatty acids and relaying that information to the hypothalmic regulatory centers.

Short-term food intake regulation refers to dietary intake in relation to the amount of food that can be assimilated in a given period of time. This regulatory process governs such activities as overeating by a nutritionally adequate subject. Regulation can occur by two methods: "metering" of food as it passes through the mouth and reflexes caused by distension of the upper GI tract.

Metering of food means that sensory receptors in the mouth and pharynx detect the amount of chewing, salivation, swallowing, and tasting and thereby quantitate the amount of food that passes through the mouth. This information then passes to the hypothalamic feeding center to inhibit hunger. Similar hunger-inhibiting signals may also be sent from the upper GI tract. Pressure sensors in the tract can detect if it is distended from overeating, and relay the appropriate signal to the hypothalamus.

In addition to the hypothalmic centers, the conscious centers of the cerebral cortex also enter into the regulation of food intake. Such regulation can effect not only the quantity of food, but also the kinds of food consumed. Poor food choice and consumption, regulated by conscious centers, can lead to the paradoxical situation of overfed yet nutritionally starved subjects. Such conditions are becoming increasingly prevalent in the more affluent nations of the world.

Psychological factors may also play a role in food intake. Compulsive overeating may be the result of a psychological "override" of the normal physiological control mechanisms for feeding behavior.

Obesity is a chronic nutritional disease which results when caloric intake exceeds expenditure and the excess calories are converted to fat. Much obesity is simply the result of poor eating habits that have upset the balance of the hunger and satiety centers of the brain. Dieting and the adoption of better eating habits in such a case may be sufficient to reverse the obesity.

Some cases of obesity may be the result of an inherited imbalance between the hunger and satiety centers. As a result, the obese person appears to have an elevated mean body weight which his hunger centers constantly maintain. If he diets, and loses weight, he may experience such acute hunger that he eats voraciously until he returns to his former body weight. His appetite will then moderate so as to maintain that weight, neither increasing nor decreasing. Like a thermostat, his biological ''hungerstat'' is set at a higher than normal value.

Obesity can be treated by four different methods. These are (1) surgical intervention to shorten the length of the intestine or wiring the jaw shut, (2) increased exercise to burn off excess calories, (3) decreased food consumption, and (4) drug control of appetite.

The first method is extremely drastic and should not be used unless the condition of obesity is life-threatening and cannot be treated otherwise. The second and third methods, when performed under the guidance of a physician, are the preferred methods. Unfortunately, the selfcontrol required is great and most patients cannot adhere to their demands. The use of the fourth method, the chemical suppression of appetite, in combination with methods two and three, may be the most effective procedure for the largest population of patients.

The development of chemical appetite suppressants requires an understanding of the neurochemical basis of the hypothalmic hunger control function.

The Neurochemical Basis of Feeding

During the 1950s it was noted that several monoamines were not evenly distributed in the brain, but rather appeared to be localized in specific areas within the brain.[96] Furthermore, different monoamines appeared to be localized in different areas. As a result of the observations of this uneven distribution it was suggested that these monoamines might function as neurotransmitters in these specific areas of the CNS. By using fluorescence histochemical techniques, investigators were able to visualize monoamines localized in specific nerve cells within the brain. Two monoamines appeared to be localized in neurons and cell bodies within the lateral and ventromedial hypothalamus, the brain areas believed to control appetite. These monoamines were norepinephrine and dopamine.[97]

The direct involvement of these monoamines in hunger regulation has been demonstrated in experiments in which specific nerves containing either norepinephrine or dopamine were destroyed using the neurotoxin, 6-hydroxydopamine (6-OH-DA).[98] On the basis of these experiments, it appears that dopaminergic neurons in the lateral hypothalamus may be intimately involved in the regulation of this feeding center, whereas norepinephrine neurons in the lateral hypothalamus do not affect feeding behavior.

Experiments in which nerve fibers in the ventromedial hypothalamus were selectively destroyed by 6-OH-DA were also performed. Destruction of the norepinephrine-containing ventral ascending nerve bundle only moderately stimulated feeding and weight gain compared to the ravenous eating observed if the ventromedial hypothalamus is lesioned. Thus, while norepinephrine levels within the ventromedial hypothalamus may be involved in the regulation of the satiety center, norepinephrine may not be the sole regulatory factor in this function.

Finally, recent studies have shown a correlation between brain serotonin levels and feeding behavior. Intraventricular injection of parachlorophenylalanine, which depletes brain serotonin levels, caused rats to overeat and gain weight. It would appear, therefore, that serotonin neurons in the brain may also be responsible for inhibiting food intake.

MONOGRAPHS

AMPHETAMINE

Mechanism of Action

CNS Activity

The first clinical observations of amphetamine' action were reported by Piness et al.[99] in 1930, but it was not until 1933 that its CNS stimulant activity was reported.[100] The neurohumoral basis of this CNS activity, however, was not elucidated until the development of the compound α-methyl-*p*-tyrosine. α-Methyl-*p*-tyrosine inhibits the enzyme tyrosine hydroxylase, which is one of the enzymes responsible for the biosynthesis of norepinephrine (Figure 31). When animals were treated with α-methyl-*p*-tyrosine prior to administering amphetamine, the behavioral effects of amphetamine were abolished. The effects could be restored by the administration of the norepinephrine precursor L-dopa. If diethyldithiocarbamate, an inhibitor of the norepinephrine biosynthetic enzyme dopamine-hydroxylase, is given prior to amphetamine, no inhibition of the behavioral effects of amphetamine is noted. Based on the sequence of enzymes and their products in the biosynthesis of norepinephrine (Figure 31) and the effect of the inhibitors on the action of amphetamine, it was concluded that the principal behavioral effects of amphetamine must be dependent on the availability and release of dopamine rather than on a direct effect by amphetamine on some biological receptor.

The predominant behavioral effects of amphetamine include increased motor activity, mental alertness, decreased fatigue, and mild euphoria.

Anorexigenic Activity

While it appears that the principal behavioral effects of amphetamine in the CNS are mediated by dopamine, the mechanism of the anorectic activity of amphetamine is less clear. Treatment of the ventral noradrenergic bundle of the hypothalamus with 6-OH-DA results in the destruction of over 90% of noradrenergic terminals, but only a few dopamine terminals. Such treatment was shown to block the anorectic activity of amphetamine.[101] This suggests that amphetamine may act through norepinephrine by mediating its activity on adrenergic receptors in the satiety center of the ventromedial hypothalamus.

Conversely, it has been demonstrated that the anorectic action of amphetamine can also be antagonized by dopamine receptor blockers such as haloperidol and pimozide.[102] Thus, it is not clear whether the anorectic activity of amphetamine is mediated via noradrenergic or dopaminergic systems. Nor is it clear whether the anorectic activity involves the lateral or ventromedial portions of the hypothalamus. Lesions of the lateral hypothalamus usually result in a decrease in the anorectic activity of amphetamine. On the other hand, lesions of the ventromedial hypothalamus may cause either a decrease or an increase in anorectic activity. The results of ventromedial lesions have been contradictory. Therefore, the precise regulatory center and neurohumoral system for the action of amphetamine has not yet been solved. It has been possible though to study the effect of amphetamine on model norepinephrine and dopamine neurons. Through these studies one can determine the mechanism of action at the cellular level, irrespective of which neurohumoral system is involved in anorexic activity.

The major action of amphetamine, at both adrenergic and dopaminergic neurons, is the release of endogenous neurotransmitter from neuronal storage sites. This increased release of neurotransmitter is reflected in an increased turnover rate for both neurotransmitters as measured by their increase in metabolite production.

A second major action of amphetamine is its apparent inhibition of the mechanism of catecholamine uptake into the presynaptic terminals. Inhibition of uptake would thus lead to high synaptic levels of catecholamines.

Several other effects of amphetamine have also been observed which may play a lesser role in its mechanism of action. These effects include inhibition of monoamine oxidase, inhibition of dopamine and norepinephrine biosynthesis, and possibly some direct action on adrenergic receptors.

Peripheral Adrenergic Activity

While amphetamine has powerful CNS activity, it should not be forgotten that it also has activity on peripheral systems. These effects are due to its ability to release norepinephrine stores, and the resulting effects of norepinephrine on peripheral α- and β-adrenergic receptors. Thus, amphetamine can cause increases in both systolic and diastolic blood pressure, reflex bradycardia, smooth muscle relaxation, and vasoconstriction.

Structure-Activity Relationships

Amphetamine was first synthesized in 1887 by Edeleanu[103] as part of a program on the manufacture of aliphatic amines. Since that time, numerous derivatives of the phenethylamine structure have been synthesized. Earlier in this chapter an extensive discussion of the structure-activity relationships for phenethylamines having direct agonist activity on adrenergic receptors was presented. Similarly, a discussion on the structure-activity relationships for indirect-, direct-, and mixed-acting adrenergic agents was also presented. The structure-activity relationship trends presented in those discussions adequately reflect the modifications possible on the structure of amphetamine which might affect peripheral adrenergic action. This discussion will therefore be concerned solely with those structural modifications which might affect anorexigenic activity of amphetamine. Houlihan and Babington[104] have recently written an excellent review of structure-activity relationship trends for anorexigenics.

There are three general areas of chemical modification on the phenethylamine structure: the aromatic ring, the amine, and the ethylamine backbone.

The Ethylamine Side Chain

As was the case with other adrenergic agents, the two-carbon distance between the aromatic ring and the amine provides optimal activity. A number of compounds, however, have been synthesized in which the ethylamine side chain is incorporated into various heterocyclic systems such as piperidines, oxazolines, thioimidazolines, and morpholines. Such compounds maintained their anorexigenic activity. These compounds, however, tend to exhibit greater CNS stimulation effects relative to the parent amphetamine.

A number of alkyl substitutions can be made on the ethyl side chain, but the presence of an α-methyl group appears to be very important. Removal of the methyl group or substitution with other alkyl groups results in marked or complete loss of anorectic activity. Strong electron withdrawing groups in the α position are also deleterious. On the other hand, compounds with an additional methyl group (i.e., dimethyl) on the α-carbon maintain activity. Such compounds include phentermine and chlorphentermine (Table 15). Larger alkyl groups, however, begin decreasing activity.

Substitution on the β-Carbon

Substitution on the β-carbon is generally detrimental. All alkyl groups decrease activity. A β-hydroxyl group, as in ephedrine, or a β-keto group, as in diethyl propion, also decreases activity, but enough remains that they can still be used as anorexigenics. Presumably, the polar oxygen decreases to some degree the ability of the compound to penetrate the CNS. In the case of ephedrine the OH also promotes direct adrenergic receptor activity.

The α-carbon in amphetamine is asymmetric. This asymmetry appears to be important for anorexigenic activity in that the S(+) isomer is much more active than the R(−) isomer.

Table 15
PHARMACEUTICAL PREPARATIONS OF ANOREXIGENICS

Generic name and structure	Proprietary name, formulation	Manufacturer	Dosage form, dose
Amphetamine 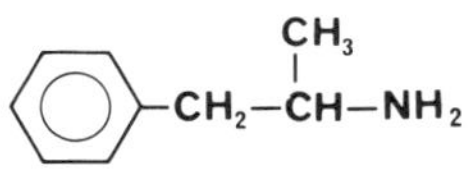	Benzedrine sulfate (± amphetamine)	Smith Kline and French	5- or 10-mg Tablets, 15 mg sustained-release capsule, 15 to 30 mg/day
	Biphetamine (dextroamphetamine and amphetamine 1:1)	Pennwalt	7.5 to 20 mg/Capsule, cation exchange resin complex, 1 capsule/day
	Delcobese (dextroamphetamine sulfate or adipate amphetamine sulfate or adipate 1:1)	Delco	5 to 20 mg/Tablet or capsule, 5 to 30 mg/day
	Dexamyl (dextroamphetamine sulfate and amobarbital)	Smith Kline and French	5 mg/Tablet; 10 to 15 mg/spansule; 30 to 40 mg/day
	Dexedrine (dextroamphetamine sulfate)	Smith Kline and French	5-, 10-, and 15-mg Spansules, sustained release, 10 to 15 mg/day
	Eskatrol (dextroamphetamine sulfate and prochlorperazine)	Smith Kline and French	15 mg/Spansule, 15 mg/day
	Obetrol (dextroamphetamine sulfate and saccharate and amphetamine sulfate and aspartate	Obetrol	10 mg and 20 mg/Tablet, 10 to 20 mg/day
	Obetan (dextroamphetamine tannate)	Mallinckrodt	5 and 7.5 mg/Tablet, 10 to 15 mg/day
Chlorphentermine 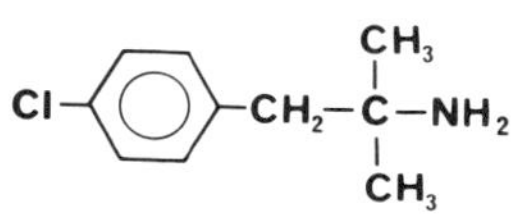	Pre-Sate (hydrochloride salt)	Warner/Chilcott	65 mg/Tablet, 1 tablet/day

Table 15 (continued)
PHARMACEUTICAL PREPARATIONS OF ANOREXIGENICS

Generic name and structure	Proprietary name, formulation	Manufacturer	Dosage form, dose
Chlortermine 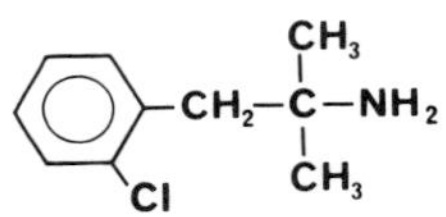	Voranil (hydrochloride salt)	USV Pharmaceutical	50 mg/Tablet, 1 tablet/day
Phendimetrazine 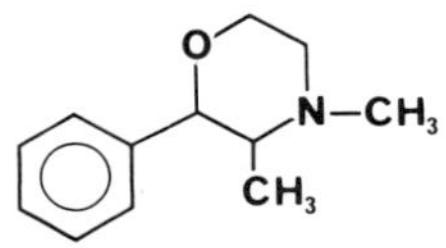	Bontril (tartrate salt) Melfiat (tartrate salt)	Carnrick Reid-Provident	35 mg/Tablet, 35 to 70 mg/day 35 mg/Tablet, 105-mg time release capsules
Phenmetrazine 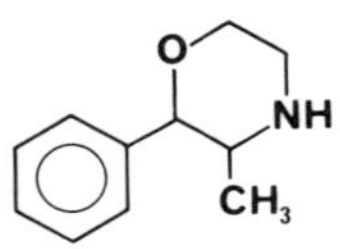	Preludin (hydrochloride salt)	Boehringer- Ingelheim	25 mg/Tablet, 50- and 75-mg time release capsules, 50 to 75 mg/day
Metamphetamine 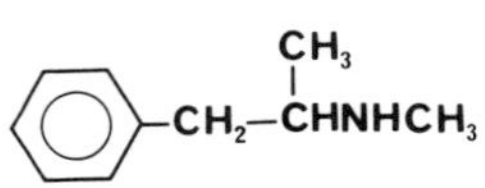	Desoxyn (metamphetamine hydrochloride) Fetamin (d-metamphetamine hydrochloride)	Abbott Mission	5 and 10 mg/Tablet, 5 to 15 mg/day 5 mg/Tablet, 5 to 15 mg/day
Benzphetamine 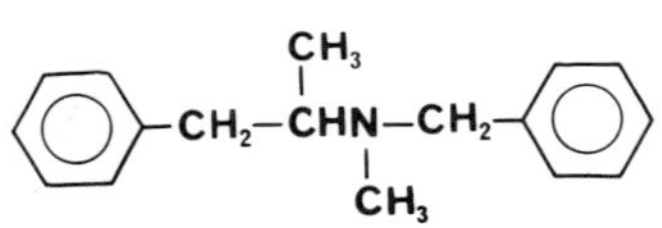	Didrex (benzphetamine hydrochloride)	Upjohn	25 and 50 mg/Tablet, 50 to 150 mg/day

Phentermine 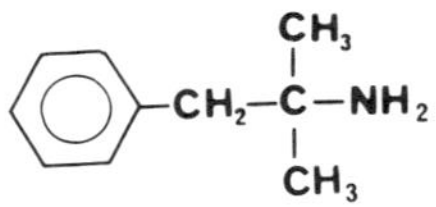	Fastion (hydrochloride salt) Ionamin (cation exchange resin complex)	Beecham Pennwalt	30 mg/Capsule, 30 to 60 mg/day 15 and 30 mg/Capsule, 15 to 30 mg/day
Mazindol 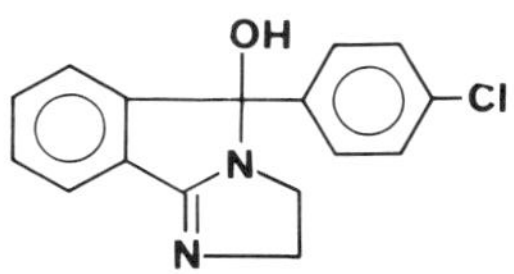	Sanorex (base)	Sandoz	1 and 2 mg/Tablet, 2 to 6 mg/day
Diethylpropion 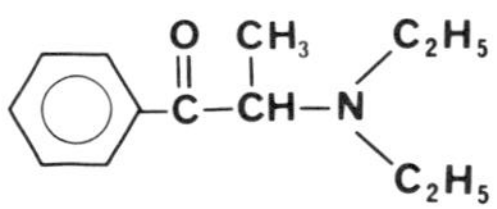	Tenuate (hydrochloride salt) Tepanil (hydrochloride salt)	Merrell Riker	25 mg/Tablet, 75-mg time-release tablets; 25 to 75 mg/day 25 mg/Tablet, 75-mg time-release tablets; 25 to 75 mg/day
Fenfluramine 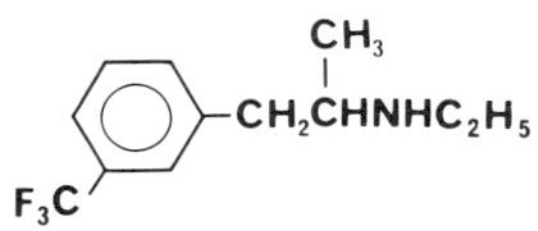	Pondimin (hydrochloride salt)	Robins	20 mg/Tablet, 20 to 60 mg/day

Aromatic Substitution

As was the case with substitution on the β-carbon, substitution on the aromatic ring of amphetamine almost always results in decreased activity. Functionalities which have the best chance of maintaining activity are lipophilic groups such as Cl, F, and CF_3 substituted in the *meta* or *para* positions of the ring.

Amine Substitution

The addition of alkyl groups to the amine of amphetamine almost always results in decreased anorexigenic activity. The exception to this rule is the presence of alkyl groups which can be easily removed from the nitrogen by metabolism, thus generating the parent amphetamine. Such substituents would include small alkyl groups such as methyl and ethyl, and aryl groups which are attached to the amine by a benzylic carbon. Benzylic and allylic carbons are particularly susceptible to P-450 oxidative *N*-dealkylation (see pharmacokinetics section).

One should be cautious in evaluating structure-activity relationship trends for anorexigenics, particularly for the purpose of drug design. Since the exact locus of activity and the receptor(s) involved in mediating these actions are not known, structure-activity relationship trends may be meaningless. As discussed in a previous section, adrenergic, dopaminergic, and even serotinergic receptors may be involved in regulating feeding behavior. Thus, increases or decreases in activity as a function of structural modification may reflect a shift in mechanism or site of action, rather than affinity for a given "anorexigenic receptor".

Pharmacokinetics

Absorption and Distribution

After a 10-mg oral dose, amphetamine is rapidly absorbed, and reaches maximal blood levels in 2 to 3 hr. Maximal plasma concentrations in man are 10 to 30 μg/mℓ. The volume of distribution is about 2 to 4 ℓ/kg body weight. When resin-bound amphetamine is administered the absorption rate was slower, and peak plasma levels were lower.

After an i.p. injection of amphetamine sulfate, the compound was quickly distributed to most tissues, but relatively high concentrations were found in the liver, brain, spleen, lung, and kidney.[105] The compound readily penetrates the blood-brain barrier. Interestingly, one of the areas of the brain where amphetamine localizes is the hypothalamus, which is associated with feeding behavior.

The plasma half-life of amphetamine in man is 6 to 8 hr, but may vary greatly depending on urine pH. Excretion is almost exclusively via the urine. Since amphetamine is a basic amine, its amount of tubular reabsorption in the kidney is dependent on the pH of the urine. If the urine is acidic, the compound is protonated, not readily reabsorbed, and rapidly excreted unchanged. If the urine is neutral or basic, amphetamine remains unionized, is extensively reabsorbed, and the duration of action prolonged. Under normal conditions, up to 50% of amphetamine is excreted unchanged. Reabsorption also provides a greater opportunity for metabolism of the compound, and thus, a greater number of amphetamine metabolites can be isolated from basic or neutral urine.

Aromatic Hydroxylation

Hydroxylation of the ring occurs exclusively at the *para* position. In man, however, this is a relatively minor pathway, accounting for only 3 to 7% of an administered dose. Oxidation can occur via a microsomal P-450 enzyme, but this may not be the only enzyme involved.

β-Carbon Hydroxylation

Amphetamine can be hydroxylated on the β-carbon via the enzyme dopamine β-hydroxylase, which is found in adrenergic nerve terminals. Like aromatic hydroxylation, β-carbon hydroxylation is a relatively minor pathway, accounting for about 3% of the administered dose.

N-Dealkylation

N-Dealkylation is a primary route of metabolism for *N*-substituted amphetamines. As mentioned in the structure-activity relationships section, the biological activity of many *N*-substituted compounds may be due to the fact that the compound is first dealkylated to generate the active amine. The amount of dealkylation which occurs appears to be dependent on both the type of substituent as well as the degree of nitrogen substitution. The larger the alkyl substituent, the more likely it will be cleaved. Thus, only 10% of *N*-methyl amphetamine is dealkylated, whereas 45% of *N*-isopropyl amphetamine is dealkylated. The degree of nitrogen substitution also influences dealkylation in that tertiary amines are dealkylated faster than secondary amines. Finally, there appears to be some stereoselectivity in the dealkylation process. The (+) isomers of amphetamine are more extensively dealkylated than are the (−) isomers.

Like aromatic hydroxylation, the primary site for *N*-dealkylation is the P-450 system of the liver. Oxidation of the carbon adjacent to the amine yields an α-carbinolamine which nonenzymatically decomposes to the free amine and corresponding aldehyde. Substituents which have carbons adjacent to the amine which are particularly susceptible to oxidation by P-450 (i.e., benzylic, allylic, tertiary), are thus more readily hydroxylated and cleaved.

Oxidative Deamination

The major route of metabolism for amphetamine is deamination. Deaminated metabolites account for 20 to 50% of the administered dose. *N*-Substituted amphetamines also undergo deamination, but usually undergo *N*-dealkylation first. This is particularly true if the alkyl substituent is large.

The mechanism for this oxidation is not clear-cut, and two major routes may be involved. One route involves a carbinolamine intermediate which arises from α-carbon hydroxylation (Figure 33). This unstable carbinolamine would then nonenzymatically decompose to the corresponding ketone. Alternatively, amphetamine could undergo *N*-oxidation to form the hydroxylamine, oxidation to the oxime, and hydrolysis to the ketone.

Evidence has been presented which supports the existence of both pathways.[106] Some discrepancies have been found, however, which suggest that other pathways are possible, including one which might require an α-hydroxy hydroxylamine intermediate (shown in brackets in Figure 33).

Regardless of the route by which the 1-phenyl-2-propanone is formed, it is known that this ketone can undergo futher oxidative metabolism to benzoic acid or reductive metabolism to the corresponding alcohol. These compounds can in turn undergo oxidation and conjugation to form other metabolites.

Uses and Dosage

Amphetamines are used for their cerebral stimulant properties in the treatment of mental disorders such as mood disturbances, chronic nervous exhaustion, psychoneurosis, and schizophrenia.[48] However, the use of amphetamines in the treatment of these disorders has been limited.

The cerebral stimulant property of amphetamine has also been used in the treatment of narcolepsy. However, amphetamine should not be used to combat fatigue or exhaustion or to replace sleep in normal persons. Amphetamine has also been found to be of value in the treatment of hyperkinesia and abnormal behavior patterns in children. The largest use of amphetamine and its analogs is in the treatment of obesity. The long-term value of such therapy, however, is doubtful due to the development of tolerance and dependence (see

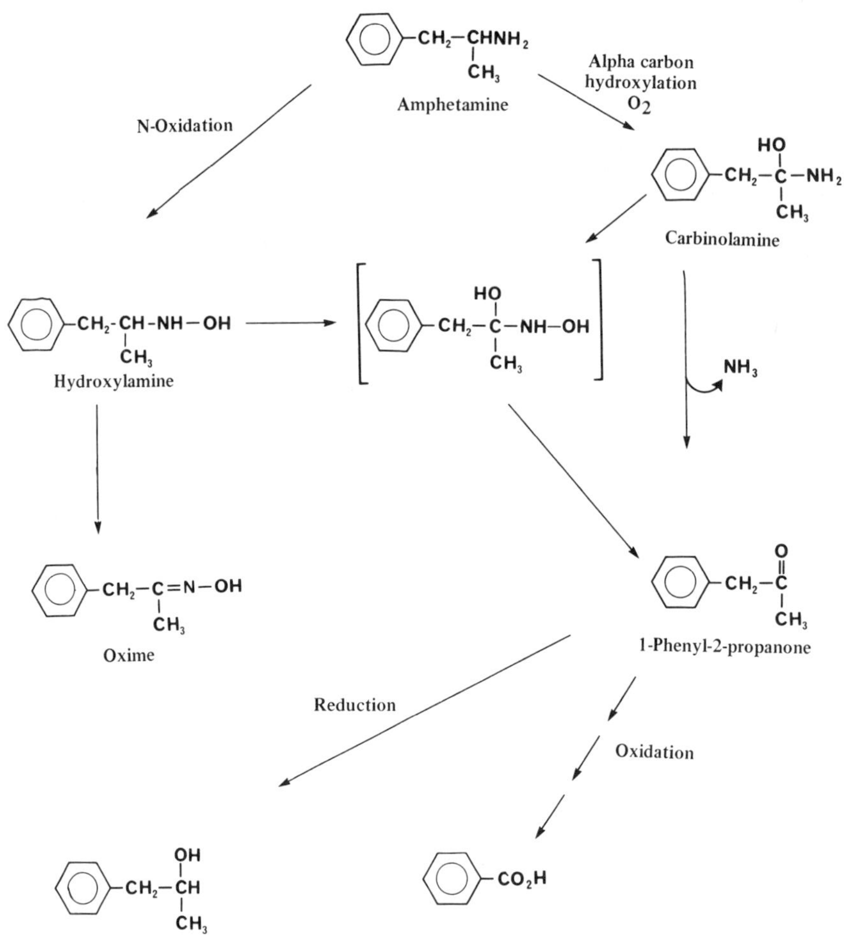

FIGURE 33. Possible route for the metabolism of amphetamine.

toxicity section). Pharmaceutical preparations and dosages for amphetamine are given in Table 15.

Toxicity

The major toxicity problem of amphetamine is the developement of tolerance. The mechanism involved in the development of this tolerance has been the subject of a great deal of research. One popular theory involves the production of a false transmitter. As discussed in the mechanism of action section, amphetamine exerts its pharmacological effects by displacing neurotransmitters (norepinephrine or dopamine) from their neuronal storage sites. It is this released neurotransmitter, acting on its receptor, which produces the pharmacological effects observed on amphetamine administration. Amphetamine has little or no (depending on the receptor) direct action.

As discussed in the pharmacokinetics section, however, amphetamine can undergo several different types of metabolic biotransformations. Two of these biotransformations are *para*-hydroxylation and β-carbon hydroxylation which produce the compounds *p*-hydroxyamphetamine and *p*-hydroxynorephedrine (Figure 34). Both of these structural modifications lead to compounds which are structurally more similar to norepinephrine than amphetamine itself. As a result of these modifications, it is believed that *p*-hydroxynorephedrine can be taken up and preferentially accumulated in the neuron at the expense of the more easily

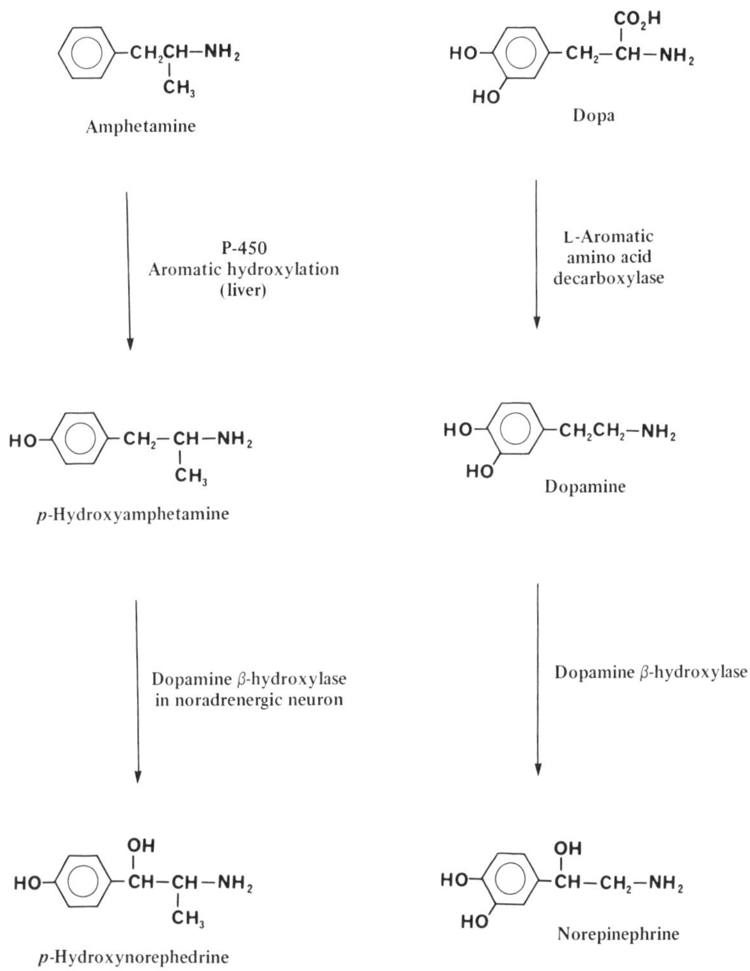

FIGURE 34. Biosynthesis of the false transmitter, *p*-hydroxy-norephedrine.

metabolized norepinephrine. While *p*-hydroxyamphetamine does have direct agonist activity on adrenergic receptors, it is only 1/500 that of norepinephrine.

Thus, on long-term therapy, the theory states that amphetamine would be metabolized to *p*-hydroxynorephedrine which would accumulate in the neuron, thus depleting norepinephrine stores. Further administration of amphetamine would thus liberate the false transmitter *p*-hydroxynorephedrine rather than norepinephrine. Since this compound is only 1/500 as active as norepinephrine at the adrenergic receptor, the pharmacological response is less. This then would appear as tolerance to amphetamine action.[107]

While the false transmitter theory for the development of amphetamine tolerance has a number of attractive features, it may not totally explain the tolerance syndrome. Part of the problem is that some amphetamine analogs, which because of their structure cannot be metabolized to false transmitters, still give rise to tolerance on prolonged therapy.

In addition to tolerance, amphetamine has a number of other toxicities as a result of its CNS and peripheral activity. Among its CNS toxicities are nervousness, insomnia, irritability, dizziness, headache, and increased motor activity. Some of its peripheral toxicities include hyper- or hypotension, tachycardia, palpitations, nausea, vomiting, and cramps. Psychic dependence or addiction is possible. Amphetamine should not be coadministered with monoamine oxidase inhibitors since they can potentiate the effects of neurotransmitters released by amphetamine. The LD_{50} (oral) in mice is 22 mg/kg and in rabbits, 85 mg/kg.[49]

Physical-Chemical Data

Structure:	Figure 34
Empirical formula:	$C_9H_{13}N$
Mol wt:	135.2
Boiling point:	200 to 203°C
Chemical name:	(±)-α-methylphenethylamine
Generic name:	amphetamine
Proprietary name:	Benzedrine, others (see Table 15)
Physical description:	colorless, mobile, slowly volatile liquid
Solubilities:	1 in 50 of water, freely soluble in ethanol, ether, and chloroform
Salt forms:	hydrochloride, sulfate, phosphate
pK_a	9.94.

OTHER COMPOUNDS OF INTEREST

As discussed in the structure-activity relationships section in the amphetamine monograph, a number of different structural types of compounds have anorexigenic activity. The structures and pharmaceutical preparations for some of the most used compounds are given in Table 15. All these compounds, however, probably act at noradrenergic or dopaminergic receptors. One compound, however, is noteworthy in that it may act via serotonergic receptors.

Fenfluramine (Table 15) is an amphetamine analog which has anorexigenic activity. Unlike amphetamine, however, the anorexigenic activity of fenfluramine is not inhibited by agents that deplete brain catecholamines or block dopamine receptors. Furthermore, lesioning of the ventromedial noradrenergic bundles of the hypothalamus, rather than reducing anorexigenic activity, as it does with amphetamine, in fact enhances the anorexigenic activity of fenfluramine.[108] Such experimental observations suggest that the anorexigenic activity of fenfluramine may be mediated through a system other than a noradrenergic or dopaminergic one. A likely candidate is the serotonergic system.

Fenfluramine, at anorectic doses, has been shown to deplete brain serotonin levels. Furthermore, it was shown to be able to both displace serotonin from neuronal storage granules as well as inhibit its uptake. The possible involvement of serotonin receptors in regulating feeding behavior has been previously discussed. The possibility that its anorexigenic activity is mediated through this system is supported by the following observations:[109]

1. Selective destruction of brain serotonin neurons antagonizes fenfluramine anorexia.
2. Serotonin receptor antagonists, such as methergoline, inhibit fenfluramine anorexia.
3. Chlorimipramine, which inhibits the effect of fenfluramine on brain serotonin, also inhibits the effect of fenfluramine on food intake.

It appears, therefore, that it may be possible to elicit anorexigenic activity through serotonin receptors. If this is so, then structure-activity studies based on fenfluramine analogs may provide new avenues of investigation toward new anorexigenic agents.

TESTING METHODS FOR PRESYNAPTIC ADRENERGIC AGENTS

The majority of experiments on agents acting presynaptically seek information not only on the type or magnitude of pharmacological response, but also on the mechanism whereby that response is elicited. For this reason the following discussion will be divided into two sections: experimental parameters for determining site and mechanism of action and pharmacological testing systems. A particularly good review of this material has been provided by Armstrong and Green.[110]

Experimental Parameters for Determining Site and Mechanism of Action

1. Ganglionic blockers can be observed by electrical stimulation of preganglionic nerves and measurement of postganglionic response. In vivo and in vitro methods are available.
2. Transmitter releasing agents can be diagnosed by measurement of both pharmacological response as well as release of norepinephrine from neuronal stores. The activity of these agents can also be characterized by the fact that continued stimulation of the system leads to decreased response as neurotransmitter stores are depleted. Similarly, uptake inhibitors tend to decrease or abolish the effects of these agents. Testing of these agents is particularly suitable in perfused organ preparations or isolated tissue baths.
3. Neuronal blocking agents are diagnosed by the fact that they inhibit electrically stimulated sympathomimetic response. Furthermore, the inhibition can be reversed by neurotransmitter releasing agents, and antagonized by neuronal uptake inhibitors. Electrically induced release of norepinephrine in perfusates is also antagonized.
4. Presynaptic receptor agonists are best measured in preparations which allow detection of neurotransmitter efflux. In order to observe activity, synaptic and extraneuronal uptake mechanisms must be inhibited by cocaine and metanephrine, respectively. Stimulation of presynaptic α-receptors leads to an inhibition of efflux of both norepinephrine and dopamine β-hydroxylase from neuronal storage sites.
5. False transmitters are characterized by the fact that they generally cause decreased responses in sympathetic systems. The false transmitter should be detectable in the tissue bath or perfusate, and it should be released on electrical stimulation of the nerve.
6. Uptake inhibitors are characterized by their ability to deplete neurotransmitter stores, and thus, inhibit sympathetic activity in in vivo preparations. In in vitro preparations, uptake inhibitors may potentiate activity of direct agonists.

Pharmacological Testing Systems

1. The cat spleen can be prepared as either a perfused organ preparation or it can be cut into strips for isolated tissue studies. Norepinephrine released by presynaptic agonists, depolarization, or releasing agents can be measured in the venous effluent from the isolated organ, or in the tissue bath. Electrical sympathetic stimulation of the preparation causes contraction of the strips, or increased vascular resistance in the perfused organ. The system is not generally suitable for testing ganglionic blockers.
2. Rabbit pulmonary artery can be dissected and cut into a spiral strip for use in isolated tissue baths. It is mostly vascular smooth muscle and is particularly sensitive to vasoactive agents. Detectable amounts of norepinephrine are released into tissue baths by sympathetic stimulation, or releasing agents. The preparation does not contain ganglionic nerves, and thus, is not suitable for testing ganglionic blockers. It also does not appear to possess presynaptic receptors.
3. The vas deferens from rat, along with its hypogastric nerve, can be dissected and mounted in an isolated tissue bath. The vas deferens will contract on sympathetic or electrical stimulation. Additional sensitivity can be obtained by longitudinally cutting the preparation to expose more surface area. The hypogastric nerve is particularly useful for testing ganglionic blockers. Norepinephrine efflux is difficult to detect.
4. The pithed rat system is useful for observation of the effects of total sympathetic outflow. The rat trachea is canulated and artificially respirated. The left carotid artery

is canulated for measurement of blood pressure and pulse. Drugs are administered through the left femoral vein. The spinal cord is pithed with a steel rod which is insulated except for the area corresponding to the thorax and lumbar vertebrae. Electrical pulses through this rod cause stimulation of sympathetic outflow from those areas of the spinal cord. Total physiological response in the presence and absence of test drug is measured.

5. In vivo responses to presynaptic agents can also be observed in typical anesthetized cat nictitating membrane preparations.

REFERENCES

1. **Oliver, G. and Schaefer, J.,** The physiological effects of extracts of the suprarenal capsules, *J. Physiol. (London),* 18, 230, 1895.
2. **Abel, J. J. and Crawford, A. C.,** On the blood pressure raising constituent of the suprarenal capsule, *Bull. Johns Hopkins Hosp.,* 8, 151, 1897.
3. **Von Furth, I.,** Zur Kenntniss der brenzcatechinahnlichen substanz der nebenneiren, *Zietschr. f. Physiol. Chem.,* 29, 105, 1900.
4. **Takamine, J.,** The isolation of the active principle of the suprarenal gland, *J. Physiol. (London),* 27, 29P, 1901.
5. **Stolz, L.,** Uber Adrenalin und Alkylaminoacetobrenzcatechin, *Ber. Dtsch. Chem. Ges.,* 37, 4149, 1904.
6. **Dakin, H. D.,** The synthesis of a substance allied to adrenalin, *Proc. R. Soc. (London),* 76, 491, 1905.
7. **Elliott, T. R.,** On the action of adrenalin, *J. Physiol. (London),* 31, xxp, 1904.
8. **Barger, G. and Dale, H. H.,** Chemical structure and sympathomimetic action of amines, *J. Physiol. (London),* 41, 19, 1910.
9. **von Euler, U. S.,** Identification of the sympathomimetic ergone in adrenergic nerve of cattle (sympathin N) with laevo-noradrenalin, *Acta Physiol. Scand.,* 16, 63, 1849.
10. **Ehrlich, P.,** Address delivered at dedication of the George-Speyer-Haus, Sept. 6, 1906, in *Collected Papers of Paul Ehrlich,* Vol. 3, Himmelweit, F., Ed., Pergamon Press, London, 1960.
11. **Dale, H. H.,** On some physiological actions of ergot, *J. Physiol. (London),* 34, 163, 1906.
12. **Cannon, W. B. and Rosenbleuth, A.,** *Autonomic Neuro-Effector Systems,* Macmillan, New York, 1937.
13. **Ahlquist, R. P.,** A study of the adrenotropic receptors, *Am. J. Physiol.,* 153, 586, 1948.
14. **Lands, A. M.,** Sympathetic receptor action, *Am. J. Physiol.,* 169, 11, 1952.
15. **Furchgott, R. F.,** The receptors for epinephrine and norepinephrine (adrenergic receptors), *Pharmacol. Rev.,* 11, 429, 1959.
16. **Konzett, H.,** Neue broncholytisch hochwirksame Korper der Adrenalinreihe, *Naunyn-Schmeid. Arch. Exp. Pathol. Pharmakol.,* 197, 27, 1940.
17. **Lands, A. M.,** The pharmacological activity of epinephrine and related dihydroxy-phenylalkylamines, *Pharmacol. Rev.,* 1, 279, 1949.
18. **Lands, A. M., Arnold, A., McAuliff, J. P., Luduena, F. P. and Brown, T. G., Jr.,** Differentiation of receptor systems activated by sympathomimetic amines, *Nature (London),* 214, 597, 1967.
19. **Arnold, A.,** Differentiation of receptors activated by catechol amines, *Farm. Sci.,* 27, 79, 1972.
20. **Phillips, D. K.,** Chemistry of alpha and beta-adrenoceptor agonists and antagonists, in *Handbook of Experimental Pharmacology,* Vol. 54, Part I, Szekeres, L., Ed., Springer-Verlag, New York, 1980, chap. 1.
21. **Uloth, R. H., Kirk, J. R., Gould, W. A., and Larsen, A. A.,** Sulfonanilides. I. Monoalkyl and arylsulfonamidophenylethanolamines, *J. Med. Chem.,* 9, 88, 1966.
22. **Patil, P. N., Miller, D. D., and Trendelenburg, U.,** Molecular geometry and adrenergic drug activity, *Pharmacol. Rev.,* 26, 323, 1974.
23. **Easson, L. H. and Stedman, E.,** CLXX, Studies on the relationship between chemical constitution and physiological action. V. Molecular dissymmetry and physiological activity, *Biochem. J.,* 27, 1257, 1933.
24. **Jen, T. and Kaiser, C.,** Adrenergic agents. V. Conformational analysis of 1-alkylamino-3-aryloxy-2-propanols by proton magnetic resonance studies. Implications relating to the steric requirements of adrenoreceptors, *J. Med. Chem.,* 20, 693, 1977.
25. **Starke, K., Endo, T., and Taube, H. D.,** Relative pre- and post-synaptic potencies of α adrenoceptor agonists in the rabbit pulmonary artery, *Naunyn-Schmidebergs Arch. Pharmacol.,* 291, 55, 1975.
26. **Langer, S. Z.,** Presynaptic regulation of catecholamine release, *Biochem. Pharmacol.,* 23, 1793, 1974.

27. **Berthelsen, S. and Pettinger, W. A.,** A functional basis for classification of adrenergic receptors, *Life Sci.,* 21, 595, 1977.
28. **Hartmann, M. and Isler, H.,** Chemische konstitution und pharamcologishe wirk-samkeit von in 2-stellung substitutierten imidazolinen, *Naunyn-Schmiedebergs Arch. Exp. Pathol. Pharmakol.,* 192, 141, 1939.
29. **Mujic, M. and Van Rossum, J. M.,** Comparative pharmacodynamics of sympathomimetic imidazolines; studies on intestinal smooth muscle of the rabbit and the cardiovascular system of the cat, *Arch. Int. Pharmacodyn.,* 155, 432, 1965.
30. **Jackman, L. M. and Jen, T.,** H^1 and C^{13} nuclear magnetic resonance studies on the tautomerism, geometrical isomerism, and conformation of some cyclic amidines, guanidines, and related systems, *J. Am. Chem. Soc.,* 97, 2811, 1975.
31. **Ruffolo, R. R., Jr., Turkowski, B. S., and Patil, P. N.,** Lack of cross-desensitization between structurally dissimilar -adrenoceptor agonists, *J. Pharm. Pharmacol.,* 29, 378, 1977.
32. **Sutherland, E. W. and Rall, T. W.,** Formation of a cyclic adenine ribonucleotide by tissue particles, *J. Biol. Chem.,* 232, 1065, 1958.
33. **Robison, G. A., Butcher, R. W., and Sutherland, E. W.,** *Cyclic AMP,* Academic Press, New York, 1971.
34. **Moss, J. and Vaughn, M.,** Activation of adenylate cyclase by choleragen, *Annu. Rev. Biochem.,* 48, 581, 1979.
35. **Rodbell, M., Krans, H. M. J., Pohl, S. L., and Birnbaumer, L.,** The glucagon-sensitive adenylate cyclase system in plasma membranes of rat liver. IV. Effects of guanyl nucleotiedes on binding of 125I-glucagon, *J. Biol. Chem.,* 246, 1872, 1971.
36. **Moss, J., Osborne, J. C., Jr., Fishman, P. H., Brewer, H. B., Vaughn, M., and Brady, R. O.,** Effect of gangliosides, and substrate analogs on the hydrolysis of nicotinamide adenine dinucleotide by choleragen, *Proc. Natl. Acad. Sci. U.S.A.,* 74, 74, 1977.
37. **Mukherjee, C. and Lefkowitz, R. J.,** Regulation of beta adrenergic receptors in isolated frog erythrocyte plasma membranes, *Mol. Pharmacol.,* 13, 291, 1977.
38. **Su, Y.-F., Harden, T. K., and Perkins, J. P.,** Catecholamine-specific desensitization of adenylate cyclase: evidence for a multistep process, *J. Biol. Chem.,* 255, 7410, 1980.
39. **Hanski, E., Seville, N., and Levitzki, A.,** The allosteric inhibition by calcium of soluble and partially purified adenylate cyclase from turkey erythrocytes, *Eur. J. Biochem.,* 76, 513, 1977.
40. **Endoh, M. and Schumann, H. J.,** Frequency-dependance of the positive inotropic effect of methoxamine and naphazoline mediated by alpha receptors in the isolated rabbit papillary muscle, *Naunyn-Schmidedebergs Arch. Pharmacol.,* 287, 377, 1975.
41. **Watanabe, A. M., Hathaway, D. R., Besch, H. R., Jr., Farner, B. B., and Harris, R. A.,** Alpha adrenergic reduction of cyclic adenosine monophosphate concentration in rat myocardium, *Circ. Res.,* 40, 596, 1977.
42. **Amer, S. M.,** Cyclic guanosine 3′,5′-monophosphate and gallbladder contraction, *Gastroenterology,* 67, 333, 1974.
43. **Aviado, D. M. and Micozzi, M. S.,** Systemic pharmacology of adrenergic activators and inhibitors: effects on the respiratory system, in *Handbook of Experimental Pharmacology,* Vol. 54, Part II, Szekeres, L., Ed., Springer-Verlag, New York, 1980, chap. 1.
44. **Bruce, R. B. and Ditts, J. E.,** The determination and excretion of phenylephrine in urine, *Biochem. Pharmacol.,* 17, 335, 1968.
45. **Troup, A. E. and Mitchner, H.,** Degradation of phenylephrine hydrochloride in tablet formulations containing aspirin, *J. Pharm. Sci.,* 53, 375, 1964.
46. Litton Industries, *Physicians Desk Reference,* Vol. 31, Medical Economics, Oradell, N. J., 1977.
47. **Griffin, J. P. and D'Arcy, P. F.,** *A Manual of Adverse Drug Interactions,* John Wright & Sons, Bristol, 1979, chap. 19.
48. *American Hospital Formulary,* American Society of Hospital Pharmacists, Bethesda, Md., 1982.
49. **Clarke, E. G. C.,** *Isolation and Identification of Drugs,* Vol. 1, Pharmaceutical Press, London, 1974.
50. **Brodie, B. B., Aronow, L., and Axelrod, J.,** The fate of benzazoline (Priscoline) in dog and man and a method for its estimation in biological material, *J. Pharmacol. Exp. Ther.,* 106, 200, 1952.
51. **Brunner, H., Hedwall, P. R., Maitre, L., and Meir, M.,** Antihypertensive effects of alpha-methylated catecholamine analogs in the rat, *Br. J. Pharmacol.,* 30, 108, 1967.
52. **Albertson, N. F., McKay, F. C., Lape, H. E., Hoppe, J. O., Selberis, W. H., and Arnold, A.,** The optical isomers of metaraminol. Synthesis and biological activity, *J. Med. Chem.,* 13, 132, 1970.
53. **Falkay, G., Herczeg, J., and Kovacs, L.,** Effect of beta-mimetic isoxuprine on prostaglandin biosynthesis in pregnant human myometrium *in vitro, Life Sci.,* 23, 2689, 1978.
54. **Kulz, F. and Schneider, M.,** Uber neue gefasserweiternde sympathicomimetica, *Klin. Wohnschr.,* 28, 535, 1950.
55. **Wiemers, K.,** Uber gefasserweiternde aralkyle der adrenalinbenzedrinreihe, *Arch. Exp. Pathol. Pharmakol.,* 213, 283, 1951.

56. **Lorenz, D.,** Untersuchungen mit beta-adrenergisch wirksamen substanz am wehenbereiten isoierten uterus der trachtigen ratte, in *Methoden der Pharmakologischen Geburtserleichterung und Uterus-Relaxation,* Jung, H., Ed., Georg Thiem Verlag, Stuttgart, 1972, 159.
57. **Bristol, J. A. and Evans, D. B.,** Cardiotonic agents for the treatment of heart failure, in *Annual Reports in Medicinal Chemistry,* Vol. 16, Hess, H.-J., Ed., Academic Press, New York, 1981, chap. 10.
58. **Campbell, S. F. and Danilewicz, J. C.,** Agents for the treatment of heart failure, in *Annual Reports in Medicinal Chemistry,* Vol. 13, Clarke, F. H., Ed., Academic Press, New York, 1978, chap. 10.

58a. **Morris, H. R., Taylor, G. W., Jones, C. M., Piper, P. J., Tippins, J. R., and Samhoun, M. N.,** Structure elucidation, biosynthesis and biodegradation of SRS-A from lung, in *SRS-A and Leukotrienes,* Piper, P. J., Ed., John Wiley & Sons, New York, 1981, 19.

58b. **Hammarström, S., Murphy, R. C., Samuelsson, B., Clark, D. A., Mioskowski, C., and Corey, E. J.,** Structure of leukotriene C. Identification of the amino acid part, *Biochem. Biophys. Res. Commun.,* 91, 1266, 1979.

59. **Szentivanyi, A.,** Beta-adrenergic theory of the atopic abnormality in bronchial asthma, *J. Allergy,* 42, 203, 1968.
60. **Venter, J. C., Fraser, C. M., and Harrison, L. C.,** Autoantibodies to $beta_2$ adrenergic receptors: a possible cause of adrenergic hyporesponsiveness in allergic rhinitis and asthma, *Science,* 207, 1361, 1980.
61. **Lowell, F. C.,** Therapeutic schedules in asthma, in *Asthma: Physiology, Immunopharmacology, and Treatment,* Lichtenstein, L. M. and Austen K. F., Eds., Academic Press, New York, 1972, chap. 47.
62. **Konzett, H.,** On the discovery of Isoprenaline, *Trends Pharmacol. Sci.,* 2, 47, 1981.
63. **Konzett, H.,** Neue broncholytisch hochwirksame korper der adrenalinreihe, *Naunyn-Schmiedebergs Arch. Pharmakol.,* 197, 27, 1940.
64. **Bennett, P., Blackwell, E., and Davis, D. S.,** Metabolism of drugs by intestinal mucosa-effect of salicylamide on bio-availability of oral isoprenaline in dogs, *Abstracts VI Int. Cong. of Pharmacology,* Pergamon Press, New York, 1975, 583.
65. **Reichl, R., Traunecker, W., and Engelhardt, A.,** Quantitative pharmacological investigations with 3-methoxy-isoprenaline, the primary metabolite of isoprenaline, *Proc. Eur. Soc. Drug Tox.,* 12, 142, 1971.
66. **Aviado, D. M.,** Potential hazards of bronchodilator aerosols. Toxicity of aerosols, *J. Clin. Pharmacol.,* 15, 86, 1975.
67. **Osol, H., Pratt, R., and Gennaro, A., Eds.,** *The Dispensatory of the United States of America,* 25th ed., J. P. Lippincott, Philadelphia, 1960, 500.
68. **Chen, K. K. and Schmidt, C. F.,** The action of ephedrine, the active principle of the chinese drug Ma Huang, *J. Pharmacol. Exp. Ther.,* 24, 339, 1924.
69. **Patil, P. N. and Jacobowitz, D.,** Steric aspects of adrenergic drugs. IX. Pharmacologic and histochemical studies on isomers of cobefrin (α-methylnorepinephrine), *J. Pharmacol. Exp. Ther.,* 161, 279, 1968.
70. **Thiercelin, J. F., Jacquot, C., Rapin, J. R., and Cohen, Y.,** Pharmacocinetique de la DL-Norephedrine ^{14}C chez le rat, *Arch. Int. Pharmacodyn. Ther.,* 220, 153, 1976.
71. **Wilkinson, G. R. and Beckett, A. H.,** Absorption metabolism, and excretion of the ephedrines in man. I. The influence of urinary pH and urine volume output, *J. Pharmacol. Exp. Ther.,* 162, 139, 1968.
72. **Hartley, D., Jack, D., Lunts, L. H. C., and Ritchie, A. C.,** A new class of selective stimulants of beta-adrenergic receptors, *Nature (London),* 219, 861, 1968.
73. **Brittain, R. T., Jack, D., and Ritchie, A. C.,** Recent beta-adrenoreceptor stimulants, *Adv. Drug Res.,* 5, 197, 1970.
74. **Bianchi, C. P.,** Cellular pharmacology of contraction of skeletal muscle, in *Cellular Pharmacology of Exitable Tissues,* Narahashi, T., Ed., Charles C Thomas, Springfield, Ill., 1975.
75. **Sattin, A. and Rall, T. W.,** The effect of adenosine and adenine nucleotides on the adenosine 3′,5′-phosphate content of guinea pig cerebral cortex slices, *Mol. Pharmacol.,* 6, 13, 1970.
76. **Wells, J. N., Baird, C. E., Wu, Y. L., and Hardman, J. G.,** Cyclic nucleotide phosphodiesterase activities of pig coronary arteries, *Biochim. Biophys. Acta,* 384, 430, 1975.
77. **Kramer, G. L., Garst, J. E., Mitchel, S. S., and Wells, J. N.,** Selective inhibition of cyclic nucleotide phosphodiesterase by analogues of 1-methyl-3-isobutylxanthine, *Biochemistry,* 16, 3316, 1977.
78. **Wells, J. N., Garst, J. E., and Kramer, G. L.,** Inhibition of separated forms of cyclic nucleotide phosphodiesterase from pig coronary arteries by 1,3-disubstituted and 1,3,8-trisubstituted xanthines, *J. Med. Chem.,* 24, 954, 1981.
79. **Williams, J. E., Lowitt, S., Polson, J. B., and Szentivanyi, A.,** Pharmacological and biochemical activities of some monomethyl xanthines and methyluric acid derivatives of theophylline and caffeine, *Biochem. Pharmacol.,* 27, 1545, 1978.
80. **Kass, I. and Mingo, T. S.,** Bitolterol mesylate (Win 32784) aerosol. A new long-acting bronchodilator with reduced chronotropic effects, *Chest,* 78, 283, 1980.
81. **Bell, S. C., Capetola, R. J., and Ritchie, D. M.,** Pulmonary and antiallergy drugs, in *Annual Reports in Medicinal Chemistry,* Vol. 14, Hess, H.-J., Ed., Academic Press, New York, 1979, chap. 6.

82. **Patil, P. N. and Ruffolo, R. R., Jr.,** Evaluation of adrenergic alpha and beta receptor activators and adrenergic alpha and beta receptor blocking agents, in *Handbook of Experimental Pharmacology,* Vol. 54, Part I, Szekeres, L., Ed., Springer-Verlag, New York, 1980, chap. 3.
83. **Liebman, J.,** Modification of the chronotropic action of sympathomietic amines by reserpine in the heart lung preparation of the dog, *J. Pharmacol. Exp. Ther.,* 133, 63, 1961.
84. **Castillo, J. C. and deBeer, E. J.,** The tracheal chain: a preparation for the study of antispasmodics with particular reference to bronchodilator drugs, *J. Pharmacol. Exp. Ther.,* 90, 104, 1947.
85. **Akeasu, A.,** The action of drugs on the isolated trachea, *J. Pharm. Pharmacol.,* 4, 671, 1952.
86. **Triggle, D. J.,** Adrenergics: catecholamines and related agents, in *Burger's Medicinal Chemistry,* Part III, 4th ed., Wolff, M. E., Ed., John Wiley & Sons, New York, 1981, 238.
87. **Tainter, M. L. and Chang, D. K.,** The antagonism of the pressor action of tyramine by cocaine, *J. Pharmacol. Exp. Ther.,* 30, 193, 1927.
88. **Fleckenstein, A. and Burn, J. H.,** The effect of denervation on the action of sympathomimetic amines on the nictitating membrane, *Br. J. Pharmacol.,* 8, 69, 1953.
89. **Burn, J. H. and Rand, M. J.,** The action of sympathomimetic amines in animals treated with reserpine, *J. Physiol. (London),* 144, 314, 1958.
90. **Carlsson, A. and Lindquist, M.,** *In-vivo* decarboxylation of α-methyldopa and α-methyl metatyrosine, *Acta Physiol. Scand.,* 54, 87, 1962.
91. **Day, M. D. and Rand, M. J.,** A hypothesis for the mode of action of α-methyldopa in relieving hypertension, *J. Pharm. Pharmacol.,* 15, 221, 1963.
92. **Kopin, I. J.,** False adrenergic transmitters, *Annu. Rev. Pharmacol.,* 8, 377, 1968.
93. **Brown, G. L. and Gillespie, J. S.,** Output of sympathin from the spleen, *Nature (London),* 178, 980, 1956.
94. **Kirpekar, S. M. and Puig, M.,** Effect of flow-stop on adrenaline release from normal spleens and spleens treated with cocaine, phentolamine, or phenoxybenzamine, *Br. J. Pharmacol.,* 43, 359, 1971.
95. **Adler-Grashinsky, E. and Langer, S. Z.,** Possible role of a beta-adrenoceptor in the regulation of noradrenalin release by nerve stimulation through a positive feedback mechanism, *Br. J. Pharmacol.,* 53, 43, 1975.
96. **Sano, I., Gamo, T., Kakimoto, Y., Taniguchi, K., Takegada, M., and Nishinuma, K.,** Distribution of catechol compounds in human brain, *Biochem. Biophys. Acta,* 32, 586, 1959.
97. **Ungerstedt, U.,** Stereotaxic mapping of the monoamine pathways in the rat brain, *Acta Physiol. Scand. Suppl.,* 367, 1, 1971.
98. **Marshall, J. F.,** Neurochemistry of central monoamine systems as related to food intake, in *Appetite and Food Intake,* Silverstone, T., Ed., Abakon Verlagsgesellschaft, Berlin, 1976.
99. **Piness, G., Miller, H., and Alles, G. A.,** Clinical observations on phenylaminoethanol sulfate, *JAMA,* 94, 790, 1930.
100. **Alles, G. A.,** The comparative physiological actions of *d,l*-beta-phenylisopropylamines. I. Pressor effect and toxicity, *J. Pharmacol. Exp. Ther.,* 47, 339, 1933.
101. **Ahlskog, J. E. and Hoebel, B. G.,** Overeating and obesity from damage to a noradrenergic system in the brain, *Science,* 182, 166, 1973.
102. **Frey, H. A. and Schulz, R.,** On the central mediation of anorexigenic drug effects, *Biochem. Pharmacol.,* 22, 3041, 1973.
103. **Edeleanu, L.,** Ueber einige derivate der phenylmethylacrylsaure und der phenylisobuttersaure, *Chem. Ber.,* 20, 616, 1887.
104. **Houlihan, W. J. and Babington, R. G.,** Anorexigenics, in *Burger's Medicinal Chemistry,* Part III, 4th ed., Wolff, M. E., Ed., John Wiley & Sons, New York, 1981, chap. 59.
105. **Axelrod, J.,** Studies on sympathomimetic amines. II. The biotransformation and physiological disposition of D-amphetamine, D-*para*-hydroxyamphetamine, and D-methamphetamine, *J. Pharmacol. Exp. Ther.,* 110, 315, 1954.
106. **Struyker-Boudier, H. A. J.,** Kinetics and biotransformation of adrenergic activators and inhibitors: adrenergic transmitter releasing agents, in *Handbook of Experimental Pharmacology,* Vol. 54, Part II, Szekeres, L., Ed., Springer-Verlag, New York, 1981.
107. **Caldwell, J., Croft, J. E., and Sever, P. S.,** Tolerance to the amphetamines: an examination of possible mechanisms, in *Amphetamines and Related Stimulants: Chemical, Biological, Clinical, and Sociological Aspects,* Caldwell, J., Ed., CRC Press, Boca Raton, 1980, chap. 9.
108. **Saminin, R., Bernasconi, S., and Garattini, S.,** The effect of selective lesioning of brain catecholamine containing neurons on the activity of various anorectics in the rat, *Eur. J. Pharmacol.,* 34, 373, 1975.
109. **Garrattini, S. and Samanin, R.,** Anorectic drugs and brain neurotransmitters, in *Appetite and Food Intake,* Silvertone, T., Ed., Abakon Verlagsgesellschaft, Berlin, 1976.
110. **Armstrong, J. M. and Green, A. F.,** Evaluation of agents that release or modify release of adrenergic transmitter, in *Handbook of Experimental Pharmacology,* Vol. 54, Part I, Szekeres, L., Ed., Springer-Verlag, New York, 1981.

α- AND β-ADRENOCEPTOR BLOCKING AGENTS

Raymond P. Alquist* and James T. Stewart

INTRODUCTION

The adrenergic receptors were originally delineated in 1948 on the basis of comparative responses to a series of closely related agonists.[1] At that time, only α-blockers were known. These included the ergot alkaloids, piperoxan and dibenamine. In 1958 the first β-adrenoceptor blocking agent, dichloroisoproterenol (DCI), was described by Powell and Slater[2] and Moran and Perkins.[3]

Clinical Uses of Adrenoceptor Blocking Agents

α-Adrenoceptor Blockers

This class of drugs has been used in a variety of vasospastic disorders because of their ability to block the vasoconstriction produced by norepinephrine activation of α-adrenoceptors. α-Block is useful in hypovolemic shock by preventing excessive reflex vasoconstriction that impedes blood flow in small vessels. In hypertension, α-blockers have generally been unsatisfactory because of the severe tachycardia produced in part by the postural hypotension. Recent studies have shown that there are selective α-blockers, for example, prazosin, that can be used to treat hypertension.

β-Adrenoceptor Blockers

These drugs produce a partial sympathectomy of the heart. This is the basis of their beneficial action in angina pectoris, obstructive cardiomyopathy, and certain tachyarrhythmias. All β-blockers are effective in hypertension, either renal or essential. They seem equally effective in preventing sudden death following a myocardial infarction. Some β-blockers prevent or reduce the tremor in essential tremor, prevent the occurrence of migraine, or diminish the peripheral effects of anxiety.

In the U.S. at the present time the FDA has approved only certain indications for each accepted β-blocker. In reality, all β-blockers are essentially equal. The principal differences are their pharmacokinetics, potency, and receptor selectivity.

Test Methods

Some general methods for detecting and quantitating adrenoceptor block will be briefly reviewed.

α-Receptor Block

The classical method is the production of epinephrine "reversal". Originally this was done using isolated uterine strips from rabbits. The addition of epinephrine to the bath caused contraction. Following the addition of an ergot alkaloid to the bath, epinephrine caused only relaxation. The same effect can be produced using the arterial pressure in anesthetized animals. Epinephrine normally produces a pressor response. Following an α-receptor blocking agent, epinephrine now produces a depressor response.

In man, α-blockers evoke postural hypotension by blocking reflex vasoconstriction. This is not a positive test since other types of drugs, such as ganglionic blockers, can also produce postural hypotension. A more usual test in man is the ability of the α-blocker to diminish the pressor response to an i.v.-infused α-agonist such as phenylephrine.

* The passing away of Dr. Ahlquist is acknowledged with sadness.

Table 1
PROTOTYPES OF α-ADRENOCEPTOR BLOCKING AGENTS[5]

Competitive blockers	Noncompetitive blockers
Imidazolines	β-Haloalkylamines
Phentolamine	Dibenamine
Tolazoline	Phenoxybenzamine
Ergot alkaloids	Tetramine disulfides
Ergotamine	BHC
Dihydroergocryptine	AOC
Others	
Prazosin	
Yohimbine	
Halperidol	

β-Receptor Block

In animals, the usual test is the inhibition of the depressor response and tachycardia produced by i.v.-administered isoproterenol. While this test can also be used in man, there is also a more physiologic procedure. Determine the response of the heart rate to graded exercise. Administer the β-blocker and repeat the graded exercise. A β-blocker will significantly reduce the heart rate increase.

Radioligand Binding Studies

The availability of radioactive agonists and blocking agents with high specific activity has made possible this type of receptor characterization. In brief, the method consists of making a suitable membrane preparation containing the receptors. This preparation is incubated with the radioligand, filtered, washed, and counted. A similar preparation is used containing an excess of unlabeled agonist or blocking agent to measure nonspecific binding. The difference between the two preparations is the specific binding.

General Pharmacology of α-Adrenoceptor Blocking Agents

This class of drugs has been known for many years.[4] The ergot alkaloids were probably the first compounds in which this type of action was recognized. As shown in Table 1, there are two types of α-blocking agents. The competitive blockers bind with the receptor and thus prevent agonists from binding with the receptor. The block can be overcome by increasing the dose of the agonist. The noncompetitive blocking agents chemically modify the receptor so it is no longer able to respond to agonists. This block cannot be completely overcome.

The α-blocking agents demonstrate a great deal of structural variation. Only the imidazolines and haloalkylamines have a structure similar to norepinephrine.

The α-adrenoceptors have been divided into two subtypes. α_1, The classical α-adrenoceptor, is located on the membrane of the effector cells. The α_2-receptor was first postulated as being a presynaptic receptor whose function was to modulate the release of norepinephrine from the nerve end. Although the α_2-receptor was needed to explain the action of the antihypertensive drug prazosin, these receptors are apparently found in many other locations.

Most of the α-adrenoceptor blocking agents have other pharmacodynamic actions in addition to their blocking effect. The α-adrenoceptor blocking agents covered in this review are shown in Table 2.

Table 2
CHEMICAL NAMES OF α-ADRENOCEPTOR BLOCKING AGENTS

Generic name	Trade name	Chemical name
	Competitive Blockers	
Phentolamine	Regitine, Rogitine, Regitin	*m* (*N*-2-Imidazoline-2-ylmethyl)-*p*-toluidino-phenol (HCl or mesylate)
Prazosin	Minipress, Sinetens, Hypovase	1-(4-Amino-6,7-dimethoxy-2-quinazolinyl)-4-(2-furoyl)piperazine HCl
	Noncompetitive Blockers	
Dibenamine	—	*N,N*-Dibenzyl-β-chloroethylamine
Phenoxybenzamine	Dibenzyline, Dibenyline	*N*-(2-Chloroethyl)-*N*-(1-methyl-2-phenoxyethyl)benzylamine HCl
	Other Agents	
Ergotamine Tartrate	Ergomar, Gynergen, Femergin, Lingraine	12′-Hydroxy-2′-methyl-5′α-(phenylmethyl)ergotaman-3′,6′,18-trione tartrate
Bromocryptine mesylate	Parlodel	2-Bromo-12′-hydroxy-2′-(1-methyllethyl)-5′α-(2-methylpropyl) ergotaman-3′,6,18-trione mesylate
Dihydroergocryptine	—	—
Yohimbine	Yobinol	17α-Hydroxyyohimban-16-carboxylic acid methyl ester
Tetramine disulfides	—	—
Haloperidol	Haldol, Serenace, Serenase	4[4-(4-chlorophenyl)-4-hydroxy-1-piperindyl]-1-(4-fluorophenyl)-1-butanone

α-ADRENOCEPTOR BLOCKING AGENTS MONOGRAPHS*

PHENTOLAMINE

Mechanism of Action

In addition to producing a competitive blockade of all α-adrenoceptors, phentolamine and tolazoline produce other responses on smooth muscle and the heart. These drugs are β-adrenoceptor agonists causing tachycardia and myocardial stimulation. They are also muscarinic agonists stimulating the smooth muscle of the intestine, an action blocked by atropine. And they have a histamine-like action, causing vasodilation, and an increase in gastric hydrochloric acid secretion. Phentolamine is the more effective α-adrenoceptor blocking agent, while tolazoline, used today only as a vasodilator, has less α-adrenoceptor blocking action but more of the other actions listed above.[6] In addition, phentolamine also blocks the receptors for 5-HT.

The principal physiologic response to an effective dose of phentolamine is postural hypotension. Since the β-adrenoceptors of the heart are not blocked, severe tachycardia and palpitations result. Thus, phentolamine is unsatisfactory as an antihypertensive drug. It has recently become apparent that the tachycardia produced by phentolamine is not entirely due to the reflex response. The α_2-receptors, which inhibit the release of norepinephrine from the axon, are blocked by phentolamine. This results in a loss of the negative feedback control so that there is more norepinephrine at the β-receptors of the heart.

* Second monographs on phentolamine, phenoxybenzamine, and prazosin may be found in the Antihypertensive Agents chapter, *Handbook of Anti-Inflammatory and Cardiovascular Agents*.

Structure-Activity Relationships

Many 2-substituted imidazolines have been synthesized during the past 40 or more years. Many different types of active compounds have been found. These include sympathomimetics, antihistamines, and histamine-like and cholinergic-stimulating substances. Phentolamine, a derivative of tolazoline, was prepared principally as a diagnostic agent for pheochromocytoma. This action is described below.

Pharmacokinetics

There is not much known about the fate of phentolamine. It is probably 90% metabolized by the liver since only about 10% of the active drug is excreted in the urine. Thus, phentolamine is a short-acting drug.

Uses and Dosages

Phentolamine is occasionally used in the diagnosis of pheochromocytoma. Following the i.v. injection of 5 mg, the arterial pressure falls. If excessive amounts of catecholamines from the tumor are present, the fall in pressure will be greater than if the catecholamines were not present. Phentolamine is also given by i.v. infusion during the surgical removal of the pheochromocytoma to prevent the response to released catecholamines produced by manipulation of the tumor. Phentolamine hydrochloride in the form of 50-mg tablets is occasionally used to treat peripheral vascular disease.[7] Phentolamine has been tried in the treatment of hypertension,[8] and in the treatment of acute myocardial infarction.[9]

Toxicity

The adverse responses to phentolamine are associated both with cardiac stimulation and intestinal stimulation. This drug can cause tachycardia, cardiac arrhythmias, and, occasionally, anginal pain. The GI stimulation may result in abdominal pain, vomiting, diarrhea, and exacerbation of peptic ulcer.

Physical-Chemical and Other Data

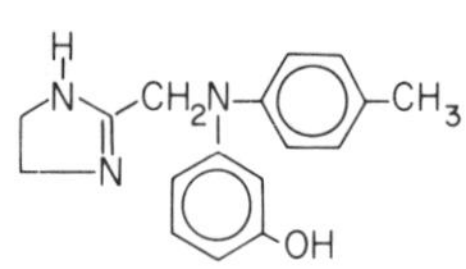

PHENTOLAMINE

The pK_a of phentolamine is 7.7. Phentolamine HCl (mp 240°C) is a white or slightly grayish, odorless, bitter powder. Its solubility in water is 1 g/50 mℓ and 1 g/70 mℓ in alcohol. It is very slightly soluble in chloroform and ether, and practically insoluble in acetone and ethyl acetate. The pH of a 1% aqueous solution lies between 4.5 and 5.5. Aqueous solutions of the drug are not stable.

Phentolamine mesylate (mp 178°C) is a white, odorless, bitter powder that is slightly hygroscopic. Its solubility in water is 1 g/50 mℓ, 1 g/23 mℓ in alcohol, and 1 g/660 mℓ in chloroform. The pH of a 1% aqueous solution lies between 4.5 and 5.5. Aqueous solutions of the salt are not stable. The mesylate salt is said to be superior to the hydrochloride, especially in the formulation of injectables.

PHENOXYBENZAMINE

Mechanism of Action

Dibenamine (see Table 2) was the first of a series of haloalkylamines described by

Nickerson and Goodman in 1947.[10] This compound was shown to act in the following way. When first administered, it acts as a competitive blocking agent. Then it gradually becomes an irreversible blocking agent over a period of about 90 min. These compounds, related to the nitrogen mustards, cyclize in the body to an active aziridinium ion. Since dibenamine has the proper conformation to bind with the receptor, this active aziridinium ion gradually alkylates some part of the receptor to form a covalent bond. This produces a long-lasting blockade of the α-adrenoceptor; an effect which may last for several days. This mechanism is illustrated below.

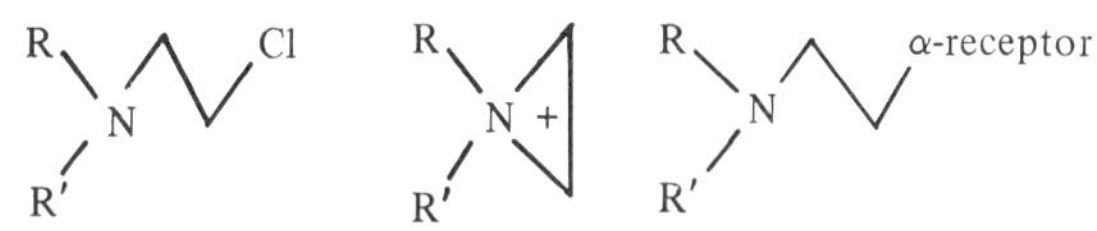

Phenoxybenzamine, the presently available derivative, is more potent and faster acting. The presence of a catecholamine or a competitive α-adrenoceptor blocking agent when the haloalkylamine is administered inhibits the development of the noncompetitive block by the latter. Phenoxybenzamine also enhances the release of norepinephrine by blocking the α_2 presynaptic receptor. It also seems to enhance the exocytosis of the granules containing norepinephrine and dopamine β-hydroxylase. Phenoxybenzamine is not specific for α-adrenoceptors. It also blocks histamine receptors, 5-HT receptors, and muscarinic receptors. The β-adrenoceptor is unaffected by the haloalkylamines.

Structure-Activity Relationships

Nickerson and Gump[11] studied 218 compounds related to dibenamine. They found that to be active a compound must meet the following requirements: (1) it must be a tertiary amine or the quaternary derivative of an active amine, (2) it must include at least one β-haloalkyl group capable of forming an intermediate ethylenimmonium (or vinyl) derivative, (3) it must include an unsaturated ring substituent attached to the amine in such a way as to allow resonance stabilization of the active intermediate, and (4) in the case of benzyl derivatives, there must be no substitution on the phenyl ring, which tends to be out of the plane of the ring.

Loew et al.[12] studied compounds that were related both to dibenamine and to diphenhydramine. They found several compounds that not only had α-adrenergic blocking activity, but also had good antihistaminic properties. For more information about β-haloalkylamines, see Graham.[13]

Pharmacokinetics

Absorption from the GI tract is incomplete and variable, with 20 to 30% being absorbed in an effective form. The haloalkylamines are very unstable in water solutions. Little is known about their metabolic fate. Most of the metabolic products are excreted in 24 hr. There is some accumulation of phenoxybenzamine in body fat.

Uses and Dosages

Phenoxybenzamine has been used to treat acute hypertension due to sympathomimetic overdosage and certain foods in the presence of MAO inhibitors.

Phenoxybenzamine can be used in the medical management of pheochromocytoma. If severe tachycardia is present, a β-blocker should be used concurrently. In hypovolemic shock, α-blockers, in conjunction with adequate volume replacement, are effective.

Toxicity

Most of the adverse responses to phenoxybenzamine are due to the α-receptor blockade.

Postural hypotension and tachycardia occur. Exercise, large meals, or alcohol consumption increase the postural response. Other effects are miosis, nasal stuffiness, and loss of ejaculatory ability. Phenoxybenzamine is a local irritant; nausea and vomiting can occur on oral administration. Local tissue irritation can occur during i.v. injection.

Physical-Chemical and Other Data

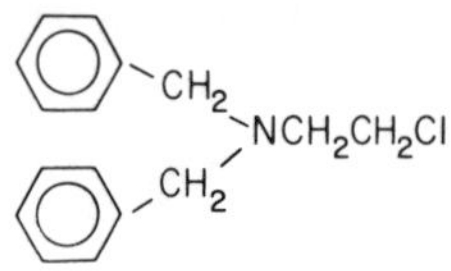

DIBENAMINE

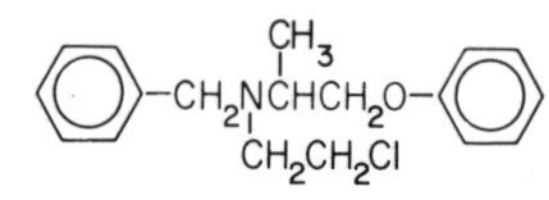

PHENOXYBENZAMINE

Phenoxybenzamine hydrochloride (mp 140°C) is a white, crystalline, odorless powder. It is freely soluble in alcohol and chloroform, soluble in water, and insoluble in ether.

PRAZOSIN*

Mechanism of Action

A quinazoline compound, prazosin was originally introduced as a vasodilator drug. However, it was later found that this substance was a good α-adrenoceptor blocking agent. Unlike phentolamine, prazosin does not cause a significant increase in heart rate. To explain this it was postulated that prazosin blocks only the α_1 adrenoceptor.[14] Because it does not block the α_2-adrenoceptor, the so-called presynaptic receptor, prazosin does not prevent the negative feedback control of the released norepinephrine. Therefore, there is not an excessive amount of norepinephrine at the sympathetic nerve terminal in the myocardium.[15] Prazosin produces both arterial and venous vasodilation. It has little effect on cardiac output, heart rate, or right atrial pressure. However, with the first dose of prazosin there may occur considerable postural hypotension. Subsequent doses, however, only lower the arterial pressure without a significant amount of postural hypotension.

Structure-Activity Relationships

In 1941 it was observed by Moe et al.[16] that some potential antimalarial aminoquinolines had an antihypertensive side effect. This led to the development of a series of 2-amino-6,7-dimethoxyquinazolines as potential antihypertensives. One of these, prazosin, was introduced as a vasodilator in 1974. Its mode of action was subsequently described as an α-receptor blocking agent.[17,18]

Pharmacokinetics

The pharmacokinetic parameters for prazosin are shown in Table 3. About half of the administered oral dose is absorbed. The half-life is about 3 hr. It has been shown by Pitterman et al.[19] that a 5-mg oral dose of prazosin in normal subjects effectively blocked the pressor response to infused phenylephrine. This blockade persisted for at least 7 hr. It was found that the α-adrenoceptor blockade did not correlate with prazosin plasma concentration. In patients with congestive failure of the circulation, the elimination of prazosin is slower than in normal patients.[20] Prazosin kinetics are unchanged by the presence of essential hypertension.[21]

* A second monograph on prazosin may be found in the Antihypertensive Agents review, *Handbook of Anti-Inflammatory and Cardiovascular Agents*.

Table 3
PRAZOSIN PHARMACOKINETICS

Absorption (% of dose)	Bioavailable (% of dose)	Effective blood conc (ng/mℓ)	$t_{1/2}$ (hr)	% Bound to plasma protein	Urinary excretion (% of dose)
50	50—57	2—50	3—6	93	0—1

Uses and Dosage

Prazosin is used in the treatment of hypertension and congestive heart failure. This drug is available in capsules containing 1, 2, or 5 mg. Prazosin is often administered in conjunction with β-adrenoceptor blocking agents and diuretics. Because prazosin is such a good vasodilator it is being tested in the treatment of heart failure by reducing afterload.[22] This use is similar to that of i.v.-administered sodium nitroprusside.

Toxicity

The adverse effects of prazosin are related to its regular pharmacodynamic action. As noted above, postural hypotension can be a problem.

Physical-Chemical and Other Data

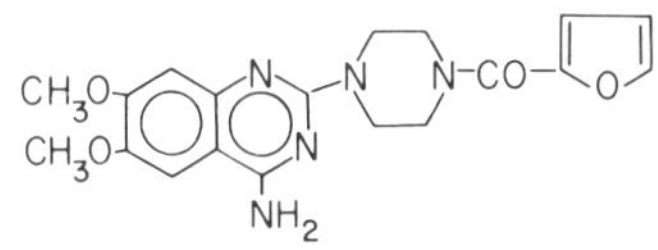

PRAZOSIN

The pK_a of prazosin is 6.5. Prazosin hydrochloride is a white, crystalline substance, slightly soluble in water and/or isotonic saline, and very slightly soluble in alcohol.

Other α-Adrenoceptor Blocking Agents

Ergot Alkaloids

The ergot alkaloids were the first known α-adrenoceptor blocking agents. Their pharmacology was described in 1906 by Dale.[23] There are many different alkaloids in ergot. Some of these are named in Table 4. Their actions are varied and complex.[24] They may be agonists, partial agonists, or blocking agents on α-adrenoceptors, 5-HT receptors, and dopaminergic receptors. None are used clinically for their α-adrenoceptor blocking action. The tritiated form of dihydroergocryptine is used extensively as a radioligand for receptor binding studies. The ergot alkaloids have long been used for their oxytocic action. Ergotamine is used in the treatment of migraine.[25] However, its vasoconstrictor effects limit the dose that can be given since gangrene of the extremities can be produced.[26] Bromocryptine is used mainly to inhibit prolactin secretion. In this case the bromocryptine is acting as a dopaminergic agonist. Bromocrytptine is also used in the treatment of Parkinsonism.

Ergotamine tartrate (mp 203°C dec) is a colorless or white, crystalline, odorless, slightly bitter powder. Its solubility in water is 1 g/500 mℓ and 1 g/500 mℓ in alcohol. Its solubility in water can be increased slightly by adding a small amount of tartaric acid. It must be protected from heat and light. Ergotamine base can be quantitatively extracted from aqueous alkaline solution (pH 8 to 11) into benzene, ether, or chloroform. The optical rotation of ergotamine base in chloroform is −125 to −155°. The pK_a of ergotamine is 6.4.

Bromocryptine mesylate (mp between 180 to 200°C dec) is a gray, tinged white or light yellow, finely crystalline powder. It is odorless or of a weak characteristic odor. The partitioning at pH 1.2 between *n*-octanol/water is 1:90. At pH 7.5, the partitioning is 1:235. The salt is relatively sensitive to autoxidation both in solid and dissolved states. The deg-

Table 4
ACTIONS OF ERGOT ALKALOIDS

Drugs	Receptors	Agonist action	Partial agonist	Blocking action
Ergotamine	α-Adrenergic	Partial	Some blood vessels Some smooth muscle	In nervous system
	5-HT	Poor in CNS	Some blood vessels	Poor in CNS
	Dopamine	None	None	None
Dihydroergotamine	α-Adrenergic	Partial	Veins	Blood vessels, smooth muscle
	5-HT	Lateral geniculate ganglion	Few smooth muscles	Few smooth muscles
	Dopamine	None	None	Sympathetic ganglia
Bromocryptine	α-Adrenergic	None	None	Less potent than dihydroergotamine
	5-HT	None	None	Weak blocker
	Dopamine	Inhibit prolactin secretion	In CNS	None
Ergotoxine (ergocryptine[a])	α-Adrenergic	Unknown	Unknown	Potent in all tissues

[a] Used as radioligand for receptor binding studies.

radation products have not been elucidated. The salt is stable up to 3 years at ambient temperature when stored in sealed polyethylene bags contained in twist-off amber glass bottles. The salt is only stable for 3 months at 50°C. It absorbs up to 6% by weight of water in tropical climates. The pK_a of bromocryptine is 4.90.

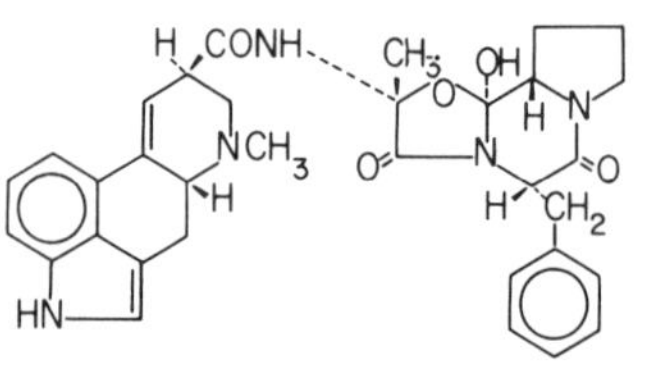
ERGOTAMINE

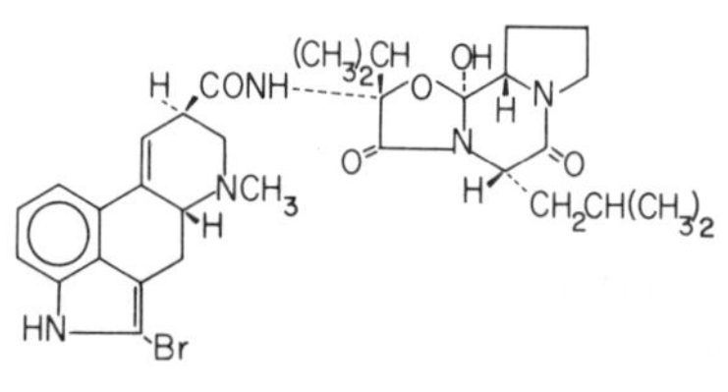
BROMOCRYPTINE

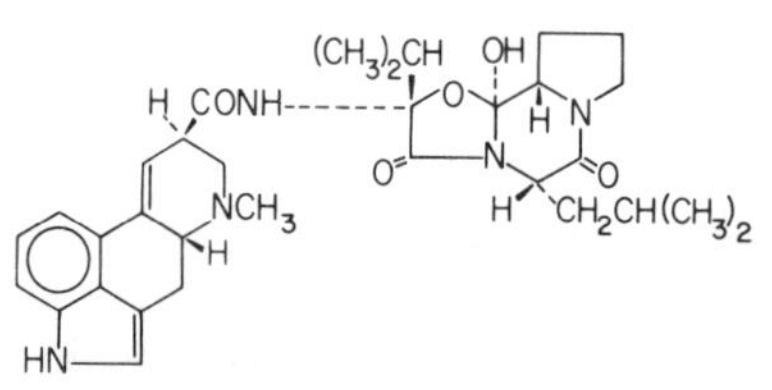
ERGOCRYPTINE

Dihydroergocryptine is one of the ergoloid alkaloids. The pK_a of dihydroergocryptine is 6.89.

Yohimbine

This oldtime alkaloid, reputed to be an aphrodisiac, has now been found to act selectively on the α_2-receptor. This compound has no clinical use but is now being used widely experimentally to study the α-receptors.

Yohimbine (mp 234°C) is the methyl ester of yohimbic acid. It crystallizes from dilute ethanol in colorless needles. It is soluble in ethanol, chloroform, and hot benzene. It is moderately soluble in ether and sparingly soluble in water. It is a monoacidic base and forms crystalline salts.

Yohimbine hydrochloride (mp 302°C dec) has a solubility of 1 g/120 mℓ in water and 1 g/400 mℓ in alcohol. Aqueous solutions of the salt give a neutral pH.

Haloperidol

This tranquilizer, and the phenothiazine tranquilizers, have some α-receptor blocking activity. This is a side effect in patients receiving these drugs; postural hypotension may occur. In addition, haloperidol is a dopamine blocking agent. The renal vasodilation produced by dopamine is blocked by haloperidol. All of the major tranquilizers block the action of dopamine in the central nervous system, resulting in the Parkinsonism-like side effects of these compounds.

Haloperidol (mp 149°C) is a white to faintly yellowish, amorphous or microcrystalline, odorless powder. It is stable for up to 5 years at ambient temperature stored in amber glass containers and up to 1 year when stored at 45°C. It discolors or degrades when exposed to natural sunlight for long periods. It is only slightly hygroscopic. Following are solubilities for haloperidol in terms of grams per 100 milliliters of solvent: acetone, 2; benzene, 1; chloroform, 6.6; alcohol, 1.7; ether, 0.5; ethyl acetate, 1.8; methanol, 1.8. The pK_a of haloperidol is 8.3.

Tetramine Disulfides

In a structure-activity relationship study of a large group of polyamine disulfides, two very active compounds have been found.[5] These compounds have the trivial designations of BHC and AOC, are noncompetitive blocking agents, and are of considerable interest because they are quite selective for α-receptors. They do not block the effects of 5-HT, histamine, or isoproterenol. Their general structure is given below.

$$(RNH(CH_2)_nNHCH_2CH_2S-)_2$$

AOC, R=H, n=8

BHC, R= 2′-METHOXYBENZYL, n=6

General Pharmacology of β-Adrenoceptor Blocking Agents

The first β-blocker was found 10 years after the α- and β-receptors were first delineated. This compound, discovered during a search for longer-acting bronchodilators, was dichloroisoproterenol. It was noted by Powell and Slater[2] that it prevented some of the inhibitory responses to epinephrine or isoproterenol. Then, Moran and Perkins[3] demonstrated that this compound was an antagonist to the cardiac-stimulating action of the catecholamines.

Dichloroisoproterenol is a strong partial agonist. Its effect as a β-blocker in man was first demonstrated by Shanks[27] in Belfast. The second β-blocker used clinically was pronetholol, a weak partial agonist. It was tested for the treatment of arrhythmia and hypertension in about 1964.[28] Unfortunately, pronetholol turned out to be carcinogenic in mice and thus was abandoned as a test compound. The next compound was propranolol. At the suggestion of Dr. James Black, propranolol was tested in the treatment of angina pectoris.[29] In 1964, Prichard and Gillam[30] showed that propranolol had an antihypertensive action. The demonstration that propranolol was an effective treatment for angina,[31,32] was a good antihypertensive agent, and was effective in treating many types of cardiac arrhythmias[33] stimulated the search for other β-blockers.

General Structure-Activity Relationships

As can be seen in Table 5, and in the structural formulas, all of the presently available β-blockers are derivatives of isoproterenol. This seems to be the necessary molecular shape and conformation for receptor binding.

Table 5
CHEMCIAL NAMES OF β-ADRENOCEPTOR BLOCKING AGENTS

Generic name	Trade name	Chemical name
Propranolol[a]	Inderal, Dociton, Kemi, Oposim	1-(Isopropylamino)-3-(1-naphthyloxy)-2-propanol HCl
Pindolol[a]	Visken, Carvisken	1-(Indol-4-yloxy)-3-(isopropylamino)-2-propanol
Timolol[a]	Timoptic, Blockadren, Betin	(−)-1-(*tert*-Butylamino)-3-((4-morpholino-1,2,5-thiadiazol-3-yl)oxy)-2-propanol maleate
Nadolol[a]	Corgard	1-(*tert*-Butylamino)-3-((5,6,7,8-tetrahydro-cis-6,7-dihydroxy-)-1-naphthyl)oxy-2-propanol
Sotalol	Sotacor, β-Cardone, Sotalex, Sotaper, Sotapor	4′-(1-Hydroxy-2-(isopropylamino)ethyl-methansulfonanilide) HCl
Alprenolol	Aptin, Gubernal, Betacard	1-(*o*-Allylphenoxy)-3-(isopropylamino)-2-propanol HCl
Atenolol[a]	Tenormin	2-(*p*-(2-Hydroxy-3-(isopropylamino)propoxy)phenyl)acetamide
Labetolol	Trandate	5-(1-Hydroxy-2-((1-methyl-3-phenylpropyl)amino)ethyl) salicylamide
Penbutolol	Betapressin	(*S*)-1-(*tert*-Butylamino)-3-(*o*-cyclopentyl-phenoxy)-2-propanol sulfate
Metoprolol[a]	Lopressor, Betaloc, Seloken	1-Isopropylamino-3-(*p*(2-methoxyethyl)phenoxy)-2-propanol tartrate
Acebutolol	Sectral	3′-Acetyl-4′-(2-hydroxy-3-(isopropylamino)propoxy)butyranilide
Practolol	Eraldin, Dalzic	4′-(2-Hydroxy-3-isopropylamino)-propoxy)acetanil
Bevantolol		1-((3,4-dimethoxyphenethyl)amino)-3-(*m*-tolyl)-2 propanol HCl
Pamatolol		Methyl-(*p*-(2-hydroxy-3-(isopropylamino)propoxy)phenethyl)-carbamate sulfate

[a] See appropriate monograph.

Ethanolamine derivatives — The original β-blockers, including DCI, pronetholol, and sotalol, are of this type. Only sotalol is clinically available today.[35]

Aryloxypropranolamine derivatives — Propranolol was the first of this type. The insertion of the $-O-CH_2-$ between the ethanolamine side chain and the aromatic group does not decrease β-blocking activity. The aromatic grouping can be phenyl, substituted phenyl, naphthalene, tetrahydronaphthalene, benzofurane, indane, indole, 3,4-dihydro-1(2H)-naphthalenone, morpholino-thiadiazole, phenanthidine, thiophene, or chromone.[35] It should be noted that when compounds of this type are synthesized, agonists as well as antagonists may result; however, most of these compounds turn out to be inactive.

Some general observations regarding structure-activity relationships can be made:

1. Most β-blockers are available as racemic mixtures. The blocking activity resides in the levo-rotatory isomer. Timolol, the most potent compound, is the levo isomer.
2. The N-isopropyl group can be substituted by *N-tert*-butyl or *N*-dimethoxyphenyl.
3. Most of the β-blockers have a substituted phenyl as the aromatic group. Substitution in the *para* position seems to impart cardioselectivity. Practolol was the first effective compound of this type.

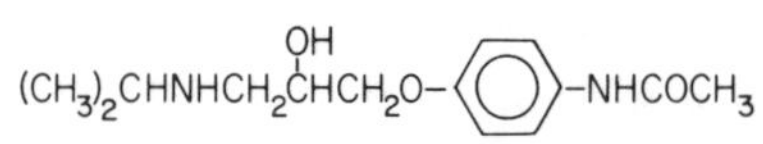

PRACTOLOL

Table 6
CHARACTERISTICS OF β-ADRENOCEPTOR BLOCKING AGENTS

Drug name	Potency ratio	ISA[a]	MSA[b]
Nonselective Blockers			
Propranolol[c]	1	None	Most
Pindolol[c]	6	Most	Some
Timolol[c]	6	Some	None
Nadolol[c]	3	None	None
Sotalol	0.3	None	None
Alprenolol	0.3	More	Some
Labetolol[d]	0.3	None	Some
Penbutolol	5—6	None	Some
Cardioselective Blockers			
Atenolol[c]	1	None	None
Metoprolol[c]	1	None	Some
Acebutolol	0.3	Some	Some
Practolol[e]	0.3	More	None
Bevantol	1	None	—
Pamatolol	0.3	—	—

[a] Intrinsic sympathomimetic action (partial agonist action).
[b] Membrane-stabilizing action (local anesthetic action).
[c] See monograph.
[d] Also blocks α-adrenoceptors.
[e] Limited use because of toxic effect (Oculomucocutaneous Syndrome).

The β-blockers can be classified into subgroups according to other properties as shown in Table 6.

Local anesthetic action — Known also as the membrane stabilizing action (MSA), at one time this was thought to be an important mechanism when using the β-blockers in the treatment of arrhythmias. However, it was found that at the serum concentrations that occur during usual dosage of the β-blockers, this effect is not apparent.

Intrinsic sympathomimetic action (ISA) — Some of the β-blockers are partial agonists. For this reason they do not produce as much bradycardia as those that do not have this action.

Cardioselectivity — At certain dose levels some of the β-blockers act more on the β_1-receptor than on the β-$_2$-receptor. This may be of some clinical importance.

Pharmacokinetics — Each β-adrenoceptor blocking agent is handled somewhat differently by the body. Some have short half-lives while others have long half-lives. Some are bound to plasma protein more than others. Their metabolism and excretion differ. The clinical relevance of these differences will be described under each individual monograph. Table 7 shows the principal pharamcokinetic parameters for some of the important β-blockers.

Table 7
COMPARATIVE PHARMACOKINETICS OF β-ADRENOCEPTOR BLOCKING AGENTS

Drug	Mol wt	Absorption (% of dose)	Bioavailable (% of dose)	Effective conc. (ng/mℓ)	Half-life (hr)	Bound HSA[a]	Urinary excretion (% of dose)
Lipid-Soluble Compounds							
Propranolol	295.81	90	30	50—100	3—6	62	0—1
Timolol	432.49	90	50	5—10	3—4	10	20
Alprenolol[b]	285.81	90	10	50—100	2—3	85	0—1
Poorly Lipid-Soluble Compounds							
Short half-life							
Metoprolol	684.82	95	50	50—100	3—4	12	3
Pindolol	248.32	90	52—86	10—25	3—4	57	5—34
Long half-life							
Atenolol	266.34		40	200—500	6—9		40
Nadolol	309.40	30	30	50—100	17—24	30	90+
Sotalol[b]	308.82		60	500—4000	5—17	54	60
Practolol[b]	266.34	95	100	150—500	6—8	32	90

[a] Human serum albumin.
[b] Not presently available in the U.S.

Table 8
METABOLITES OF PROPRANOLOL — COMPARATIVE PHARMACOKINETICS

Metabolite	Propranolol route of administration	AUC[a]	$t_{1/2}$[b]	Ref.
Propranolol glucuronide[c]	IV	Equal		37
	PO	7-Fold increase	Equal	
Naphthoxylacetic acid	IV	2-Fold increase		38
	PO	10-Fold increase	Equal	
4-Hydroxy-propranolol	IV	9-Fold decrease		39
	PO	3-Fold decrease	Equal	

[a] Area under the curve.
[b] Half-life.
[c] This metabolite has been proposed as a storage pool for propranolol.[37]

β-ADRENOCEPTOR BLOCKING DRUG MONOGRAPHS*

PROPRANOLOL

Mechanism of Action

Propranolol acts by binding competitively with all β-receptors, thus preventing the action of agonists on the receptor. The general effects produced by an effective dose include the following: bradycardia, prolongation of the P-R interval, a small increase in peripheral resistance, and a decrease in plasma renin activity.

Pharmacokinetics

Propranolol is more or less rapidly metabolized in the liver. This is known as the "first pass" effect since the drug, when taken by mouth, first goes to the liver. The principal metabolites of propranolol are shown in Table 8. The 4-hydroxypropranolol is almost as potent as propranolol: however, unlike propranolol, this metabolite has some partial agonist activity.[40]

It is difficult to predict the blood level from the oral dose when using propranolol since liver metabolism varies considerably among patients. It has been reported by one group[41] that the first pass effect may show up in patients with up to a fivefold variation. However, another group,[42] using great care in all aspects of collecting the blood, giving the drug, and determining the propranolol content, shows that there is only a 1.3-fold variation at doses of 160 mg/day and up. At 40 mg/day the variation may be as great as threefold. Some of the variability between these results may be due to differences in methods for measuring the plasma concentration of propranolol. At the present time there are at least two excellent methods. The first uses gas chromatography-mass spectrometry. The other method uses a radio receptor assay which works for both propranolol and for hydroxypropranolol.[43,44] Radioactive iodohydroxybenzyl pindolol is used as the ligand.

Uses and Dosages

Propranolol is used in the treatment of either renal or essential hypertension. It is also indicated in the treatment of most forms of angina pectoris, in fast cardiac arrhythmias, in obstructive cardiomyopathy, in hyperthyroidism, in the suppressive treatment of migraine, and in essential tremor. Propranolol is usually given by mouth in a dose sufficient to produce the desired effect. It is only the first dose given to a patient that may be dangerous; the

* Additional monograph information on propranolol, metoprolol, nadolol, atenolol, and timolol may be found in the Antihypertensive Agents review, *Handbook of Anti-Inflammatory and Cardiovascular Agents*.

Table 9
DRUG INTERACTIONS FOR β-ADRENOCEPTOR BLOCKING AGENTS

Drug	Possible effect
Aminophylline	Mutual inhibition
Isoproterenol	Mutual inhibition
Digitalis	Excessive bradycardia, heart block
Epinephrine	Hypertension, bradycardia, heart block
Tubocurarine and similar	Prolonged neuromuscular block
Insulin	Mask symptoms of hypoglycemia
Oral hypoglycemics	Mask symptoms of hypoglycemia
MAO inhibitors	Hypertension

blood level is not predictable, and the degree of sympathetic activity is unknown. Therefore, it is usual to start with a relatively small dose of propranolol (about 20 mg two or three times per day) and increase this dose daily or weekly until the desired effect is obtained. In a series of patients with essential hypertension who were on continuous diuretic therapy, there was no difference in the response whether the propranolol was given as a single dose or given three times per day.[45] Using the reduction in exercise tachycardia as the criterion, it was found that propranolol and several other β-blocking agents were effective for at least 24 hr after a single dose of the drug.[55] In the same study it was found that β-blockers with long half-lives were not superior to other β-blockers. Neither were slow-release preparations of β-blockers with short half-lives more effective. In patients that do not have any contraindications for propranolol there seems to be no upper limit to the dose of propranolol that may be given. The average maximal dose in the U.S. is between 320 and 480 mg/day. In other parts of the world the maximal dose may be as high as 960 mg/day.

Toxicity

There are certain definite contraindications to the use of propranolol. The first is uncompensated congestive failure of the circulation. In these patients their heart is being controlled mainly by sympathetic activity. The administration of β-blocker in this type of a patient can have a fatal effect. Therefore, patients with congestive failure should be treated with digitalis before propranolol is given. There is no interaction between propranolol or any other β-blocker and the digitalis inotropic effect on the heart. Obstructive pulmonary disease is another contraindication for the use of propranolol. In these cases the β-blocker can prevent the effective action of the bronchodilators being used and can also prevent the bronchodilation produced by sympathetic nerve activity. In patients with incipient asthma, propranolol can often induce an asthmatic attack. Certain types of peripheral vascular disease are a contraindication to the use of propranolol. In these cases propranolol blocks the small amount of adrenergic vasodilation that is always present. In diabetes propranolol should be used with caution. The β-blockers in this condition can have a variable effect on blood sugar. In addition, the symptoms of hypoglycemia are masked by the β-blocking effect on the cardiovascular system.

Table 9 lists some of the more obvious drug interactions that can occur when β-adrenoceptor blocking agents are used.

Physical-Chemical and Other Data

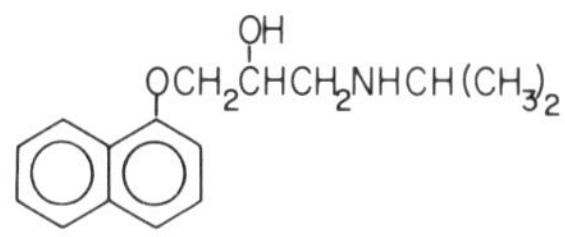

PROPRANOLOL

The pK_a of propranolol is 9.45. Propranolol hydrochloride (mp 163°C) is a white to off-white crystalline solid that is nonhygroscopic. It is odorless and bitter-tasting. It is soluble in water or ethanol (1 g/20 mℓ), slightly soluble in chloroform, and practically insoluble in ether, benzene, and/or ethyl acetate. It is unstable in light. There is an asymmetric carbon atom in the propanolamine side chain. The clinical material is marketed as the racemate. The partitioning of propranolol base between chloroform/water is 34.5:1 and between *n*-octanol/water is 17:1. The pH of a 1% w/v propranolol hydrochloride solution in water is 5 to 6.

Propranolol HCl is available in tablets of 10, 20, 40, and 80 mg. It is also available as an injection containing 1 mg/mℓ.

TIMOLOL

Mechanism of Action

Timolol is a nonselective β-adrenoceptor blocking agent. It has been shown that all of the hemodynamic effects produced by timolol are due to β-blockade.[46] Timolol lowers pressure in a dose that blocks tachycardia produced by isoproterenol.[47] Timolol is used to lower intraocular pressure by an unknown mechanism.

Pharmacokinetics

The pharmacokinetic parameters for timolol are shown in Table 7. This drug is well absorbed from the GI tract, and about 50% is bioavailable due to the first pass metabolism in the liver. At least nine metabolites have been identified for timolol but none of these shows any activity. Timolol has been shown to be eight times more potent than propranolol so that the daily dosage and effective blood level are considerably lower than those for propranolol. The effect of a single dose lasts for more than 12 hr.[48] The half-life has been found to be not significantly different between normal patients and patients with moderate chronic renal failure.[49]

Uses and Doses

Timolol is approved for the treatment of hypertension.[50] It may be used alone or in combination with other antihypertensive agents such as diuretics. The starting dose is usually 10 mg twice a day. This can be increased at weekly intervals up to 20 to 40 mg/day. The maximum dose is about 60 mg/day, usually given in two doses. Timolol is the only β-adrenoceptor blocking agent currently approved for use in patients who have survived the acute phase of a myocardial infarction. The objective of this treatment is to reduce cardiovascular mortality and the risk of reinfarction. Since cardiac arrhythmias are a common cause of deaths in these patients it is apparent that any β-adrenoceptor blocking agent should provide some protection. Some patients with angina need take timolol only once per day.[51] In patients with hypertension it has also been found that once a day dosage is all that is needed.[52]

Timolol (Timoptic) is used in the treatment of glaucoma in the form of eyedrops containing either 0.25 or 0.5% timolol. Even when carefully administered to prevent absorption of

Table 10
COMPARATIVE POTENCIES OF β-ADRENOCEPTOR BLOCKING DRUGS[55]

Drug	Sympathetic nerve stimulation Heart rate[a]	Isoproterenol Heart rate[a]	Resistance[a]	Bronchi[b]
Metoprolol	1	4	50	30
Propranolol	1	1	1	1
Practolol	1	5	350	40

[a] Determined in anesthetized cats.
[b] Determined in guinea pig exposed to histamine mist.

timolol, some of the drug can always be found in the urine and the exercise tachycardia is lower following the use of ophthalmic timolol.[53]

Toxicity

The adverse effects of timolol are identical to those of propranolol.

Physical-Chemical and Other Data

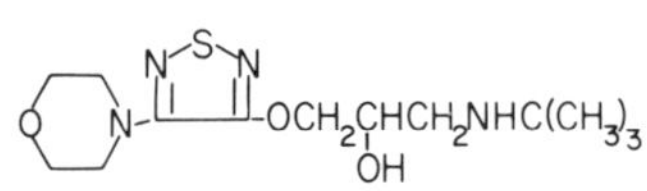

TIMOLOL

The pK_a of timolol is 9.40. Timolol hydrogen maleate (mp 202°C) occurs as a white, or practically white, odorless powder. It is freely soluble in water, soluble in alcohol and methanol, sparingly soluble in chloroform and propylene glycol, and insoluble in ether and cyclohexane. Aqueous solutions are stable up to pH 12. It must be protected from light.

Timolol is available as the (−) isomer and not as the racemate (specific rotation is −11.7 to −12.5°). The partitioning of the drug between *n*-octanol/water is 0.7 at pH 7.4.

Timolol is available in tablets of 10 and 20 mg, and as an ophthalmic solution containing either 0.25 or 0.5%.

METOPROLOL

Mechanism of Action

Metoprolol is a cardioselective β-adrenoceptor blocking agent. As shown in Table 10, metoprolol is only slightly less potent than propranolol in blocking the effect of isoproterenol on heart rate. It is much less effective in blocking the effects of isoproterenol on peripheral vascular resistance and on bronchial relaxation. The same cardioselectivity can be shown in man. In a series of patients suffering from chronic obstructive lung disease, metoprolol was compared to propranolol as to its effects on lung function.[54] It was found that, although metoprolol did induce increased obstruction, its effect was significantly less than that of propranolol.

Pharmacokinetics

As shown in Table 7, metoprolol has very similar pharamacokinetic parameters compared to propranolol,[56] the major difference being the percent bound by plasma protein. The elimination half-life of metoprolol becomes prolonged after chronic dosing.[57] When taken after meals the bioavailability of both metoprolol and propranolol is increased.[58]

Uses and Dosages

Metoprolol is indicated in the treatment of essential hypertension. The daily dose of metoprolol is in the same range as that for propranolol. However, since the former is somewhat more bioavailable, the dosage may be a little less. The daily dose of metoprolol can be given as a single dose, the effect of which lasts for at least 26 hr. The antihypertensive and side effects of metoprolol do not correlate with plasma concentration.[59]

Toxicity

The adverse effects of metoprolol are identical to those of propranolol.

Physical-Chemical and Other Data

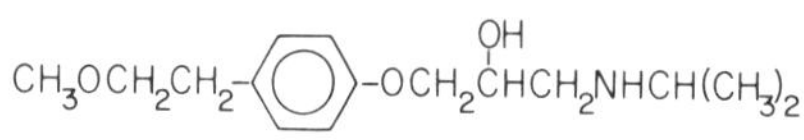

METOPROLOL

The pK_a of metoprolol is 9.78. Metoprolol tartrate (mp 121 to 122°C) is a white, odorless, crystalline powder. It is freely soluble in water, chloroform, and methylene chloride and soluble in alcohol, but sparingly soluble in acetone and ether. It must be protected from light.

ATENOLOL

Mechanism of Action

Atenolol is a potent cardioselective β-adrenoceptor blocking agent. Studies in asthmatic patients[60] show that atenolol has little antagonistic action against β_2-adrenoceptor agonists and does not significantly worsen the respiratory function. The effect of atenolol in hypertension was compared to that of three selective β-adrenoceptor blocking agents.[61] There were no differences between the drugs as far as their action on blood pressure was concerned and on glucose tolerance and insulin levels.

Pharmacokinetics

As shown in Table 7, atenolol has a half-life of between 6 and 9 hr. Atenolol is not metabolized in the liver; however, only about 50% of the drug is absorbed;[62] 95% of the drug in the plasma is excreted unchanged in the urine. In renal failure the half-life of atenolol can increase up to 100 hr.[63] The blood pressure response does not correlate closely with serum atenolol levels.

Uses and Dosage

Atenolol is indicated in the treatment of hypertension. The usual starting dose is 50 mg/day. After 1 week, if the pressure has not been suitably reduced, the dose may be doubled. The dose can be doubled at weekly intervals to a maximum of about 400 mg.[64] In other parts of the world atenolol has been used successfully for all of the indications for propranolol.

Toxicity

The adverse responses to atenolol are identical to those for propranolol. However, since atenolol is only poorly lipid soluble, it probably does not enter the central nervous system. For this reason some of the central adverse effects of propranolol may be less with atenolol.

Physical-Chemical and Other Data

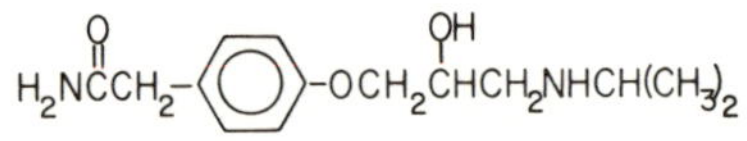

ATENOLOL

Atenolol (mp 146°C) is less soluble in water than alcohol.

NADOLOL

Mechanism of Action

The action of nadolol is the same as that of propranolol. In animals it appears to be less depressant to the myocardium than propranolol.[65] However, in humans the actions of the two drugs seem to be identical.[66]

Pharmacokinetics

As shown in Table 7, about 30% of nadolol given orally is absorbed. The drug is excreted unchanged by the kidney. The half-life of nadolol is very long, ranging between 17 and 24 hr. Nadolol need be administered only once per day.[67]

Uses and Dosages

Nadolol is effective in hypertension, angina pectoris, and arrhythmias.[68] The starting dose is about 40 mg/day. As with other β-blockers, the dose can be doubled at weekly intervals until an effective level is achieved. The average daily dose is 100 mg.

Toxicity

The adverse effects of nadolol are identical to those of propranolol. In case an overdose is given, hemodialysis is the only way to remove the drug from the body.

Physical-Chemical and Other Data

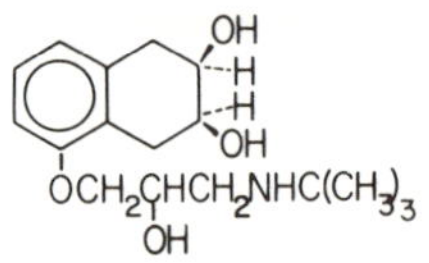

NADOLOL

The pK_a of nadolol is 9.50. Nadolol is a white to off-white, crystalline, odorless powder. It is insoluble in acetone, benzene, ether, and hexane. It is slightly soluble in chloroform and freely soluble in ethanol. The drug possesses excellent stability as a solid. Lyophilized sterile solutions in 0.1 *M*, pH 7.4, sodium phosphate buffer show no decomposition when stored at ambient temperature for 50 days. It is available in tablets containing 40, 80, 120, or 160 mg.

PINDOLOL

Mechanism of Action

Pindolol is a potent β-adrenoceptor blocking agent that possesses a good deal of intrinsic sympathomimetic action (ISA). Because of the ISA, pindolol may be less active in inducing asthma in patients with obstructive lung disease. In spite of the ISA, pindolol is effective in the treatment of angina pectoris. This drug produces less bradycardia than does propranolol, but some bradycardia is always induced.[49]

Pharmacokinetics

Pindolol is mostly excreted unchanged by the kidney. However, as shown in Table 7, about 86% of the administered dose of pindolol appears in the blood. In a study in hypertensive patients with renal impairment, it was found that the urinary excretion decreases from about 34% in patients without renal impairment to about 5% with renal impairment. At the same time, the bioavailability falls from about 86 to 52%.[68,69] Since hepatic clearance was unchanged it is presumed that the fall in bioavailability is due to a decrease in intestinal absorption.

Uses and Dosages

Pindolol is indicated in the treatment of hypertension, angina pectoris, and various cardiac arrhthymias. As noted above, it may be useful in patients with obstructive lung disease. The usual starting dose is 5 to 10 mg/day and, as with the other β-blockers, this dose may be doubled at weekly intervals until the desired therapeutic response is obtained.

Toxicity

The adverse effects of pindolol are the same as those for propranolol.

Physical-Chemical and Other Data

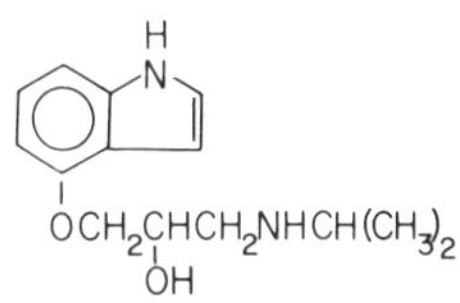

PINDOLOL

The pK_a of pindolol (mp 171°C) is 9.46.

OTHER β-ADRENOCEPTOR BLOCKING AGENTS

The chemical structures are shown in Table 11.

Labetolol is a unique β-blocker. It not only blocks β-receptors, it also blocks α-receptors. Labetolol is about one third as potent as propranolol on β-receptors. The ratio of β-block to α-block is approximately 1:4. In a comparative study in man it has been found that labetolol blocks the tachycardia produced by isoproterenol and the pressor response of norepinephrine in a similar manner.[70] In theory, a blocking agent of this type should be more effective in the treatment of hypertension. Unlike agents that do not block the α-receptor, labetolol should directly decrease peripheral resistance. Labetolol has been found to be effective in the treatment of hypertension and angina.

Labetalol has two asymmetric centers and exists as a mixture of equal proportions of four stereomers.

Acebutolol is about one third as potent as propranolol, is cardioselective, has some partial agonistic effect, and has some membrane-stabilizing action. Acebutolol is effective in hypertension but it has been found that it does not reduce the cardiac output and does not reduce plasma renin activity.[71] Acebutolol is metabolized by acetylation. The acetyl metabolite has been found to be active and to have a slightly longer half-life than for the parent compound.[72] In supraventricular arrhythmias in patients with obstructive lung disease, acebutolol has been shown to be effective against the arrhythmia without exacerbating the lung disease.[73] Antinuclear antibodies develop in many patients receiving acebutolol; however, there has been no evidence of any autoimmune disease found.[74]

Acebutolol hydrochloride (mp 141 to 144°C) is a white or slightly cream-colored powder

Table 11
STRUCTURES OF OTHER β-ADRENOCEPTOR BLOCKING AGENTS

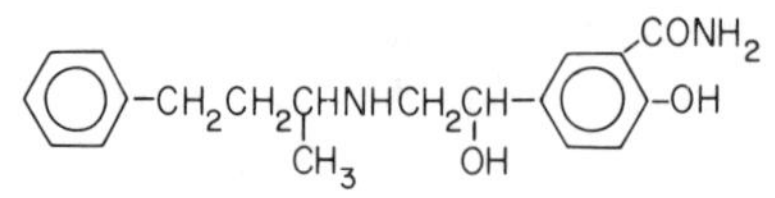

Labetolol

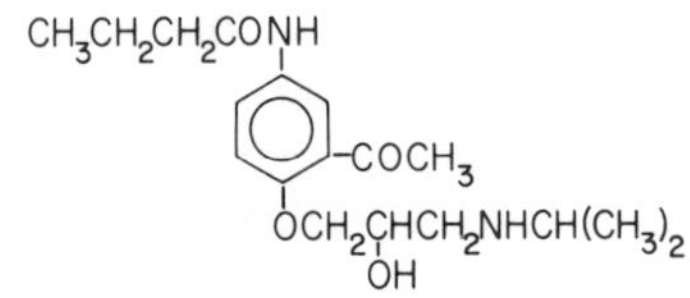

Acebutolol

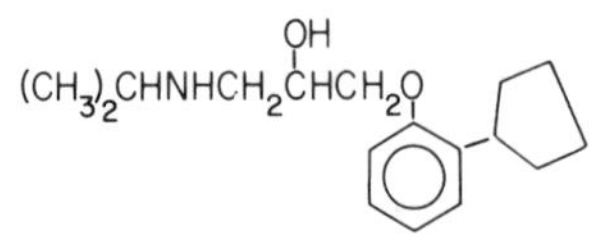

Penbutolol

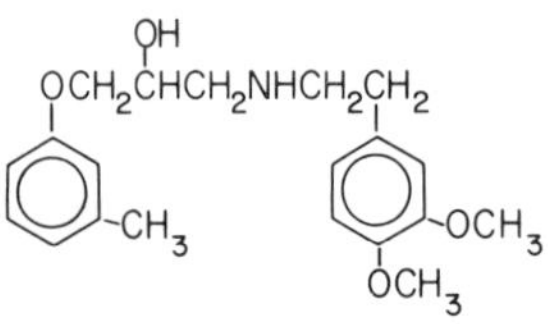

Bevantolol

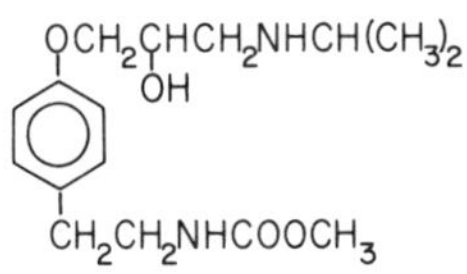

Pamatolol

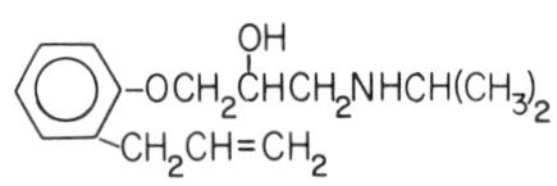

Alprenolol

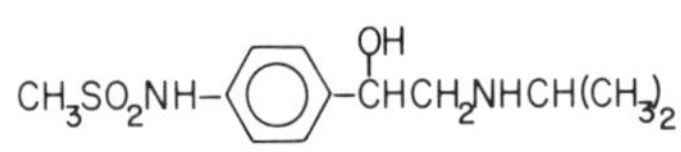

Sotalol

that is water soluble and remains in solution at pH 4.0. It has weak lipophilicity. The partitioning between *n*-octanol/water is 0.62:1.

Penbutolol is a nonselective β-adrenoceptor blocking agent. It is the levo-rotatory isomer, and is five to six times more potent than propranolol. In doses of 25 to 75 mg/day, penbutolol is just as effective as propranolol in the treatment of hypertension.[75]

Bevantolol is a newer cardioselective β-adrenoceptor blocking agent that has not been studied extensively. It has no intrinsic sympathomimetic action. In normal subjects it has been found that bevantolol shifts the albuterol dose response curve on airway resistance to the right. Doses of 75 to 150 mg were found to be similar to 100 mg of practolol.[76]

Pamatolol is a cardioselective β-adrenoceptor blocking agent. The half-life ranges between 3 and 5 hr.[77] The duration of action is directly related to the declining plasma levels. At a serum concentration of 50 ng/mℓ there is a 10% reduction in exercise tachycardia. To increase the duration of action it is necessary to give a large dose. For example, 600 mg has a 24-hr duration.[78]

Alprenolol is very similar in action to propranolol. It has been used extensively to treat hypertension and angina in Europe. Like propranolol, alprenolol is metabolized to a 4-hydroxy derivative, which has the same potency as alprenolol. However, it has a much shorter half-life: 0.8 hr as compared to 2.5 hr for alprenolol.[79]

The pK_a of alprenolol is 9.63. Alprenolol hydrochloride (mp 108 to 111°C) is a colorless or white, crystalline, odorless, or almost odorless powder that has a bitter, numbing taste. It is very soluble in water, alcohol, and chloroform and insoluble in ether. A 5% aqueous solution has a pH of 5.5 to 6.5. It is incompatible with alkali. Dry solutions are sterilized by autoclaving. The powder should be stored in airtight containers and protected from light.

Sotalol was first synthesized in 1960.[80] Its pharmacology and toxicology were first described in detail by Lish et al.[35] in 1965. Sotalol was the first pure β-adrenoceptor blocking agent having no ISA and no membrane-stabilizing action. It is effective in the treatment of hypertension, angina pectoris, and arrhythmias.[81] Sotalol has a long half-life (about 17 hr), is not bound to plasma protein, is not metabolized in the liver, and is excreted in the kidney in unchanged form.[82] Sotalol need be administered only once a day.[83]

The pK_a of sotalol is 9.80. Sotalol hydrochloride (mp 206.5°C dec) is an off-white to pale cream-colored crystalline solid. It is freely soluble in water and only slightly soluble in chloroform. It has low lipophilicity and should be protected from light.

REFERENCES

1. **Ahlquist, R. P.,** A study of the adrenotropic receptors, *Am. J. Physiol.,* 153, 586, 1948.
2. **Powell, C. E. and Slater, I. H.,** Blocking of inhibitory adrenergic receptors by a dichloro analog of isoproterenol, *J. Pharmacol. Exp. Ther.,* 122, 480, 1958.
3. **Moran, N. C. and Perkins, M. E.,** Adrenergic blockade of the mammalian heart by a dichloro analogue of isoproterenol, *J. Pharmacol. Exp. Ther.,* 124, 223, 1958.
4. **Nickerson, M.,** The pharmacology of adrenergic blockade, *Pharmacol. Rev.,* 1, 27, 1949.
5. **Melchiorre, C. and Belleau, B.,** Recent developments in structure-activity relationships among inhibitors of the adrenergic alpha-receptor, in *Adrenoceptors and Catechlolamine Action,* Kunos, G., Ed., John Wiley & Sons, New York, 1981, chap. 6.
6. **Ahlquist, R. P., Huggins, R. A., and Woodbury, R. A.,** The pharmacology of benzyl-imidazoline (Priscol), *J. Pharmacol. Exp. Ther.,* 89, 271, 1947.
7. **Taylor, S. H., Sutherland, G. R., MacKenzie, G. J., Staunton, H. P., and Donald, K. W.,** The circulatory effects of phentolamine in man with particular respect to changes in forearm blood flow, *Clin. Sci.,* 28, 265, 1965.
8. **Kelly, D. T., Delgado, C. E., Taylor, D. R., Pitt, B., and Ross, R. S.,** Use of phentolamine in acute myocardial infarction associated with hypertension and left ventricular failure, *Circulation,* 47, 729, 1973.
9. **Moyer, J. H. and Caplovitz, C.,** The clinical results of oral and parenteral administration of 2-(*N-p*-tolyl-*N-m*-hydroxyphenylaminomethyl) imidazoline hydrochloride (Regitine) in the treatment of hypertension and an evaluation of cerebral hemodynamics, *Am. Heart J.,* 45, 602, 1953.
10. **Nickerson, M. and Goodman, L. S.,** Pharmacological properties of a new adrenergic blocking agent; *N*,*N*-dibenzyl-*b*-chloroethylamine (dibenamine), *J. Pharmacol. Exp. Ther.,* 89, 167, 1947.
11. **Nickerson, M. and Gump, W. S.,** The chemical basis for adrenergic blocking activity in compounds related to dibenamine, *J. Pharmacol. Exp. Ther.,* 97, 25, 1949.
12. **Loew, E. R., Achenbach, P., and Micetich, A.,** Adrenergic blocking drugs. V. Blocking of excitatory responses to epinephrine and adrenergic nerve stimulation with *N*-alkyl-*N*-(2-chloroethyl)-benzhydrylamines, *J. Pharmacol. Exp. Ther.,* 97, 441, 1949.
13. **Graham, J. D. P.,** 2-Halogenoalkylamines, *Prog. Med. Chem.,* 2, 132, 1962.
14. **Langer, S. Z.,** Presynaptic receptors and their role in the regulation of transmitter release, *Br. J. Pharmacol.,* 60, 481, 1977.
15. **Weiner, N.,** Drugs that inhibit adrenergic nerves and block adrenergic receptors, in *The Pharmacological Basis of Therapeutics,* Gilman, A. G., Goodman, L. S., and Gilman, A., Eds., Macmillan, New York, 1980, chap. 9.
16. **Moe, G. K., Peralta, B., and Seevers, M. H.,** Central impairment of sympathetic reflexes by 8-aminoquinolines, *J. Pharmacol. Exp. Ther.,* 95, 407, 1949.
17. **Massingham, R. and Hayden, M. L.,** A comparison of the effects of prazosin and hydrallazine on blood pressure, heart rate and plasma renin activity in conscious renal hypertensive dogs, *Eur. J. Pharmacol.,* 30, 121, 1975.
18. **Scriabine, A., Constantine, J. W., Hess, H. J., and McShane, W. K.,** Pharmacological studies with some new antihypertensive aminoquinazolines, *Experientia,* 24, 1150, 1968.
19. **Pitterman, A. B., Rollins, D. E., Shen, D. D., Hurwitz, A., and Hassanein, K. M.,** Alpha adrenoceptor blockade with oral prazosin, *Clin. Pharm. Ther.,* 29, 142, 1981.
20. **Jaillon, P., Rubin, P., Yee, Y. G., Ball, R., Kates, R., Harrison, D., and Blaschke, T.,** Influence of congestive heart failure on prazosin kinetics, *Clin. Pharm. Ther.,* 25, 790, 1979.

21. **Chau, N. P., Flouvat, B. L., LeRoux, E., and Safar, M. E.,** Prazosin kinetics in essential hypertension, *Clin. Pharm. Ther.*, 28, 6, 1980.
22. **Pierpont, G. L., Franciosa, J. A., and Cohn, J. N.,** Effect of prazosin on renal function in congestive heart failure, *Clin. Pharm. Ther.*, 28, 335, 1980.
23. **Rall, T. W. and Schleifer, L. S.,** Oxytocin, prostaglandins, ergot-alkaloids, and other agents, in *The Pharmacological Basis of Therapeutics*, Gilman, A. C., Goodman, L. S., and Gilman, A., Eds., Macmillan, New York, 1980, chap. 39.
24. **Berde, B. and Sturmer, E.,** Introduction to the pharmacology of ergot alkaloids and related compounds as a basis of their therapeutic application, in *Ergot Alkaloids and Related Compounds,* Berde, B. and Schild, H. O., Eds., *Handbuch der Experimentellen Pharmakologie,* Vol. 49, Springer-Verlag, Basel, 1978.
25. **Saper, J. R.,** Migraine, II. Treatment, *JAMA*, 239, 2480, 1978.
26. **Barger, G.,** *Ergot and Ergotism,* Gurney & Jackson, Edinburgh, 1931.
27. **Shanks, R. G.,** personal communication, 1961.
28. **Prichard, B. N. C.,** Hypotensive action of pronetholol, *Br. Med. J.*, 1, 1227, 1964.
29. **Black, J. W., Crowther, A. F., Shanks, R. G., Smith, L. H., and Dornhorst, A. C.,** A new adrenergic beta receptor antagonist, *Lancet,* 1, 1080, 1964.
30. **Prichard, B. N. C. and Gillam, P. M. S.,** Use of propranolol (Inderal) in the treatment of hypertension, *Br. Med. J.*, 2, 725, 1964.
31. **Hamer, J., Grandjean, T., Melendez, L., and Sowton, G. E.,** Effect of propranolol (Inderal) in angina pectoris: preliminary report, *Br. Med. J.*, 2, 720, 1964.
32. **Gillam, P. M. S. and Prichard, B. N. C.,** Use of propranolol in angina pectoris, *Br. Med. J.*, 2, 337, 1965.
33. **Besterman, E. M. M. and Friedlander, D. H.,** Clinical experiences with propranolol, *Postgrad. Med. J.*, 41, 526, 1965.
34. **Ahlquist, R. P.,** Adrenergic beta-blocking agents, in *Progress in Drug Research,* Vol. 20, Jucker, E., Ed., Birkhauser Verlag, Basel, 1976, 27.
35. **Lish, P. M., Weikel, J. H., and Dungan, K. W.,** Pharmacological toxicological properties of two new beta adrenergic receptor antagonists, *J. Pharmacol. Exp. Ther.*, 149, 161, 1965.
36. **Walle, T., Conradi, E., Walle, U., Fagan, T., and Gaffney, T.,** Propranolol glucuronide cumulation during long-term propranolol therapy: a proposed storage mechanism for propranolol, *Clin. Pharm. Ther.*, 26, 686, 1979.
37. **Walle, T., Fagan, T., Conradi, E., Walle, U., and Gaffney, T.,** Presystemic and systemic glucuronidation of propranolol, *Clin. Pharm. Ther.*, 26, 167, 1979.
38. **Walle, T., Conradi, E. C., Walle, U. K., Fagan, T. C., and Gaffney, T. E.,** Naphthoxylactic acid after single and long-term doses of propranolol, *Clin. Pharm. Ther.*, 26, 548, 1979.
39. **Walle, T., Conradi, E. C., Walle, U. K., Fagan, T. C., and Gaffney, T. E.,** 4-Hydroxypropranolol and its glucuronide after single and long-term doses of propranolol, *Clin. Pharm. Ther.*, 27, 22, 1980.
40. **Fitzgerald, J. D. and O'Donnell, S. R.,** Pharmacology of 4-hydroxypropranolol, a metabolite of propranolol, *Br. J. Pharm.*, 43, 222, 1971.
41. **Kornhauser, D. M., Wood, A. J. J., Vestal, R. E., Wilkinson, G. R., Branch, R. A., and Shand, D. G.,** Biological determinants of propranolol disposition in man, *Clin. Pharm. Ther.*, 23, 165, 1978.
42. **Walle, T., Conradi, E. C., Walle, U. K., Fagan, T. C., and Gaffney, T. E.,** The predictable relationship between plasma levels and dose during chronic propranolol therapy, *Clin. Pharm. Ther.*, 24, 668, 1978.
43. **Bilezikian, J., Gammon, D., Rochester, C., and Shand, D.,** A radioreceptor assay for propranolol, *Clin. Pharm. Ther.*, 26, 173, 1979.
44. **Rochester, C. L., Gammon, D. E., Shane, E., and Bilezikian, J. P.,** A radioreceptor assay for propranolol and 4-hydroxypropranolol, *Clin. Pharm. Ther.*, 28, 32, 1980.
45. **van den Brink, G., Boer, P., van Asten, P., Mees, E. J., and Geyskes, G. G.,** One and three doses of propranolol a day in hypertension, *Clin. Pharm. Ther.*, 27, 9, 1980.
46. **Chew, C., Collett, J., Shah, P., and Singh, B.,** Hemodynamic effects of intravenous timolol in coronary artery disease, *Clin. Pharm. Ther.*, 26, 330, 1979.
47. **Simon, G., Kiowski, W., and Julius, S.,** Antihypertensive and b-adrenoceptor antagonist action of timolol, *Clin. Pharm. Ther.*, 23, 152, 1978.
48. **Singh, B. N., Williams, F. M., Whitlock, R. M., Collett, J., and Chew, C.,** Plasma timolol levels and systolic time intervals, *Clin. Pharm. Ther.*, 28, 159, 1980.
49. **Lowenthal, D. T., Pitone, J. M., Affrime, M. B., Shirk, J., Busby, P., Kim, K. E., Nancarrow, J., Swartz, C. D., and Onesti, G.,** Timolol kinetics in chronic renal insufficiency, *Clin. Pharm. Ther.*, 23, 606, 1978.
50. **Brogden, R. N., Speight, T. M., and Avery, G. S.,** Timolol, a preliminary report of its pharmacological properties and therapeutic efficacy in angina and hypertension, *Drugs,* 9, 164, 1975.
51. **Aronow, W. S., Plasencia, G., Wong, R., and Landa, D.,** Exercise duration to angina at two and twelve hours after timolol, *Clin. Pharm. Ther.*, 29, 155, 1981.

52. **Edwards, I. R. and Else, O.,** Once-daily dosing of timolol in hypertension. A controlled study, *Eur. J. Clin. Pharmacol.*, 17, 239, 1980.
53. **Affrime, M. B., Lowenthal, D. T., Tobert, J. A., Shirk, J., Eidelson, B., Cook, T., and Onesti, G.,** Dynamics and kinetics of ophthalmic timolol, *Clin. Pharm. Ther.*, 27, 471, 1980.
54. **Wunderlich, J., Macha, H., Wudicke, H., and Huckauf, H.,** Beta-adrenoceptor blockers and terbutaline in patients with chronic obstructive lung disease, *Chest*, 78, 714, 1980.
55. **Ablad, B., Borg, K., Carlsson, E., Ek, L., Johnsson, G., Malmfors, T., and Regardh, C.,** A study of the pharmacological properties of metoprolol in animals and man, *Acta Pharmacol. Toxicol.*, 36, 7, 1975.
56. **Borg, K., Carlsson, E., Hoffmann, K., Jonsson, T., Thorin, H., and Wallin, B.,** Metabolism of metoprolol - ^3H in man, the dog and the rat, *Acta Pharmacol. Toxicol.*, 36, 125, 1975.
57. **Myers, M. G. and Thiessen, J. J.,** Metoprolol kinetics and dose response in hypertensive patients, *Clin. Pharm. Ther.*, 27, 756, 1980.
58. **Melander, A., Danielson, K., Schersten, B., and Wahlin, E.,** Enhancement of the bioavailability of propranolol and metoprolol by food, *Clin. Pharm. Ther.*, 22, 108, 1977.
59. **Karlberg, B., Nilsson, O., Tolagen, K., Nitelius, E., and Waern, U.,** Once-daily metoprolol in primary hypertension, *Clin. Pharm. Ther.*, 25, 399, 1979.
60. **Vilsvik, J. S. and Schaanning, J.,** Effect of atenolol on cardiac and ventilatory function in patients with chronic asthma, *Postgrad. Med. J.*, 53, 142, 1977.
61. **Waal-Manning, H. J.,** Atenolol and three nonselective β-blockers in hypertension, *Clin. Pharm. Ther.*, 25, 8, 1979.
62. **Mason, W. K., Winer, N., Kochak, G., Cohen, I., and Bell, R.,** Kinetics and absolute bioavailability of atenolol, *Clin. Pharm. Ther.*, 25, 408, 1979.
63. **McAinsh, J., Holmes, B. F., Smith, S., Hood, D., and Warren, D.,** Atenolol kinetics in renal failure, *Clin. Pharm. Ther.*, 28, 302, 1980.
64. **van Rooijen, G., Boer, P., Mees, E., and Geyskes, G.,** Effects of atenolol and propranolol when added to long-term antihypertensive diuretic therapy, *Clin. Pharm. Ther.*, 26, 420, 1979.
65. **Frishman, W. and Wilverman, R.,** Clinical pharmacology of the new beta adrenergic blocking drugs. II. Physiologic and metabolic effects, *Am. Heart J.*, 97, 797, 1979.
66. **Frishman, W. H.,** *Clinical Pharmacology of the Beta Adrenoceptor Blocking Drugs*, Appleton-Century-Crofts, New York, 1980.
67. **Dreyfuss, J., Brannick, L. J., Vukovich, R. A., Shaw, J. M., and Willard, D. A.,** Metabolic studies in patients with nadolol: oral and intravenous administration, *J. Clin. Pharmacol.*, 17, 300, 1977.
68. **Lavene, D., Weiss, Y. A., Safar, M. E., Loria, Y., Agorus, N., Georges, D., and Milliez, P. L.,** Pharmacokinetics and hepatic extraction ratio of pindolol in hypertensive patients with normal and impaired renal function, *J. Clin. Pharm.*, 17, 501, 1977.
69. **Wainer, E., Boner, G., and Rosenfeld, J. B.,** Effects of pindolol on renal function, *Clin. Pharm. Ther.*, 28, 575, 1980.
70. **Richards, D. A. and Prichard, B. N. C.,** Concurrent antagonism of isoproterenol and norepinephrine after labetalol, *Clin. Pharm. Ther.*, 23, 253, 1978.
71. **Dreslinski, G. R., Aristimuno, G. G., Hesserli, F. H., Suarez, D. H., and Frohlich, E. D.,** Effects of beta blockade with acebutolol on hypertension, hemodynamics, and fluid volume, *Clin. Pharm. Ther.*, 26, 562, 1979.
72. **Meffin, P. J., Winkle, R. A., Peters, F. A., and Harrison, D. C.,** Dose-dependent acebutolol disposition after oral administration, *Clin. Pharm. Ther.*, 24, 542, 1978.
73. **Aronow, W. S., Van Camp, S., Turbow, M., Whittaker, K., and Lurie, M.,** Acebutolol in supraventricular arrhythmias, *Clin. Pharm. Ther.*, 29, 149, 1981.
74. **Cody, R. J., Calabrese, L. H., Clough, J. D., Tarazi, R. C., and Bravo, E. L.,** Development of antinuclear antibodies during acebutolol therapy, *Clin. Pharm. Ther.*, 25, 800, 1979.
75. **Cangiano, J. L., Ayala, J. M., Boksay, J. E., and Martinez-Maldonado, M.,** Penbutolol: a new beta blocker in the treatment of moderate essential hypertension, *J. Clin. Pharm.*, 19, 384, 1979.
76. **Mackay, A. D., Gribbin, H. R., Baldwin, C. J., and Tattersfield, A. E.,** Assessment of bronchial beta blockade after oral beventolol, *Clin. Pharm. Ther.*, 29, 1, 1981.
77. **Carruthers, S., Pentikainen, P., Hosler, J., and Azarnoff, D.,** Kinetics of pamatolol, a cardioselective beta adrenoreceptor blocker, *Clin. Pharm. Ther.*, 26, 682, 1979.
78. **Carruthers, S. G., Hosler, J. P., Pentikainen, P., and Azarnoff, D. L.,** Pamatolol: phase I evaluation of the pharmacodynamics of a cardioselective beta adrenoceptor blocking drug, *Clin. Pharm. Ther.*, 24, 168, 1978.
79. **Collste, P., Borg, K. O., Astrom, H., and von Bahr, C.,** Contribution of 4-hydroxy-alprenolol to adrenergic beta receptor blockage of alprenolol, *Clin. Pharm. Ther.*, 25, 416, 1979.
80. Mead Johnson Research Center Archives, Evansville, Ind., 1960.

81. **Aviado, D. M.,** Sotalol, a pure beta adrenergic blocking agent for treatment of essential hypertension and other cardiovascular diseases, *J. Clin. Pharm.*, 19, 497, 1979.
82. **Schnelle, K., Klein, G., and Schinz, A.,** Studies on the pharmacokinetics and pharmacodynamics of the beta adrenergic blocking agent sotalol in normal man, *J. Clin. Pharm.*, 19, 516, 1979.
83. **Pasquel, R.,** Pharmacologic considerations in determining efficacy of once-daily sotalol administration to hypertensive patients, *J. Clin. Pharm.*, 19, 523, 1979.

DIRECT AND INDIRECT CHOLINERGICS

Emmanuel B. Thompson and James E. Gearien

INTRODUCTION

The Cholinergic Nervous System

The branch of the autonomic nervous system in which the acetylcholine (Structure 1) is the neurotransmitter carrying the nerve impulse from the nerve to the muscle is commonly referred to as the "cholinergic" or "parasympathetic nervous system". Cholinergic agonists are drugs whose pharmacological profiles resemble the effects of stimulation of the parasympathetic division of the autonomic nervous system. However, because of the ubiquitous nature of this system, the responses to parasympathetic stimulation and the drugs which affect it involve several organs, such as the eyes, glands, smooth muscles of the bronchioles, GI tract, blood vessels, urinary bladder, skeletal muscle, and the heart.

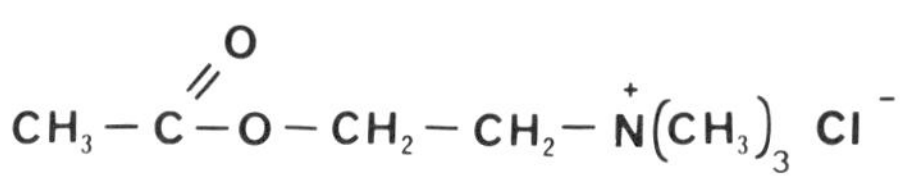

1

This system innervates both striated and smooth muscle. Stimulation of the cholinergic nervous system establishes the "tone" of striated muscle. In smooth muscle it is responsible for the peristaltic action of the GI and genitourinary tracts. Stimulation of the cholinergic nervous system increases the secretion of digestive enzymes and hydrochloric acid. It also causes constriction of the pupils of the eye.

Muscles of the cardiovascular system are innervated by the cholinergic nervous system. Acetylcholine or stimulation of the cholinergic nervous system induces a decrease in cardiac rate, the force of cardiac contraction, and vasodilation. Since cardiac muscle is also innervated by the adrenergic nervous system and since acetylcholine may liberate norepinephrine, the value of cholinergic drugs in cardiovascular disease is limited.

The Role of Neurotransmitters in Muscle Contraction

In order to understand the mechanism of action of cholinergic drugs, it is necessary to examine the role played by chemical neurotransmitters in carrying nerve impulses from the nerve to the muscle.

Nerve impulses travel down a nerve axon by the creation of a potential difference. The impulses pass from one nerve axon to another, across a synapse, and from the nerve to the effector muscle by means of a chemical neurotransmitter. This chemical is biosynthesized and stored at the nerve endings. A nerve impulse triggers its liberation. Following liberation, the neurotransmitter is distributed to a receptor on the ending of another nerve or on the membrane surrounding the effector muscle. Interaction with this receptor initiates the flow of ions required for muscle contraction. The last step in this series of events is the removal of the neurotransmitter from the area of the receptor. In the cholinergic nervous system, acetylcholine is enzymatically hydrolyzed to relatively inactive compounds.

The role of chemical neurotransmitters in nerve transmission is well documented in the literature.[1,2] An "epinephrine-like" substance was postulated to be a chemical neurotransmitter in the adrenergic nervous system,[3] and in 1967 the role of a muscarine or an acetylcholine-like substance acting as a neurotransmitter was demonstrated in the cholinergic nervous system.[4] This was not confirmed until publication of the results of the investigation of Loew[5] in 1921 and those of Dale in 1926.[6]

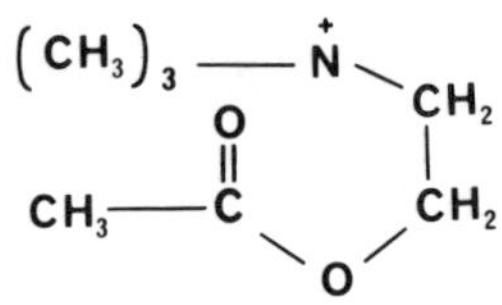

FIGURE 1. Quasi-ring conformation.

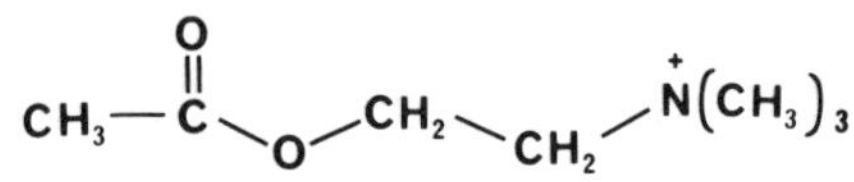

FIGURE 2. Extended conformation.

In the cholinergic nervous system, acetylcholine interacts with two structurally different receptors: (1) the nicotinic receptor on striated muscle and (2) the muscarinic receptor on smooth muscle. It was possible to identify these receptors, both capable of interacting with acetylcholine, because of the specific biological activity of two structurally dissimilar alkaloids: nicotine and muscarine.[7]

In Structure 2, nicotine interacts specifically with the receptors on striated muscle (nicotinic receptor), while muscarine shows a similarly specific interaction with the receptor present on smooth muscle (muscarinic receptor [Structure 3]).

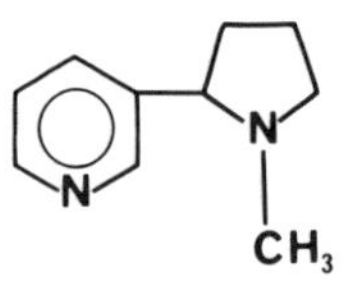

2

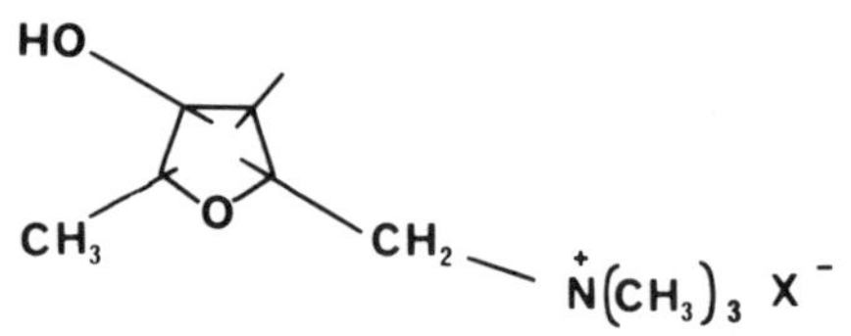

3

Subsequent to its interaction with the receptor, acetylcholine is hydrolyzed to acetic acid and choline. This extremely rapid reaction is catalyzed by the enzyme acetylcholinesterase.

The Muscarinic Receptor

The acetylcholine-mediated response (muscarinic) on smooth muscle has been intensively studied and while the structure of the receptor has not been totally elucidated, a number of amino acids present in the receptor site have been identified. Early structure-activity studies established the importance to potency of the quaternary nitrogen and the ester (or similarly, hydrogen bonding group). This suggested a receptor that contains a negative charge, referred to as the ''anionic site'' and an area containing a group capable of hydrogen bonding, called the ''esteratic site''. From an examination of possible conformations of the acetylcholine molecule, these sites are separated by a distance of (3 to 7 Å).

This distance between these two binding sites cannot be determined from models of acetylcholine since it may bind in any one of a number of conformations ranging from the ''quasi-ring'' structure (Figure 1) or ''extended'' conformation (Figure 2).

A study of the isomers of acetoxycyclopropyltrimethylammonium iodide (Structure 4) suggests that (+) *trans* isomer (2S-acetoxycyclopropyltrimethylammonium iodide) is the most potent isomer. This suggests that the anionic and esteratic sites of the receptor are located from 5 to 7 Å apart and that acetylcholine binds in the extended conformation.

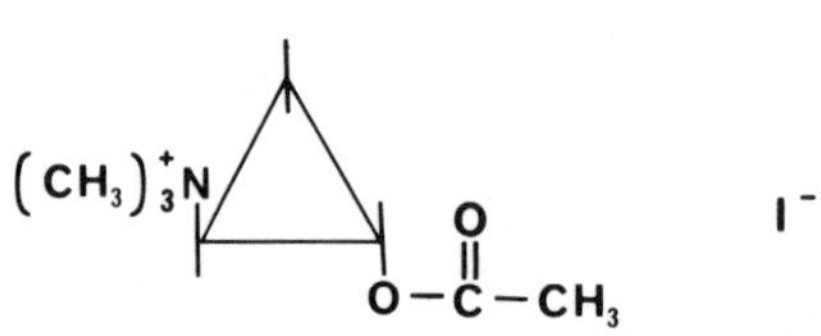

4

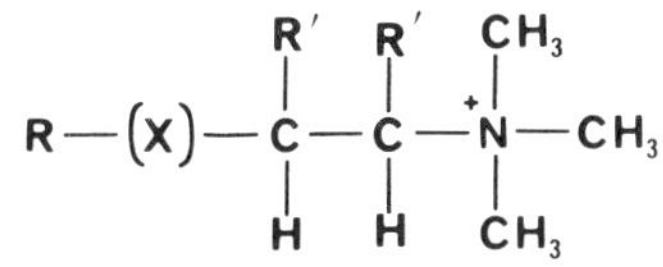

FIGURE 3. General structure for cholinergic agonist. R = CH_3 or H_2N; X = hydrogen bonding groups usually $-C(=O)-O-$; R′ = H or CH3.

Functional groups present in the receptor include the carboxyl groups of glutamic and aspartic acids, one or more hydroxyl groups from serines, and an imidazole ring from histidine.

Studies also suggest the presence of an SH group in the active site and at least one other in the vicinity of the active site.[10] These latter groups do not appear to participate in the binding of clinically useful agonists or antagonists.

The Nicotinic Receptor

Little is known about the molecular makeup of the nicotinic receptor except that it also contains an anionic site which reacts with a positive charged atom on agonist or antagonist. The nicotinic activity of *cis*-2-acetoxycyclopropyltrimethylammonium iodide (Structure 5) and other compounds suggest that acetylcholine binds to the receptor in a conformation resembling that in Figure 1. Nicotinic receptors occur on striated muscle and at the ganglia.

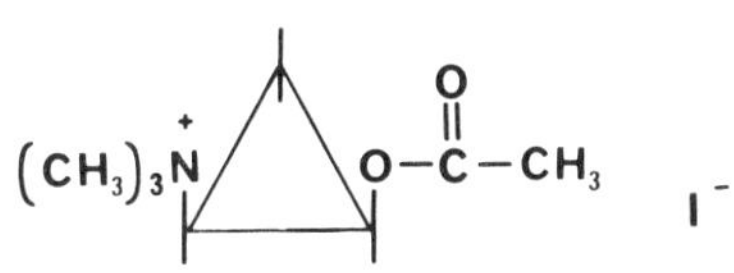

5

STRUCTURE-ACTIVITY RELATIONSHIPS OF DIRECT-ACTING CHOLINERGIC AGONISTS

In most cases, muscarinic agents resemble acetylcholine in chemical structure and interact with the muscarinic receptor through bonding similar to that exhibited by acetylcholine. The following can be viewed as a general structure for the clinically useful cholinergic agonist (Figure 3).

Both binding studies and structure-activity relationships suggest agonists are attracted to the receptor by an ionic attraction between the positive charge on the quaternary nitrogen atom and an ionized carboxyl group on the receptor (anionic site). Binding is accompanied by a conformational change of the receptor so that two of the methyl groups on the quaternary nitrogen are held by both hydrophobic and van der Waals forces. The ester group of acetylcholine is hydrogen bonded to a hydroxyl group (serine) on the receptor and the ethylene bridge separating the quaternary ammonium group for the ester group is hydrophobically bonded to the receptor. The methyl of the acetyl group is also held to the receptor by both hydrophobic and van der Waals forces.

The ability of agonists to activate the muscarinic receptor (intrinsic activity) has been related to the size and shape of the molecule. Belleau[11] has postulated that the change in conformation of receptors required for the previously described binding of agonists places the receptor in its "active state" in which it permits the flow of ions required for muscle contraction.

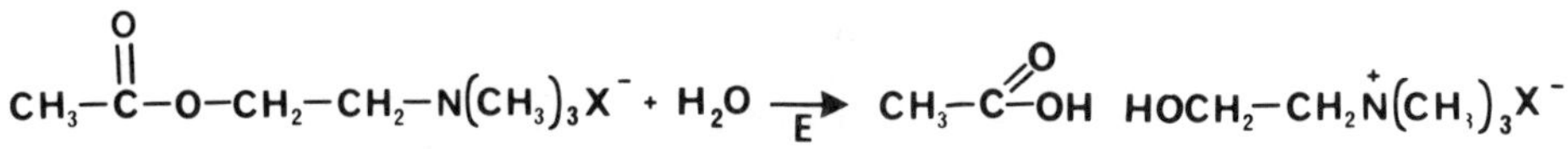

FIGURE 4. E = Acetylcholinesterase.

The requirements for binding to the muscarinic receptor along with the size requirement for intrinsic activity define the structural prerequisite for cholinergic agonists. These requirements include:

1. A positively charged atom (nitrogen in acetylcholine) for ionic attraction to the receptor.
2. Two of the groups on the quaternary nitrogen should be methyl.
3. A group capable of hydrogen bonding is required for high potency. While high potency is achieved with an ester group separated by two atoms from the quaternary ammonium group, it may be substituted by a carbonyl, ether, or amide without loss of intrinsic activity.
4. For optimum receptor fit the hydrogen-bonding group should be separated from the quaternary ammonium group by two carbons.
5. For pure agonist activity a molecule should not exceed the length of acetylcholine. Ing[12] formulated a rule that stated that for a cholinergic molecule to have maximum activity, there must be five atoms present between the quaternary nitrogen and the terminal hydrogen.

Substitution of the ethylene bridge separating the quaternary ammonium group in acetylcholine gave compounds with interesting pharmacological activity. When a methyl group is α-substituted to the nitrogen atom of acetylcholine, a compound with high nicotinic activity and low muscarinic activity is obtained. However, if a methyl group is placed β to the nitrogen, the resulting compound has high muscarinic and low nicotinic activity.[13,14]

A number of cyclic compounds which have structures comparable to that required for receptor binding have high activity or cholinergic agonists. In addition to muscarine (Structure 3), a structurally similar dioxalone (Structure 6) is one of the most potent muscarinic agonists known.[15]

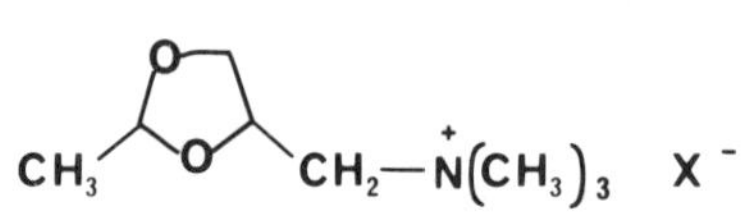

6

Of the stereochemical isomers of dioxalane, the most potent is the (−) *cis*-2R-methyl-4-*S*-trimethylammonium-1,3-dioxalene iodide. High stereospecificity of the receptor for the binding for a similar isomer of acetyl β-methyl choline and muscarine has been noted. The potency ratio of the *S* and *R* isomers may be as great as 500.

INDIRECT-ACTING CHOLINERGIC AGONISTS (INHIBITORS OF ACETYLCHOLINESTERASE)

Acetylcholine, after liberation from its storage vessels and reaction with the receptor, is removed from the receptor area by hydrolysis (Figure 4). The reaction is efficiently catalyzed by the enzyme acetylcholinesterase. Compounds capable of inhibiting this enzyme are called "indirect cholinergic agonists". This inhibition causes an accumulation of acetylcholine in the receptor area, which increases stimulation of muscles innervated by acetylcholine.

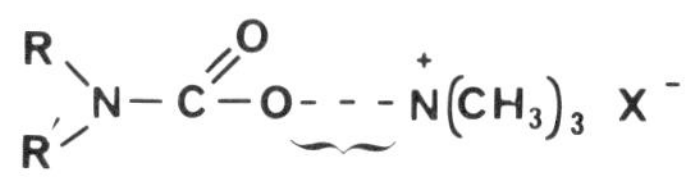

FIGURE 5. General structure for an anticholinesterase. R′ = CH_3; R = CH_3 or H.

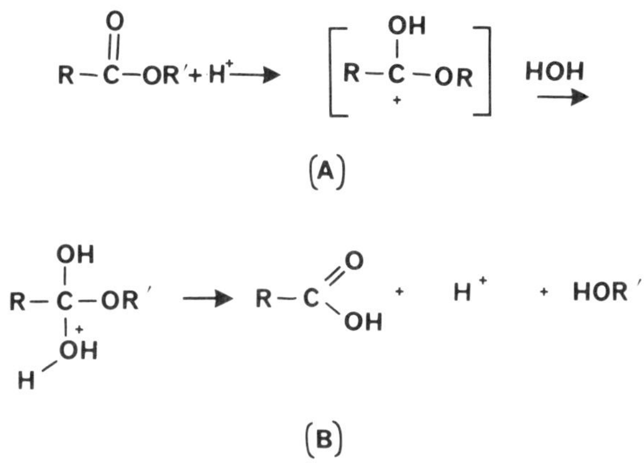

FIGURE 6. Hydrolysis of acetylcholine by acetylcholinesterase.

An early clue to the design of antiacetylcholinesterases arose from the discovery that the cholinergic activity of the alkaloid physostigmine (Structure 7) resulted from its ability to inhibit acetylcholinesterase.

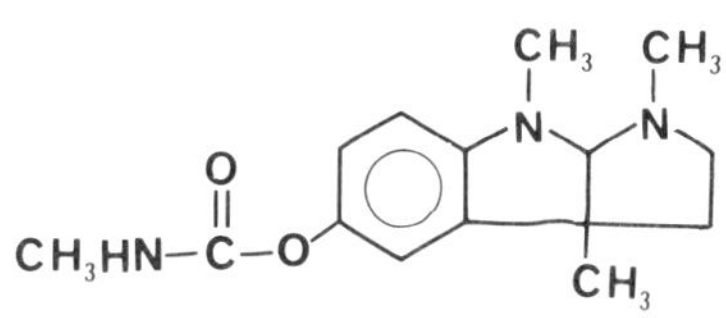

7

This discovery prompted the biological evaluation of N-substituted carbamate esters of choline-like molecules as inhibitors.

Further research has shown that two structural types of compounds may provide effective inhibition of acetylcholinesterase:

1. *N*-alkyl or *N,N*-dialkyl carbamate esters structurally related to acetylcholine are effective reversible acetylcholinesterase inhibitors. While it is difficult to describe all members of this class by a general formula, the structure depicted in Figure 5 serves as a useful generalization.
2. Esters and amides of phosphoric, phosphonic, and pyrophosphoric acids also inhibit acetylcholinesterase. Since the phosphorylation of the enzyme is very slowly reversed they are often referred to as ''irreversible inhibitors''.

Mechanism of Action of Anticholinesterases

Acetylcholinesterase hydrolyzes acetylcholine by a mechanism similar to the acid hydrolysis of esters (Figure 6). The binding of acetylcholine to the enzyme is assumed to be similar to its binding to the muscarinic receptor.[1] A number of mechanisms have been postulated for the protonation of the ester group. These include protonation by a carboxyl group in the active site, protonation by a hydrogen from an amide group of the backbone

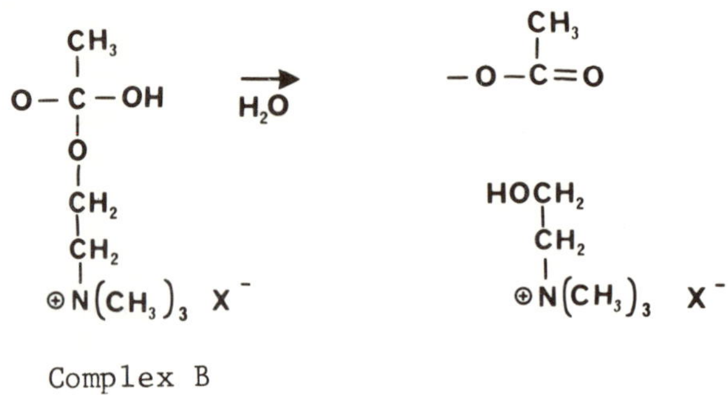

FIGURE 7. Enzyme substrate complex where one hydrogen atom of the water molecule is replaced by the enzyme (Complex B).

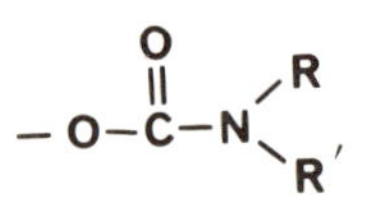

FIGURE 8. Carbamoylated enzyme.

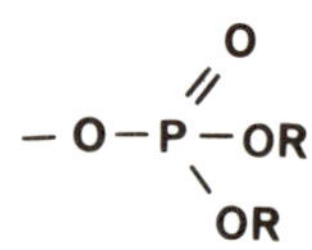

FIGURE 9. Phosphorylated enzyme.

protein structure, and a conserted protonation by the imidazole ring and a carboxyl group. This provides a structure related to intermediate A in Figure 6. Protonation is followed by a nucleophilic attack at the ester carbon by a hydroxyl group from a serine residue to produce an enzyme substrate complex [B] depicted in Figure 7 in which one hydrogen atom of the water molecule is replaced by the enzyme (Complex B). This intermediate is stabilized when choline departs as a leaving group. It should be noted that this leaves the enzyme acetylated and therefore inactive. Regeneration of the enzyme by hydrolysis of the acetyl proceeds at a rate measurable in milliseconds.

Carbamate and phosphate esters are similarly hydrolyzed by acetylcholinesterase and therefore form an inactive carbamoylated or phosphorylated enzyme (Figures 8 and 9).

The carbamoylated enzyme hydrolyzes to the active enzyme in minutes, but the reaction is much slower than in the case of the acetylated enzyme. The rate of the hydrolysis in the phosphorylated enzyme is measured in hours.[16] If untreated, the phosphorylated enzyme undergoes further reaction with the enzyme and the product formed is not hydrolyzed.

In the case of the phosphorylated enzyme, compounds which are capable of increasing the rate of hydrolysis to the active enzyme have been discovered. The most widely used of these, 2-PAM, will be discussed later.

Structure-Activity Relationships of Anticholinesterases

The requirements for binding to the active site of acetylcholinesterase is similar to that prerequisite for cholinomimetics with the exception that larger molecules may bind, undergo hydrolysis, and acylate the enzyme. The formulas shown in the monographs illustrate the importance of an aromatic system in the molecules. The presence of the aromatic system and the failure of the molecules to obey the "Rule of Five" distinguish them from muscarinic agonists.

THERAPEUTIC USES OF CHOLINOMIMETICS AND ANTIACETYLCHOLINESTERASES

A number of health problems may result from or may disrupt the regulation of the contraction of muscles innervated by the cholinergic nervous system.

Myasthenia gravis is characterized by a loss of tone of striated muscle. The disease may result from the destruction of nicotinic receptors in these muscles. Agents which increase the concentration of acetylcholine or similar substances in the receptor area may alleviate the symptoms of this disease.

Postoperative distention of the abdomen or absence of peristalsis in the gastrointestinal and/or genital urinary tract can be eliminated by cholinergic agonists which stimulate the cholinergic nervous system.

Administration of agonists into the eye reduces intraocular pressure and is used in certain types of glaucoma.

TOXICITIES OF CHOLINERGIC AGONISTS

Drugs which activate cholinergic receptors following stimulation of the parasympathetic nervous system, whether directly (the choline-esters, cholinomimetic alkaloids) or indirectly (the reversible and irreversible cholinesterase inhibitors) all produce side effects (toxicities) which are an extension of their pharmacological actions. While the possible source of variation in pharmacological activity of these agonists is based on their duration of effect, the major toxicities can nevertheless be summarized as follows.

Muscarinic Effects

Effects on smooth muscle — Increased GI tract peristalsis resulting in diarrhea, relaxed sphincters with increased urination, bronchoconstriction, vasodilitation with a fall in blood pressure, miosis (constriction of the pupil of the eye), and cycloplegia.

Effects on cardiac muscle — Bradycardia (decreased heart rate) which, together with vasodilitation, produces a fall in arterial pressure. Excessive decrease in heart rate may affect A-V node conduction to the point of heart block.

Effects on exocrine glands — Increased lacrimation, salivation, and sweating. All of these effects can be attenuated or abolished by atropine.

Nicotinic Effects

These effects include generalized muscular twitching followed by muscle weakness. Most of the other side effects are similar to those produced by nicotine.

Central Effects

These effects vary in severity according to the drug used. They are more persistent with the irreversible cholinesterase inhibitors like diisopropylfluorophosphate (DFP), tetraethylpyrophosphate (TEPP), parathion, and malathion than those of the reversible inhibitors such as physostigmine (ESERINE), neostigmine (Pro-Stigmine), or pyridostigmine (MESTINON). Side effects include restlessness, dizziness, tremor, aphasia, disorientation, and hallucination. In cases of severe poisoning, coma, convulsions, and death may occur from neuromuscular paralysis of the intercostal muscles.

Some organophosphorus compounds produce a demyelination of the nervous tissue; others become more toxic as a result of enzymatic metabolism within the body. An example is parathion, which is oxidized to paraoxon, which itself is a cholinesterase inhibitor. However, it should be noted that some sulfur-containing organophosphorus compounds are converted into toxic substances (sulfones and sulfoxides) with no anticholinesterase activity. Drug interactions of cholinomimetics are shown in Table 1.

METABOLISM OF DIRECT AND INDIRECT CHOLINERGIC AGONISTS

Metabolites of cholinergic agonists are listed in Table 2. Since most of these compounds are derivatives of acids, products resulting from hydrolysis are the predominant metabolites.

Table 1
DRUG INTERACTIONS OF CHOLINOMIMETICS

Drug	May react with	To produce
	Direct-Acting Cholinergics	
Pilocarpine	Ethyl alcohol	Prolonged action of alcohol
	Cholinesterase Inhibitors	
Neostigmine	Kanamycin, neomycin, streptomycin	Reduces neuromuscular blockade toxicity of antibiotics
	Organo-Phosphate Cholinesterase Inhibitors	
PAM	Nondepolarizing muscle relaxants (e.g., curare)	Antagonizes relevant effect of curare
Echothiophate	Succinylcholine depolarizing muscle relaxant	Potentiates the relaxant effect
Cholinergics	Sympathomimetic amines	Decreased mydriatic effect
Cholinergics	Antihistamines	Enhanced cholinergic effects
Cholinergics	Corticosteroids	Decreased cholinergic effect

Table 2
METABOLITES OF CHOLINOMIMETICS

Excreted unchanged	Hydrolysis products	Other
Bethanechol	Acetylcholine	Isofluorophate®
Carbachol	Physostigmine	Echothiophate[a]
	Neostigmine[b]	
	Pyridostigmine	
	Ambenonium	

[a] Metabolized to a number of phosphorous-containing compounds.
[b] After hydrolysis some of the phenolic portion is conjugated. The unchanged compound has also been isolated (50%) from the urine.

BIOLOGICAL EVALUATION OF DIRECT AND INDIRECT CHOLINERGIC AGONISTS

The qualitative and quantitative aspects of the effects of cholinergic agonists (known as the pharmacological activity) are usually assessed in biological systems. In order to study the pharmacological activity of cholinergic agonists and antagonists, both isolated tissue or organ preparations (in vitro) and intact animal (in vivo) tests have been devised. These tests will be described briefly as to model, purpose, and execution. For greater detail of these procedures, the reader is referred to the references cited in this section.

The following portion of this review discusses some of the procedures used to study the activity of chemical compounds on the parasympathetic system. The authors have selected preparations that are used quite widely and those with which they have had personal experience. However, because of the wide variety of test models available, it is necessary to clearly understand the uses and limitations of each.

The entire GI tract is innervated by both adrenergic and cholinergic divisions of the autonomic nervous system. The parasympathetic supply from the esophagus to the ileum is furnished by the vagus (Xth cranial nerve). Parasympathetic stimulation generally augments

Table 3
METHODS FOR ASSAY OF CHOLINERGIC ACTIVITY IN VITRO

Test system	Technique	Characteristic of assay method	Ref.
Frog rectus abdominus muscle preparation	Organ bath	Develops slow contraction in response to Ach	26
Perfused frog heart	Organ bath	Perfused constant pressure, quite sensitive to Ach	27
Leech dorsal muscle preparation	Organ bath	Not suitable because of length of response time involved, although highly sensitive to Ach	29
Chick semispanalis cervics muscle preparation	Organ bath	Preparation produces slow contraction to Ach, and Ach-like substances	30,31,32

the contractions or tone of both coats of the muscular tube (i.e., the longitudinal and circular muscles), while they are relaxed by sympathetic (adrenergic) stimulation.

Physiologically, the intestines exhibit two fundamental types of movement: (1) pendular rhythm of the longitudinal muscle of the "resting" intestines and (2) localized contraction rings or circular muscles which produce peristaltic movements. Both pendular and peristaltic movements are conspicuous through their regularity and well maintained by the excised intestine immersed in oxygenated Ringer's solution. Similar movements are observed in the living animal as reported by Trendelenburg.[17]

Parasympathetic agonists, such as choline-esters, physostigmine, pilocarpine, and muscarine, all increase the tonus, the pendular movement, and the peristalsis in mammalian intestine. These facts have been used in the development of the assay procedures described.

Cholinergic Agonists Assay Procedures

Pieces of intestinal tissue from any small animal will continue to give responses to drugs or electrical stimulation for many hours if kept in a suitable solution, usually in Tyrode solution, through which air is bubbled. As a result, the isolated ileum preparation from mice, rats, guinea pigs, and rabbits has been utilized for assay of cholinergic agonist activity.

In Vitro Test Systems — Choline Esters

Isolated Guinea Pig Preparation

Methods for performing tests which measure contraction of an organ or tissue are very similar regardless of preparation employed (see Table 3). The isolated tissue method which utilizes the organ bath is widely popular. For assay purposes, the guinea pig ileum is chosen because there is relatively little spontaneous activity, whereas with the rabbit or rat intestines, there is regular pendular movement (i.e., continuous contraction and relaxation). As with tissues from other animals, the guinea pig ileum has been used to evaluate drugs which act at the parasympathetic neuroeffector junction.[18] Briefly, the assay method consists of killing a guinea pig by a blow on the head and bleeding the animal. The abdomen is opened by a midline incision and the cecum lifted forward to reveal the ileum joined to it at the back. A length of ileum about 2 cm long is cut and removed and suspended in an organ bath containing Tyrole solution maintained at a constant temperature of 37 $\pm$ 0.5°C. Acetylcholine solution is added sequentially to the bath in various concentrations. Each dose is allowed to make contact with the tissue for at least 3 min or until maximal contraction is obtained. Graphically, the effective dose (ED_{50}) of acetylcholine may be determined as that dose required to cause a 50% increase in contractions. Alternatively, the assay can be performed with a solution of other cholinergic agonists.

Other anatomical preparations obtained from the guinea pig have been used as in vitro test systems for cholinergic activity.

Longitudinal Smooth Muscle From the Guinea Pig Intestine

Paton and Zar[19] used the longitudinal smooth muscle from the guinea pig ileum as another model to study parasympathetic activity. The longitudinal muscle stripped from the intestinal tissue containing Auerback's plexus was maintained in a tissue bath containing Kreb's solution at 37°C and gassed with 95% oxygen and 5% carbon dioxide. Electrical stimulation of the muscle strip with platinum electrodes (stimuli 6 to 600 Hz, 1-msec duration) or application of drugs such as acetylcholine, arecoline, or muscarine for 20 sec caused maximal contraction in the plexus-containing strip. Such a preparation is useful for studying agents which can stimulate or inhibit directly receptors at the parasympathetic neuroeffector junction.

Miscellaneous In Vitro Test Systems

The study of peristaltic movements, longitudinal contractions, intraluminal pressure, and the volume of fluid mobilized was described in detail by Bülbring et al.[20] They had used the guinea pig ileum to investigate the influence of acetylcholine and other spasmogens. Castillo and de Beer[21] first reported the use of the guinea pig "tracheal chain" for the study of cholinergic activity. It was later modified by Constantine,[22] who introduced the "spiral tracheal chain". However, in addition to the guinea pig, preparations from other animal sources have been used in drug evaluation procedures. Christensen and Daniel[23] used the circular muscle of the "cat esophagus" to determine the effects on smooth muscle of drugs active at the parasympathetic neuroeffector junction. Paton and Zar[14] also had used the "detrusor" muscle obtained from the rabbit's urinary bladder to study the effects of acetylcholine. A major drawback with the use of this preparation is its thickness, which slows down the diffusion of gases. The isolated "guinea pig eye", as well as eyes obtained from C57 mice and hooded rats, have been used to evaluate drugs with parasympathetic activity.[24] The pigmentation of the eyes of these animals facilitates visualization and measurement of the pupil. Since miosis is the characteristic response of the eye to parasympathetic stimulation, constriction of the pupil of the eye is used as the index of activity for the above preparations.[25]

In Vitro Test Systems (Anticholinesterases or Cholinesterase Inhibitors)

A practical method of ascertaining whether a drug acts by a mechanism involving cholinesterase inhibition is by biological assay using the isolated heart or similar preparation. The basis of this is as follows: a test dose of acetylcholine is administered that causes a standard response. Then the test drug, e.g., physostigmine, is administered. If it inhibits cholinesterase, the action of acetylcholine is intensified and greatly prolonged. A test model using the isolated right atrial preparation from a rabbit or guinea pig, to which the sinus node is still attached, was described by Ursillo and modified by Vincenzi and West.[33] Essentially, the test utilizes the hypothesis that subthreshold electrical stimulation of the intrinsic nerves within the sinus node causes release of the neurotransmitters acetylcholine or norepinephrine, depending on which division of the autonomic nervous system was activated. The "electro-release" phenomenon, as it was called by Vincenzi and West,[33] has been used to assay for anticholinesterase activity.

In Vivo Test Systems (Choline Esters)

The use of intact animals to establish the effects of chemical compounds is the basis of most pharmacological evaluation of drug action. There are several in vivo preparations available for demonstrating the actions of drugs which stimulate the parasympathetic system. As a result of their actions, most of these agents are used clinically as stimulants of GI motility in postoperative abdominal distention, postoperative nonobstructive urinary retention, in myasthenia gravis, and open angle glaucoma.

Measurement of the Pupil Size

Mice are widely used in this procedure for evaluation of drug action, but dogs, cats, and rabbits can also be utilized. The pupil size is controlled primarily by tonic activity of the parasympathetic nervous system. An increase in parasympathetic activity or blockade of the sympathetic system produces miosis. The pupil size of mice or rats can be measured with a dissecting microscope using a calibrated eye piece micrometer.[34] Diameters are measured prior to and at various times after the injection of the test drugs. The change in pupil diameter after drug or saline is compared.[35,36] The drugs being tested are administered parenterally or topically on the cornea.

Anesthetized Dog Preparation

The anesthetized dog preparation is often used to monitor the effects of vagal stimulation and of autonomic drugs on blood pressure and heart rate. It is widely used to determine dose-response curves and pharmacological activities of drugs. The species of animals, the routes of administration, the procedure for cannulation, and the methods of recording may all be varied to meet the needs of the experimenter. This evaluation procedure can also be carried out in anesthetized rats, cats, and rabbits. The arterial blood is monitored by using a pressure transducer.[28,37] Injection of acetylcholine or any other choline ester and/or stimulation of the peripheral stump of left vagus nerve usually produces a depressor response (i.e., lowered blood pressure) and a slowing of heart rate. Depressor responses that are blocked by atropine sulfate are regarded as "muscarinic" and pressor responses after 1 or 2 mg/kg of atropine sulfate are regarded as "nicotinic". The eviscerated cat blood pressure preparation is quite sensitive to acetylcholine and can detect amounts as small as 0.002 μg/kg. The rat preparation is also widely used for bioassay of minute amounts of acetylcholine and the analogs from tissue extracts.[31,41-43]

Assessment of Salivation (Use of Dog Chorda Tympani-Wharton's Duct Preparation)

Luduena and Lands[44] described the use of Wharton's duct preparation in the anesthetized dog as a means of evaluating the actions of drugs which stimulate or inhibit salivary flow. Salivation from this duct can be induced by drugs such as acetylcholine, acetyl-β-methylcholine, and pilocarpine injected i.v. into a femoral vein or by nerve stimulation and inhibited by Hemicholinium. After the administration of Hemocholinium, there is a decrease in saliva. This suggests that the blocking effect is due to the inhibition of acetylcholine synthesis.[45]

Assessment of Gastric Acid Secretion

Parasympathetic agonists activate many functions in the GI tract. These compounds are able to stimulate gastric acid secretion either by i.v. injection or direct application to the mucosal surface of the stomach lining. Among the many techniques available for determination of this activity, the most commonly used methods are those described by Visschner et al.[46-48] and Ghosh and Schild.[47,48] In 1961, Tsukamoto[49] studied drug-induced gastric secretion by visualizing changes in the Congo Red dye-painted mucosa of rats. A different method was described by Thorbjarnarson et al.[50] in 1967. It involved surgical preparation of the Heidenhain pouch in rats. The major drawback of this approach is the extent of surgery involved in the setup of this preparation.

Superior Cervical Ganglia Preparation

This is an extremely useful preparation for detecting release of acetylcholine from autonomic ganglia either by preganglionic nerve stimulation, or by close arterial injection of the drugs.[51-54] In the anesthetized cat (using urethane and choralose), the superior cervical ganglion is isolated. Arrangement is made to stimulate electrically the pre- and postganglionic fibers. The effects of nerve stimulation of drug administration can be detected from responses of the "nictitating" membrane.

In Vivo Test Systems — Cholinesterase Inhibitors

Measurement of cholinesterase enzyme levels in the intact animal before and after administration of the drug diisofluorophosphate (DFP) is the most practical and feasible method of determining whether a given compound inhibits cholinesterase. In the presence of an inhibitor of this enzyme system, a standard response to a test dose of acetylcholine is intensified. In 1949, Michel[55] determined cholinesterase activity of red blood cell and plasma by an electrometric method. Similarly, Smith et al.[56] described both an in vivo and in vitro manometric technique for determination of cholinesterase activity using pyridostigmine, neostigmine, and edrophonium as the inhibitors.

ACETYLCHOLINE

Mechanism of Action

Acetylcholine, the neurotransmitter in the cholinergic nervous system, acts at synapses or at the neuromuscular junction. It interacts readily with both nicotinic and muscarinic receptors.

Pharmacokinetics

Acetylcholine is not absorbed from the GI tract. This results partly from its ionic character. It is easily hydrolyzed and also ineffective due to poor penetration of biological membranes and rapid hydrolysis by acetylcholinesterase. While active when administered parenterally, its short duration of action (hydrolysis by acetylcholinesterase and other esterases) seriously limits its usefulness.

Uses and Dosage

Because of its poor absorption and short duration of action, acetylcholine is seldom used for systemic effects; however, a 1% solution may be employed to provide a rapid miosis of the pupil for ophthalmic procedures. It may also be used diagnostically to detect atropine or other anticholinergic poisoning, since its usual responses (flushing, sweating, lacrimation) will be absent if the receptors are occupied by an anticholinergic agent.

Acetylcholine is used in ophthalmic solution (Miochol). This solution contains acetylcholine (17%) and mannitol (5%).

Physical-Chemical Data

Acetylcholine chloride (Structure 1) is a white, crystalline, hygroscopic compound. It has a melting point of 151°C. The bromide melts at 143°C. It is unstable in aqueous solution.

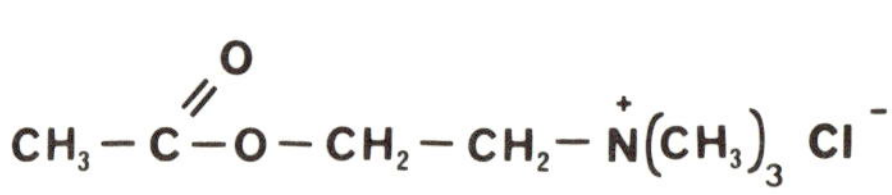

1

Its chemical name is 2-acetoxyethyltrimethylammonium chloride, choline acetate ($C_7H_{16}NO_2Cl$). Its mol wt is 181.60. Acetylcholine chloride is very soluble in water and ethyl alcohol but is insoluble in ether.

BETHANECHOL CHLORIDE

Mechanism of Action

Bethanechol acts as an agonist on the muscarinic receptor. Its fit on the receptor has been previously described.

Pharmacokinetics

As previously described, the carbamate group and the β-methyl group reduce the rate of hydrolysis of the compound sufficiently to permit oral administration. It is poorly absorbed, however, from the GI tract. It may be 10 to 20 times more effective when given s.c. than orally.

Because it is not distributed to the CNS, higher centers are not affected when it is administered in normally used doses. It has a more prolonged action than acetylcholine. Its action on the GI and GU tract may be noted in about 30 min after oral administration. In small doses (100 mg) the duration of action may be as short as 1 hr, but with larger doses (300 to 400 mg) its actions may last as long as 6 hr.

Uses and Dosage

The compound is available in 5-, 10-, and 25-mg tablets. Because of a fairly selective action on the urinary bladder, it is often used in the treatment of postoperative urinary retention or atony of the urinary bladder. It has been investigated as a diagnostic test for infantile cystic fibrosis. When used to alleviate urinary retention, 5 mg is often injected s.c. with repeated injection as required. When employed for postoperative abdominal distention, an oral dose of 10 to 20 mg several times a day is recommended.

Physical-Chemical Data

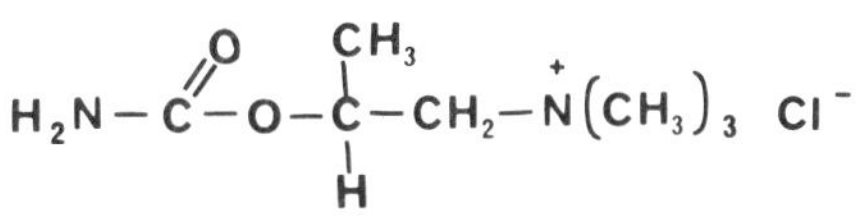

8

Bethanechol chloride is a white amorphous powder with a mp of 218 to 219°C (dec). Its chemical name is carbamoyloxy-2-propyltrimethylammonium chloride ($C_7H_{17}ClN_2O_2$). Its trade names are bethanechol chloride, Myotonachol®, Urecholine®, Miotnol®, Mictone®, and Unocarb®. One gram is soluble in 0.6 mℓ of water and 12.5 mℓ of alcohol. It is insoluble in nonpolar solvents.

PILOCARPINE

Mechanism of Action

Pilocarpine is believed to act on smooth muscle as a direct cholinergic agonist. It possesses pharmacological properties similar to acetylcholine.

Pharmacokinetics

The miotic action of pilocarpine, when applied as a 1% solution to the eye, can be noted within 30 min of application and persists for 4 to 8 hr. A reduction in intraocular pressure occurs within 1 hr.

Uses and Dosage

Pilocarpine is used chiefly as a miotic and to reduce intraocular pressure in glaucoma. Usually, a 1% solution is applied every 4 to 12 hr, but the frequency of application must be adapted to the patient. Controlled-release preparations are designed to release pilocarpine at varying rates (20 or 40 μg/hr).

Physical-Chemical Data

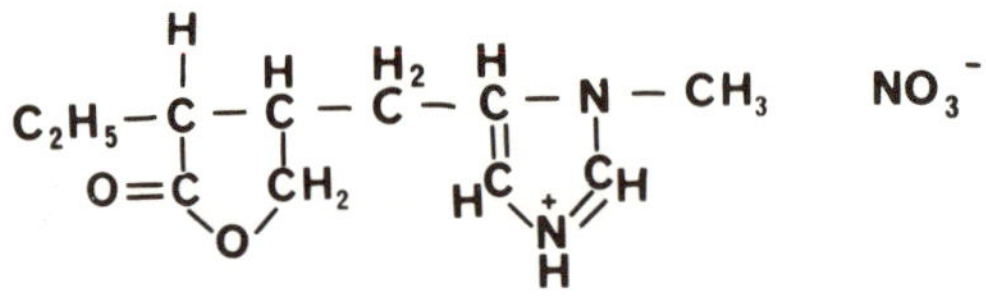

9

mp base	340°C
pK_a	7.15, 12.57
HCl	204 to 205°C
HNO_3	173 to 174°C
Base	$C_{11}H_{16}N_2O_2$ — 208.25
HCl	$C_{11}H_{17}ClN_2O_2$ — 244.72
HNO_3	$C_{11}H_{17}N_3O_5$ — 271.26

Pilocarpine is an alkaloid isolated from *Pilocarpus microphyllas* or *P. jaborandi*. The chemical name is (3-*S*-*cis*)-3-ethyldihydro-4-[(1-methyl-1H-imadazol-5yl)]-2-(3H)furanone. Trade names for preparations are Isopto carpine®, Pilocel®, Almocarpine®, Pilomiotin®, and Piloptic®. The alkaloid is soluble in water, alcohol, or chloroform and is sparingly soluble in ether and benzene. The salts are soluble in water and alcohol, but less soluble in chloroform. B-HCl — 1 g is soluble in 0.3 mℓ water, 3 mℓ alcohol, and 366 mℓ chloroform; B-NO_3 — 1 g is soluble in 4 mℓ water, 75 mℓ alcohol, and is insoluble in chloroform.

CARBACHOL

Mechanism of Action

The compound acts as an agonist at the cholinergic receptor.

Pharmacokinetics

When used as an ophthalmic solution its duration of action is from 2 to 4 hr. It is stable in aqueous solutions even when heated.

Uses and Dosage

Carbachol is used chiefly as miotic when applied as a solution to the eye. Its ability to lower intraocular pressure has led to its use in primary open angle glaucoma. While pilocarpine is often considered the drug of choice, carbachol is useful in the treatment of patients refractory or sensitive to pilocarpine. Its effectiveness is increased when formulated with a wetting agent.

Carbachol is available as a powder and in solutions of several strengths: 0.75, 1.5, 2.25, and 5% in plain solution or solutions with increased viscosity. The dosage regimen usually consists of the application of two drops of a suitable strength solution in the eye every 4 to 8 hr.

Physical-Chemical Data

$$H_2N{-}\overset{\overset{\displaystyle O}{\|}}{C}{-}O{-}CH_2{-}CH_2{-}\overset{+}{N}(CH_3)_3\ Cl^-$$

Carbachol chloride

Carbachol is a hygroscopic white, crystalline solid, with a mp of 200 to 203°C (dec). Its chemical formula is $C_6H_{15}ClN_2O_2$. Its mol wt is 182.65. Its chemical name is cholinecarbamate chloride; 2-(carbamoylethyl) trimethylammonium chloride. Trade names are Miostat® and Carbacel®. It is soluble in water and alcohol, but insoluble in ether and chloroform.

PHYSOSTIGMINE

Mechanism of Action

This alkaloid is an indirect stimulant of the cholinergic nervous system, which exerts its action by inhibition of acetylcholinesterase. Inhibition occurs through its ability to carbamoylate the enzyme.

Pharmacokinetics

Physostigmine, unlike many other cholinergic drugs, is distributed to the central nervous system. When applied to the eye as an ointment or solution, its miotic action can be noted within 30 min and can persist for 12 to 48 hr.

Uses and Dosage

While the alkaloid produces the expected responses related to a stimulation of the cholinergic nervous system, it is chiefly employed as a miotic in the treatment of glaucoma. It is an effective antidote for atropine poisoning and has been used in the treatment of the toxic effects of the tricyclic antidepressants. Doses of 1 to 4 mg given i.m. or s.c. are employed.

When employed in glaucoma, four drops of an ophthalmic solution are applied to the eye every 4 to 6 hr. An ophthalmic ointment is recommended for continuous treatment.

Physical-Chemical Data

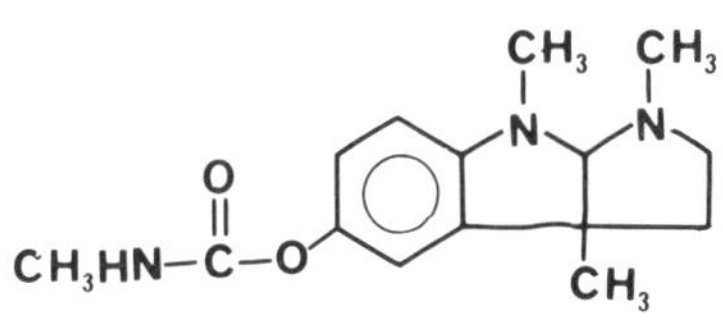

7

Physostigmine is a white crystalline solid obtained from the dried ripe seeds of *Physostigma venenosum*. The salicylate and sulfate are commonly found in pharmaceutical preparations. It has a mp of 105 to 106°C; 1 g will dissolve in 75 mℓ of water, 10 mℓ of alcohol, 30 mℓ of ether, and 1 mℓ of chloroform.

The salicylate salt melts at 184 to 186°C and has a solubility of 1 g in 90 mℓ of water, 25 mℓ of alcohol, 250 mℓ of ether, and 6 mℓ of chloroform; its pk_a is 2 and 8.1.

The sulfate has a melting point of 144 to 147°C and 1 g is soluble in 1 mℓ of water, 1 mℓ of alcohol, and 1200 mℓ of ether.

The development of colored solutions indicates decomposition to pharmacologically inactive products. Its chemical name is (3a-*S-cis*)-1,2,3,3a,8,8a-hexahydro-1,3a,8-trimethylpyrrolo(2,3-b)indol-5-ol-methylcarbamate. The chemical formula for physostigmine is $C_{15}H_{21}N_3O_2$ with a wt of 275.4; the chemical formula for physostigmine salicylate is $C_{22}H_{27}N_3O_5$, with a mol wt 413.5. Trade names are Eserine and Antilirium®.

NEOSTIGMINE

Mechanism of Action

The compound stimulates the cholinergic nervous system by inhibiting the enzyme acetylcholinesterase.

Pharmacokinetics

While rather poorly absorbed from the GI tract, neostigmine salts may be administered orally. It is not likely to cross the placenta or be present in human milk in significant quantities. It has a more rapid onset of action than pyridostigmine or ambenonium. Its effects on peristalsis or skeletal muscle are apparent within 10 to 30 min after injection or 2 to 4 hr after oral administration. After injection, its duration of action is from 2 to 4 hr. Hydrolysis to the substituted phenol occurs chiefly in the liver.

Uses and Dosage

The drug is useful in improving muscle tone in myasthenia gravis and as a diagnostic agent in detecting the disease. Because of its relative short duration of action, pyridostigmine is often preferred. It is also employed to reverse the effects of tubocurarine and other nondepolarizing neuromuscular blocking agents.

When used orally for the symptomatic treatment of myasthenia gravis, a 15- to 30-mg dose is employed. Repeated dosage is adjusted to the needs of the patient. The methylsulfate of neostigmine is administered by i.v., i.m., or s.c. injection in 1- to 2-mg doses.

Physical-Chemical Data

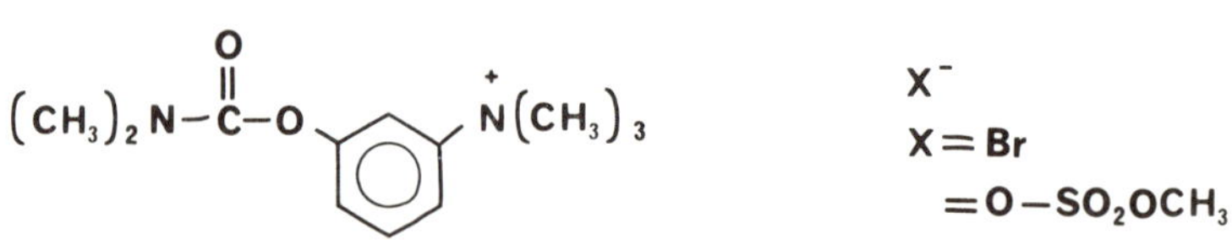

10

Neostigmine bromide is a white crystalline powder with a mp of 167 to 169°C (dec). The methylsulfate has a mp of 142 to 145°C. The bromide is soluble in 5 mℓ of water, 8 mℓ of alcohol, and 5 mℓ of chloroform, while 1 g of the methylsulfate is soluble in 0.5 mℓ of water and 6 mℓ of alcohol. Its chemical name is 3-dimethylcarbamoyloxyphenyltrimethylammonium bromide; 3-trimethylammoniumphenyl-*N*,*N*-dimethylcarbamate bromide ($C_{12}H_{19}BrN_2O_2$). Its mol wt is 303.2.

The chemical formula of methylsulfate is $C_{13}H_{22}SN_2O_6$, with a mol wt of 343.4. Trade names are Prostigmine® and Synstigmine®.

PYRIDOSTIGMINE BROMIDE

Mechanism of Action

The compound is an inhibitor of acetylcholinesterase and thereby provides an increased concentration of acetylcholine at its receptors. It also is reported to have a direct effect on skeletal muscle.

Pharmacokinetics

Pyridostigmine is intermediate in action between neostigmine and ambenonium chloride. While poorly absorbed from the GI tract, it can successfully be administered orally (60 mg).

Its effects can be seen in 30 min to 1 hr and its duration of action, although variable, may last up to 6 hr. A sustained-release tablet may extend its effective action up to 12 hr. After injection it is effective within a few minutes.

In ordinary doses it does not cross the blood-brain barrier, but the CNS is affected by higher doses. However, it may pass into the placenta.

Uses and Dosage

Pyridostigmine bromide is used chiefly in the symptomatic treatment of myasthenia gravis. Its duration of action and lower incidence of muscarinic effect have made it the drug of choice for many clinicians. It is usually administered as a 60-mg tablet three times daily or as a 100-mg extended-release tablet altered to the needs of the patient. It may be administered i.m. or s.c.

Physical-Chemical Data

Pyridostigmine bromide (Structure 11) is a white hygroscopic powder with a mp of 153 to 155°C. In terms of solubility, 1 g is soluble in 1 mℓ of water, 1 mℓ of alcohol, or 1 mℓ of chloroform.

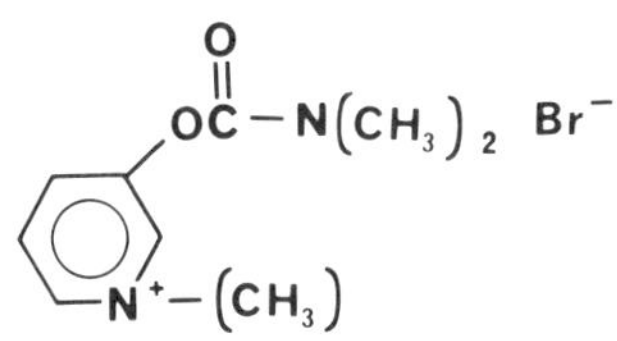

11

Its chemical name is *N,N*-3-dimethylcarbamoyloxypyridinium bromide ($C_9H_{13}B_2N_2O_2$). Its mol wt is 261.1. Trade names are Mestinon® and Regonal®.

DEMECARIUM BROMIDE

Mechanism of Action

The stimulation of the cholinergic nervous system occurs through inhibition of acetylcholinesterase.

Pharmacokinetics

The compound is a long-acting reversible inhibitor of acetylcholinesterase. When applied to the eye as an ophthalmic solution, it produces a miotic effect within 15 to 60 min, which persists for several days. Intraocular pressure is relieved within 24 hr.

Uses and Dosage

It is used in the treatment of glaucoma through relief of intraocular pressure. In such treatment an ophthalmic solution of one drop of 0.125 or 0.25% solution is applied approximately twice daily as indicated by patient response.

Physical-Chemical Data

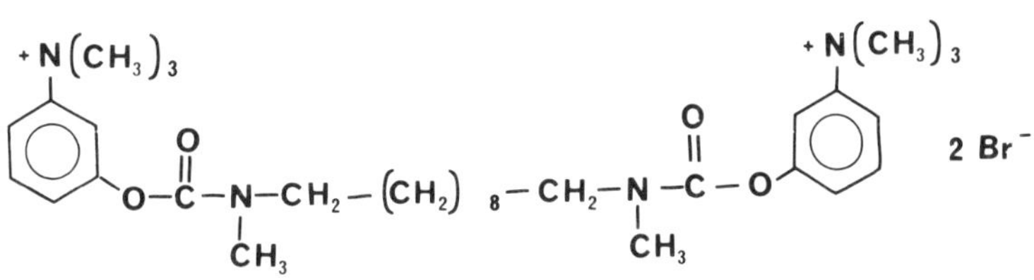

12

Its chemical name is *N,N*[1]-*bis*-3-trimethylammoniumphenoxymethylcarbamyl *N,N*[1]-dimethyldecamethylenediamine dibromide ($C_{32}H_{52}Br_2N_2O_4$; 716.61 mol wt). The compound has a mp of 268 to 270°C and is freely soluble in water and alcohol. It is insoluble in ether.

AMBENONIUM CHLORIDE

Mechanism of Action

Stimulation of the CNS occurs through a slowly reversible inhibition of acetylcholinesterase.

Pharmacokinetics

While ambenonium chloride (Structure 13), like most quaternary ammonium compounds, is poorly absorbed from the GI tract if it is administered orally, it is longer acting than neostigmine or pyridostigmine and its action persists for 4 to 8 hr. Aqueous solutions are sufficiently stable to prevent sterilization by autoclave.

Uses and Dosage

It is used for the alleviation of muscle weakness in myasthenia gravis. Tablets (10 mg) are available. The usual dose of 10.0 mg is administered three to four times daily. However, like most indirect-acting cholinergics, the regimen is usually altered to suit patient need.

Physical-Chemical Data

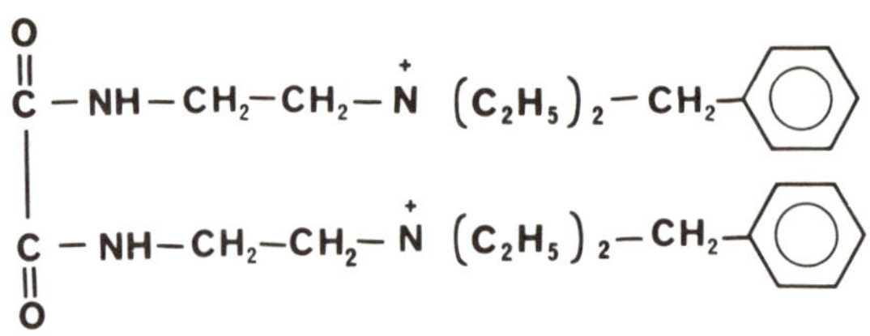

13

The compound is a white powder with a mp of 183 to 184°C; 1 g dissolves in 5 mℓ of water. Its chemical name is 2,2-oximidobis(2-chlorobenzyl)triethylammonium chloride ($C_{28}H_{42}Cl_2N_4O_2$). Other names are Ambestigmine chloride, Oxayl®, and Mytelase®.

EDROPHONIUM CHLORIDE

Mechanism of Action

Edrophonium chloride (Structure 14) is a cholinesterase inhibitor with a very short duration of action. It is therefore capable of stimulating practically all actions of the cholinergic nervous system. It is believed to interact with the anionic site of the enzyme, thereby blocking the interaction of acetylcholine.

Structure-Activity Relationship

The structure resembles those of other reversible cholinesterase inhibitors in that it contains the aromatic ring and quaternary ammonium group. However, it lacks the carbamate group. Thus, it inhibits acetylcholinesterase only by a physical competition with acetylcholine for the active site of the enzyme. This accounts for its short duration of action.

Pharmacokinetics

The duration of action of this drug is far shorter than other clinically useful anticholinesterases. When given i.v., effects on skeletal muscle can be seen in 30 sec. Its duration of action is about 5 to 10 min. In usual doses it does not pass the blood-brain barrier.

Uses and Dosage

Because of its very short duration of action, edrophonium chloride, unlike most cholinesterase inhibitors, is not useful in the treatment of myasthenia gravis. However, it is useful in both the diagnosis of the disease and in the determination of the dosage of anticholinergics required for alleviation of the symptoms.

The drug is usually administered i.v. Doses of 2 mg are injected in attempts to regulate antiacetylcholinesterase therapy. An increase of muscle strength suggests the need for increased doses of longer-acting anticholinesterase. Persons with myasthenia gravis show a rapid increase in muscle strength when edrophonium chloride is injected.

An undesirable cholinergic response can be corrected by the administration of atropine sulfate.

Physical-Chemical Data

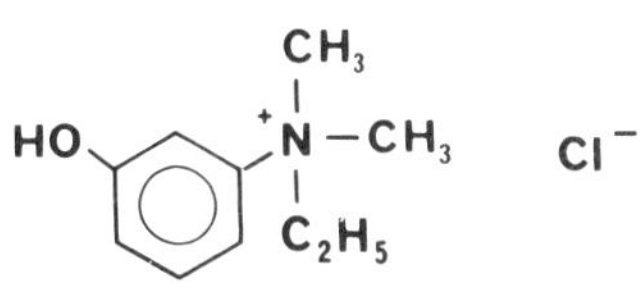

14

The compound is a white crystalline solid with a mp of 162 to 163°C (dec); 1 g will dissolve in less than 10 mℓ of water. It is insoluble in ether and chloroform. Aqueous solutions have a pH of 4. Its chemical name is 3-hydroxyphenyl-*N*,*N*-dimethyl *N*-ethylammonium chloride ($C_{10}H_{16}ClNO$).

ISOFLUOROPHATE

Mechanism of Action

The compound provides cholinergic stimulation and miosis through its action as a very slowly reversible (irreversible) inhibitor of acetylcholinesterases.

Pharmacokinetics

When applied to the eye as an ointment, miosis occurs in approximately 5 min and persists for 8 to 72 hr. The drug is unstable in aqueous solution.

Toxicities

See the discussion on toxicity of irreversible anticholinesterases. It induces a neurotoxicity probably not totally dependent on its antiacetylcholinesterase activity.

Uses and Dosage

Isofluorophate (Structure 15) is usually employed as a 0.025% ophthalmic ointment for the relief of intraocular pressure. The ointment is reapplied in 8 to 72 hr as required. It is also employed for diagnosis and treatment of convergent strabismus.

Physical-Chemical Data

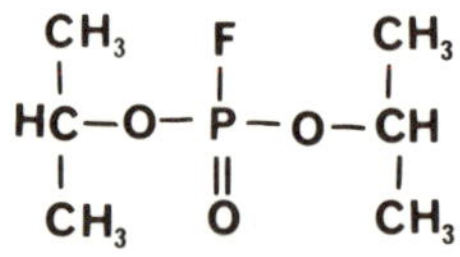

15

The compound exists as a liquid with limited solubility in water. It is soluble in alcohol, but is rapidly decomposed by moisture. It boils at 62°C/9 mm. Its chemical names are diisopropylfluorophosphate and diisopropylphosphofluorodate ($C_6H_4O_3PF$; 184.15 mol wt). Other names are DFP, Dyflor®, Fluorostigmine®, Isofluorophate, and Floropyl®.

ECHOTHIOPHATE IODIDE

Mechanism of Action

The compound stimulates the cholinergic nervous system by irreversibly inhibiting acetylcholinesterase.

Pharmacokinetics

After administration to the eye, miosis develops within 30 min and may persist up to 4 weeks. A maximum reduction in intraocular pressure occurs within 24 hr and may persist for 2 to 4 weeks.

Uses and Dosage

The compound is instilled into the eye as a 0.25% solution, one to two times daily to reduce intraocular pressure accompanying glaucoma. Dosage must be adjusted to patient requirement.

Physical-Chemical Data

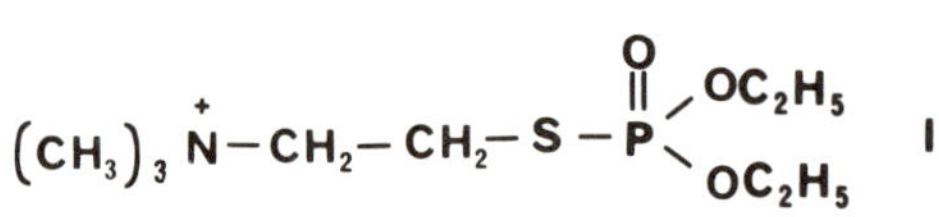

16

Echothiophate iodide is a white crystalline powder. It is hygroscopic and melts at 136 to 138°C. The powder is soluble in water and absolute ethanol. It should be stored below 8°C. Its chemical name is 2-diethoxyphosphalinylthio and *N,N,N*-trimethylethylammonium iodide ($C_9H_{23}INO_3PS$; 383.22 mol wt). Other names are Echodide® and Phospholine Iodide®.

PRALIDOXIME CHLORIDE

Mechanism of Action

Pralidoxime chloride (2-PAM) (Structure 17) acts as an antidote for poisoning resulting from the ingestion of phosphorous containing irreversible inhibitors of acetylcholinesterase. The compound fits on the phosphorylate active site of the enzyme in such a manner that the oxygen of the oxime is close to the phosphorous atom. Nucleophilic attack of the phosphorous by this oxygen results in a transition state that is stabilized when the rejuvenated enzyme departs as a leaving group.

Pharmacokinetics

Pralidoxime is metabolized in the liver. It may be administered orally (500-mg tablets) and reaches a peak blood level of 4 μg/mℓ. The half-life is relatively short (1 to 3 hr) when administered i.v. Peak plasma levels are reached within minutes of administration. The compound is rapidly excreted in the urine. About 80% of the dose will be excreted unchanged within 3 to 4 hr. 2-PAM does not pass readily through the blood-brain barrier.

Toxicities

Although it is well tolerated, a number of side effects may accompany its administration. They include dizziness, headaches, drowsiness, and blurred vision. Tachycardia and hyperventilation have been reported.

Use and Dosage

In order to react rapidly in cases of phosphate intoxication, 2-PAM is usually administered i.v. A 1-g vial is diluted with 20 mℓ of sterile water for injection and is administered over a few hours. When administered by infusion the solution is further diluted.

Physical-Chemical Data

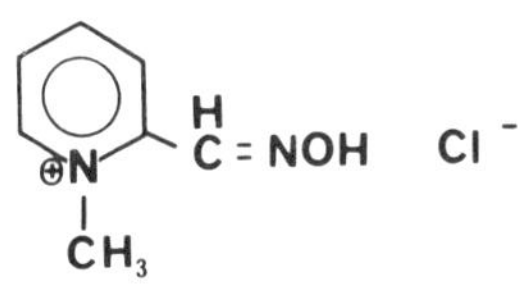

17

The chloride is a yellowish powder with a mp of 220 to 222°C. It is soluble as 1 g in 2 mℓ of water and 100 mℓ of alcohol. It is insoluble in nonpolar solvent.

The iodide melts at 212 to 214°C (dec). It is soluble as 1 g in 20 mℓ of water. It is insoluble in chloroform and ether. Its chemical name is 2-hydroxyiminomethylpyridinium chloride; 2-pyridylaldoxime methochloride (C_7H_9ClNO). Other names are 2-PAM and 2-PAMC.

REFERENCES

1. **Gearien, J. E. and Mede, K.,** Cholinergics, anticholinesterases and antispasmodics, in *Principles of Medicinal Chemistry,* Foye, W., Ed., Lea & Febiger, Philadelphia, 1981, chap 15.
2. **Taylor, P.,** Cholinergic agonists, in *The Pharmacological Basis of Therapeutics,* Gilman, A., Goodman, L., and Gilman, A., Eds., Macmillan, New York, 1980, chap. 5.
3. **Elliot, J. R.,** The action of adrenalin, *J. Physiol. (London),* 32, 401, 1905.
4. **Dixon, W. E.,** On the mode of action of drugs, *Med. Mag. (London),* 16, 454, 1907.
5. **Loewi, O.,** Ubër humorale ubertraybarkeil der therzehwirkang, *Pfug. Arch. Ges. Physiol.,* 189, 239, 1921.
6. **Dale, H.,** The action of certain esters and ethers of choline and their relation to muscarine, *J. Pharmacol. Exp. Ther.,* 6, 147, 1914.
7. **Cocoles, G.,** Cholinergic drugs and related agents, in *Wilson and Gisolds Testbook of Organic Medicinal and Pharmaceutical Chemistry,* Doerge, R., Ed., J.B. Lippincott, Philadelphia, 1980, chap. 12.
8. **Armstrong, P., Cannon, J., and Long, J.,** Conformationally rigid analogues of acetylcholine, *Nature (London),* 220, 65, 1976.
9. **Belleau, B. and Lacasse, G.,** Aspects of the chemical mechanism of complex formation between acetylcholinesterase and acetylcholine related compounds, *J. Med. Chem.,* 7, 768, 1964.
10. **Birdsell, N. and Hame, E.,** Biochemical studies on muscarinic acetylcholine receptors, *J. Neurochem.,* 27, 7, 1976.

11. **Belleau, B.,** Conformational perturbation in relation to the regulation of enzyme and receptor behavior, in *Advances in Drug Research,* Vol. 2, Harper, N. and Simmons, A., Eds., Academic Press, New York, 1965, 89.
12. **Ing, H.,** The structure action relationship of the choline group, *Science,* 109, 264, 1949.
13. **Simmenart, A.,** The action of certain derivatives of choline, *J. Pharmacol. Exp. Ther.,* 46, 157, 1932.
14. **Werzel, M.,** A suggested mechanism for the action of choline esters on animal organs inferred from a study of the effect of choline, β-methylcholine and thiocholine esters, *Experientia,* 15, 430, 1959.
15. **Belleau, B. and Paranen, J.,** Stereochemistry of the interactions of enantiomeric 1,3-dioxolan analogs of muscarone with cholinergic receptors, *J. Med. Chem.,* 6, 325, 1963.
16. **Triggle, D., Moren, J., and Bernardi, E.,** *Cholinergic Ligand Interactions,* Academic Press, New York, 1971.
17. **Trendelenburg, P.,** Physiologische und pharmacokologische versuche über die dunndarmperistaltik, *Arch. Exp. Pathol. Pharmacokol.,* 81, 55, 1917.
18. **Paton, W. D. M.,** The response of the guinea-pig ileum to electrical stimulation by coaxial electrodes, *J. Physiol.,* 127, 40P, 1955.
19. **Paton, W. D. M. and Zar, M. A.,** The origin of acetylcholine released from guinea-pig intestine and longitudinal muscle strips, *J. Physiol.,* 194, 13, 1968.
20. **Bülbring, E., Crema, A., and Saxby, O. B.,** A method for recording peristalsis in isolated intestine, *Br. J. Pharmacol.,* 13, 440, 1958.
21. **Castillo, J. C. and deBeer, E. J.,** The tracheal chain. I. A preparation for the study of antispasmodics with particular reference to bronchodilator drugs, *J. Pharmacol. Exp. Ther.,* 90, 104, 1947.
22. **Constantine, J. W.,** The spirally cut tracheal strip preparation, *J. Pharm. Pharmacol.,* 17, 384, 1965.
23. **Christensen, J. and Daniel, E. E.,** Effect of some autonomic drugs on circular esophageal smooth muscle, *J. Pharm. Exp. Ther.,* 159, 243, 1968.
24. **Beaver, W. T. and Riker, W. F.,** The quantitative evaluation of autonomic drugs on the isolated eye, *J. Pharm. Exp. Ther.,* 138, 48, 1962.
25. **Ursillo, R. C.,** Excitation of nerves in isolated auricles, *J. Pharmacol.,* 122, 78A, 1958.
26. **Chang, H. C. and Gaddum, J. H.,** Choline esters in tissue extracts, *J. Physiol.,* 79, 255, 1934.
27. **Boyd, I. H. and Pathak, C. L.,** The response of perfused frog hearts to minute quantities of acetylcholine and the variation in sensitivity with season, *Physiology,* 176, 191, 1965.
28. **Chiou, C. Y., Long, J. P., Cannon, J. G., and Armstrong, P. D.,** The cholinergic effects and rates of hydrolysis of conformationally rigid analogs of acetylcholine, *J. Pharmacol. Exp. Ther.,* 166, 243, 1969.
29. **MacIntosh, F. C. and Perry, W. L. M.,** Biological estimation of acetylcholine, *Meth. Med. Res.,* 3, 78, 1950.
30. **Child, K. J. and Zaimis, E. A.,** A new biological method for the assay of depolarizing substances using the isolated semispinalis muscle of the chick, *Br. J. Pharmacol. Chemother.,* 15, 412, 1960.
31. **Tyler, C.,** The assay of acetylcholine with the isolated semispinalis cervicis muscle of the chick, *Analyst,* 85, 8, 1960.
32. **Ginsborg, B. L. and Warriner, J.,** The isolate chick biventer cervicis nerve muscle preparation, *Br. J. Pharmacol. Chemother.,* 15, 410, 1960.
33. **Vincenzi, F. F. and West, T. C.,** Release of autonomic mediators in cardiac tissue by direct subtreshold electrical stimulation, *J. Pharmacol.,* 141, 185, 1963.
34. **Long, J. P. and Armaly, M. F.,** Studies of the mydriatic properties of anticholinergic agents in mice, *J. Pharmacol. Exp. Ther.,* 150, 389, 1965.
35. **Armaly, M. F. and Long, J. P.,** Factors influencing the activity of topically applied mydriatic agents in mice, *Arch. Intern. Pharmacodyn.,* 161, 423, 1966.
36. **Edge, N. D.,** Mydriasis in the mouse: a quantitative method of estimating parasympathetic ganglion block, *Br. J. Pharmacol.,* 8, 10, 1953.
37. **Chiou, C. Y. and Long, J. P.,** Effects of α,α′-bis-(dimethylammonium acetaldehyde diethylacetal)-β,β′-diacetylbiphenyl bromide (DMAE) on neuromuscular transmission, *J. Pharmacol. Exp. Ther.,* 167, 344, 1969.
38. **Shibayama, F., Sokabe, H., and Sakai, F.,** Determination of the blood pressure-response curve, *Arch. Intern. Pharmacodyn.,* 153, 334, 1965.
39. **Darko, L. L., Cannon, J. G., Long, J. P., and Burks, T. F.,** Synthesis and cholinergic effects of certain N-methoxylated quaternary compounds, *J. Med. Chem.,* 8, 841, 1965.
40. **Knapp, F. M., Mitchell, W. K., and Himurich, W. A.,** Measurement of basilar and vertebral arterial pressures in the dog, *Life Sci.,* 4, 251, 1965.
41. **MacIntosh, F. C. and Perry, W. L. M.,** Biological estimation of acetylcholine, *Meth. Med. Res.,* 3, 78, 1950.
42. **Chrusciel, M. and Varagic, V.,** The effect of galantamine on the blood pressure of the rat, *Br. J. Pharmacol.,* 26, 295, 1966.

43. **Dauda, G., Szokol, M., and Devenyi, I.,** Effect of renal medullary extracts on blood pressure of the rat, *Acta Physiol.*, 28, 273, 1965.
44. **Luduena, F. P. and Lands, A. M.,** An investigation of the pharmacological actions of three potent antispasmodic compounds and their corresponding methosalt, *J. Pharmacol. Exp. Ther.*, 110, 282, 1954.
45. **Wilson, H. and Long, J. P.,** The effect of hemocholinium (HC-3) at various peripheral cholinergic transmitting sites, *Arch. Intern. Pharmacodyn.*, 120, 343, 1959.
46. **Visscher, F. E., Seay, P. H., Tazelaar, A. P., Jr., Veldkamp, W., and Vander Brook, M. J.,** Pharmacology of pamine bromide, *Br. J. Pharmacol.*, 13, 118, 1958.
47. **Ghosh, M. N.,** A quantitative method for the assay of inhibitors of acid gastric secretion in the rat, *Br. J. Pharmacol.*, 13, 118, 1958.
48. **Ghosh, M. N. and Schild, H. O.,** Continuous recording of acid gastric secretion in the rat, *Br. J. Pharmacol.*, 13, 54, 1958.
49. **Tsukamoto, M.,** Effects of histamine, compound 48/80, atropine, and diamox, administered to gastric submucosa, on acid secretion in the rat, *Gastroenterology*, 41, 572, 1961.
50. **Thorbjarnarson, B., Reas, R., Bjornsson, S., Watkins, G. M., and Martin, K. A.,** The denervated gastric pouch of the rat, *Surg. Res.*, 7, 499, 1967.
51. **Perry, W. L. and Talesnik, J.,** The role of acetylcholine in synaptic transmission at parasympathetic ganglia, *J. Physiol.*, 119, 455, 1953.
52. **Craincross, K. D., McCulloch, M. W., Story, D. F., and Trinker, F.,** Modification of synaptic transmission in the superior cervical ganglion by epinephrine, norepinephrine and nortriptyline, *Int. J. Neuropharmacol.*, 6, 293, 1967.
53. **Desiraju, T.,** An analysis of the action of succinylcholine on transmission in the superior cervical ganglion of the cat, *J. Neurochem.*, 13, 323, 1966.
54. **McKinstry, D. N., Koenig, E., Koelle, W. A., and Koille, G. B.,** The release of acetylcholine from a sympathetic ganglion by carbachol, *Can. J. Biochem. Physiol.*, 41, 2599, 1963.
55. **Michel, H. O.,** Electrometric method for determination of red blood cell and plasma cholinesterase activity, *J. Lab. Clin. Med.*, 34, 1564, 1949.
56. **Smith, C. M., Mead, J. C., and Cunna, K. R.,** Antagonism of tubocurarine. III. Time course of action of pyridostigmine, neostigmine and edrophonium *in vivo* and *in vitro*, *J. Pharmacol. Exp. Ther.*, 120, 215, 1957.

CHOLINERGIC ANTAGONISTS

Emmanuel B. Thompson and James E. Gearien

INTRODUCTION

Cholinergic Antagonists

Excess activity of the cholinergic nervous system results in increased smooth muscle contraction, which may cause muscle spasm and increased secretion by the exocrine glands. Excessive stimulation is usually alleviated by the administration of antagonists competitive to acetylcholine. These act on receptors at the neuromuscular junction. The structure and function of these receptors were described in the previous review. Such inhibitors are referred to as "cholinergic antagonists" or "cholinergic blocking agents". In general, they are used chiefly for their action on the muscarinic receptor on smooth muscle.

The recognition that atropine (Structure 1) and scopolamine (Structure 2), two alkaloids isolated from *Solanaceus* plants, have the ability to block the action of acetylcholine at the muscarinic receptor, has provided a lead to the synthesis of simpler compounds having similar activity. The action of atropine on muscarinic receptors in various tissue can be classified as: (1) antispasmodic, (2) antisecretory, and (3) mydriatic. Investigation of the biological activity of structurally similar compounds was stimulated by the hope of separating these activities and finding therapeutic agents capable of exerting only a single response. While this has not been completely achieved, Table 1 indicates the feasibility of this approach.

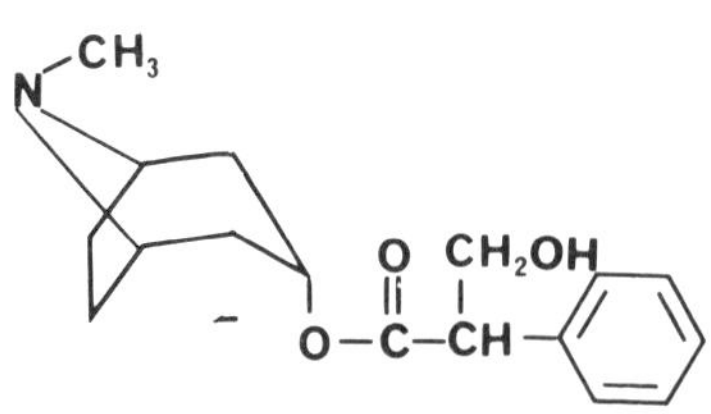

1

Atropine

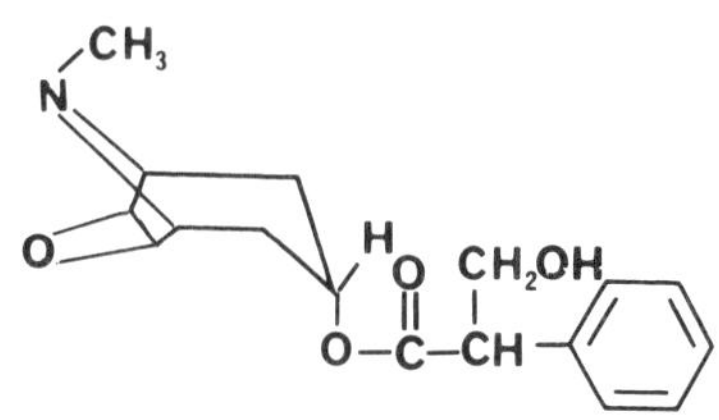

2

Scopolamine

Since acetylcholine is the neurotransmitter at synapses, it has been possible to design competitive acetylcholine antagonists which act at the ganglia. Acetylcholine is the neurotransmitter at synapses for both the sympathetic (adrenergic) and parasympathetic (cholinergic) nervous system. These compounds differ in biological responses and clinical usefulness from those acting at the muscarinic receptor. Compounds competing with acetylcholine for receptors on the ganglia are commonly referred to as "ganglionic blocking agents" and are used chiefly to control hypertension.

Compounds resembling curare[1] block the action of acetylcholine at the nicotinic receptor. Such compounds are administered during surgical procedures to relax striated muscles. While these are referred to as muscle relaxants, they act at the periphery and should not be confused with that class of minor tranquilizers acting on the central nervous system (CNS).

Structure-Activity Relationships for Anticholinergics

Most compounds acting as antagonists at the muscarinic receptor are competitive antagonists of acetylcholine and bear a resemblance to those structural features of that compound which are important to receptor binding. The encircled portion of the atropine molecule in Structure 1 illustrates this relationship. The structural requirements for acetylcholine binding

Table 1
ANTISECRETORY, ANTISPASMODIC, AND MYDRIATIC ACTIVITIES OF SOME ANTICHOLINERGICS

	Equipotent molar ratios to atropine		
Drug	**Antisecretory**	**Antispasmodic**	**Mydriatic**
Homatropine	0.44	8.5	7.7
Methantheline	0.37	0.48	3.0
Propantheline		0.40	1.6
Benzactyzine	5.6	3.5	17
Dicyclomine	60	8.0	8.8
Propivone	40	29	5000

are described in the previous review and are described in the literature.[2] Atropine and its synthetic analogs with cholinergic antagonist activity, while differing widely in structure, share two common features — they resemble acetylcholine in structure and are appreciably larger molecules. Belleau[3-5] suggests that this increased size produces a conformational change in the receptor that renders it biologically inactive.

The earliest synthetic or semisynthetic cholinergic antagonists were analogs of atropine, usually with somewhat simplified structures. These include homatropine hydrobromide (Structure 3) and homatropine methobromide (Structure 4).

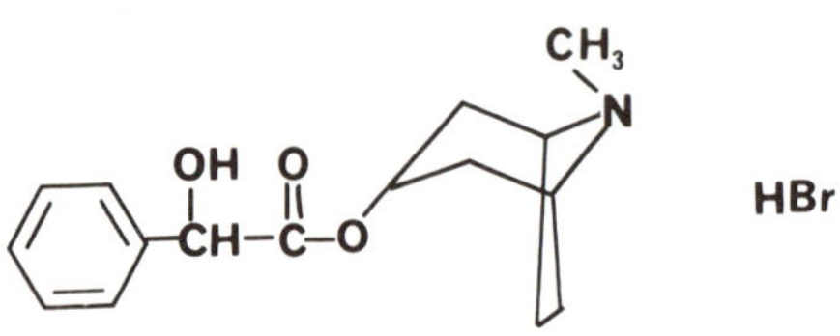

3

Homatropine Hydrobromide

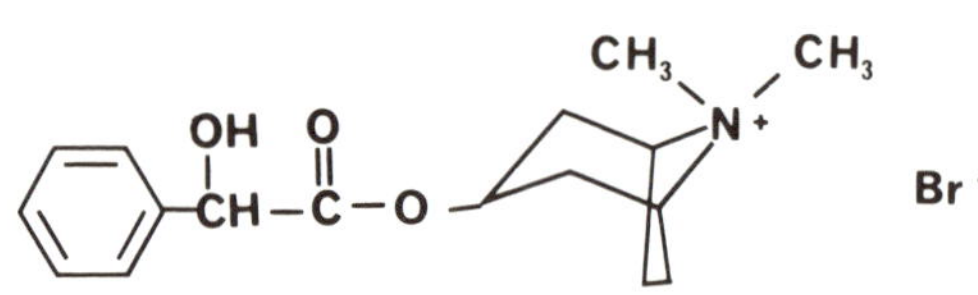

4

Homatropine Methobromide

These compounds have an altered acyl group. One of the first compounds in which the heterocyclic ring was substituted by a trialkylamine was amprotropine (Structure 5). Like atropine, it is an ester of tropic acid, specifically the diethylaminoethyl ester, thus demonstrating that the heterocyclic ring is not necessary for anticholinergic activity.

Variations of the tropic acid portion of the atropine demonstrated that the hydroxyl group of tropic acid, although contributing to the potency of the compound, is not essential to their intrinsic activity. Dialkylaminoalkyl esters of diphenyl acetic acid (Structure 6) demonstrated clinically useful anticholinergic activity.

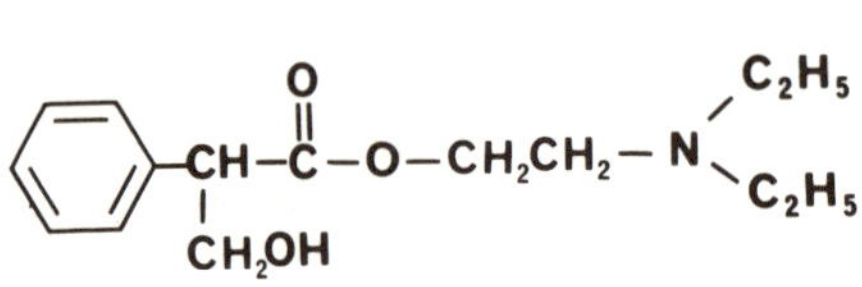

5

Amprotropine

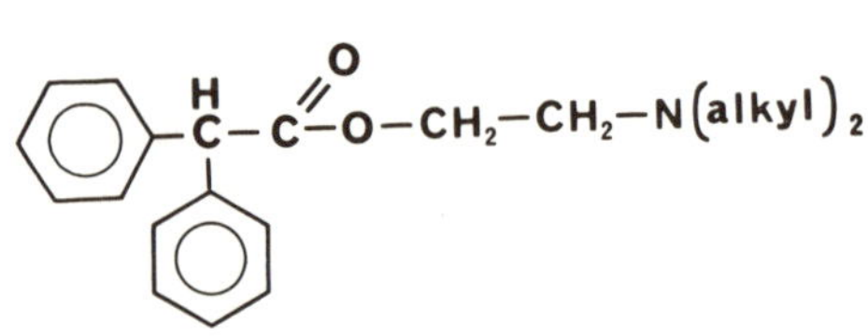

6

Diphenyl Acetic Acid

Substitution of one phenyl by a cyclohexyl group provided Trasentin H (Structure 7), which appeared to be more useful clinically. This suggested that other groups could be substituted for phenyl with retention of activity. It is important to note that dialkylaminoalkyl esters of phenyl acetic acid are partial agonists and that esters of triphenyl acetic acid are relatively inactive. A general structure for anticholinergic activity is depicted in Structure 8.

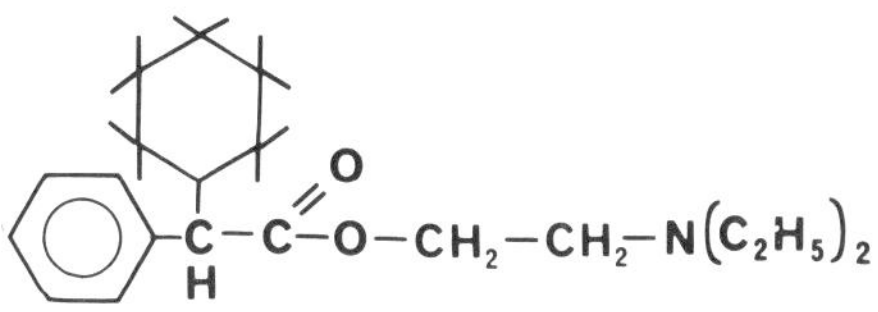

7

Trasentin H

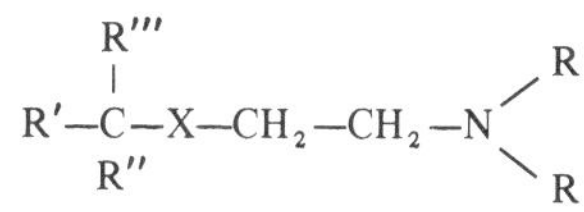

R	=	alkyl or heterocyclic
R′	=	usually aryl
R″	=	group larger than H
X	=	a hydrogen bonding group
R‴	=	hydrogen

8

Anticholinergic Activity

Structural features required for a high-potency anticholinergic compound are

1. A tertiary alkylamino or quaternary ammonium group separated from a hydrogen-bonding group by two or three carbon atoms. It is assumed that the amino group is protonated in the vicinity of the receptor and forms an ionic bond at the anionic site of the receptor.[6] Alkyl substituents may vary from methyl to butyl.
2. A hydrogen bonding group which is capable of interaction with the esteratic site.[7] This group may vary in structure as long as the hydrogen-bonding property is retained. Such groups as alcoholic hydroxyls, amines, and amides have satisfactorily replaced the ester group. Diphemanil methylsulfate is one of the few clinically used anticholinergics that does not have a hydrogen-bonding group.
3. Groups at R″ which are larger than hydrogen. A highly potent anticholinergic can be achieved if this and R′ are diphenyl, phenyl cyclohexyl, or a larger ring stem as found in propantheline (Structure 9).

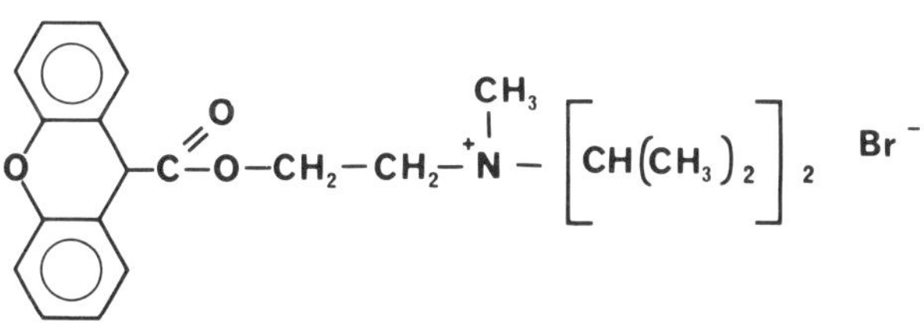

9

Propantheline

Some limitations exist in the size of the R″ groups. Dialkylaminoalkyl esters of the acids in Structures 10 to 12 are relatively inactive.[7]

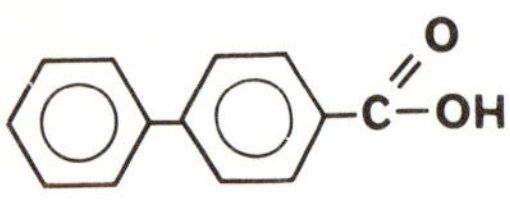

10

Biphenylcarboxylic Acid

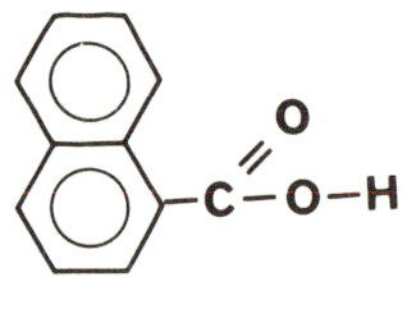

11

1-Naphthylcarboxylic Acid

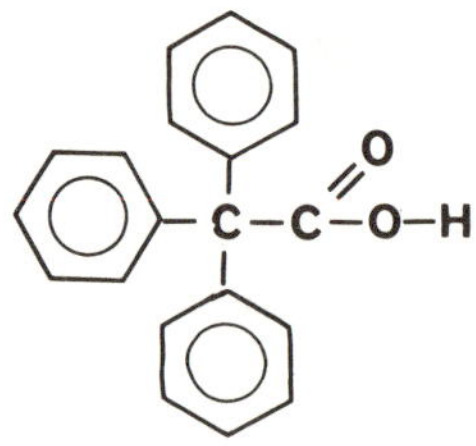

12

Triphenylacetic Acid

It has also been noted that the potency of optically active cholinergic blocking agents is associated with their stereochemistry. Atropine is a racemic compound. The l-isomer (hyoscyamine) is a potent anticholinergic while the d-compound has only weak activity.[8] Optically active substituted acetic acid esters show a similar specificity.

Mechanism of Action of Cholinergic Antagonists

Cholinergic antagonists compete with acetylcholine for the muscarinic receptor and those clinically useful compounds can be considered to be reversible competitive inhibitors. Since they compete favorably for the receptor in low concentration, it can be assumed that those portions of the molecule which are present in addition to the acetylcholine structure also participate in binding (R″ and R′).

If the three-dimensional structures of these compounds are examined, it would indicate that the amino group, the ethylene bridge, the hydrogen-bonding group, and *R′* and *R″* groups could fit on a planar receptor with a third substituent (R‴) extending out from the receptor. If it is assumed that the binding site of acetylcholine is depressed, this would suggest that a third R‴ substituent might bind at right angles to the plane of the receptor and might bind to a receptor found in a cavity.[9] This type of binding accounts for the fact that in vitro experiments show that esters of 2,2-diphenylpropionic acid (Structure 13) are more potent than esters of diphenylacetic acid (Structure 14).[10]

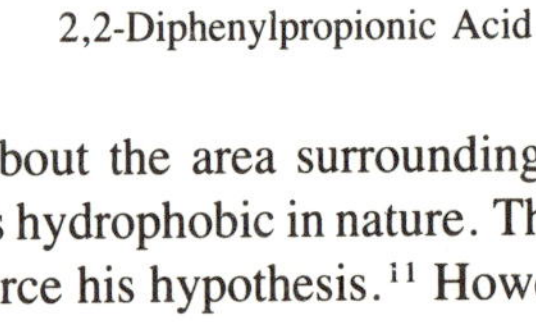

13

2,2-Diphenylpropionic Acid

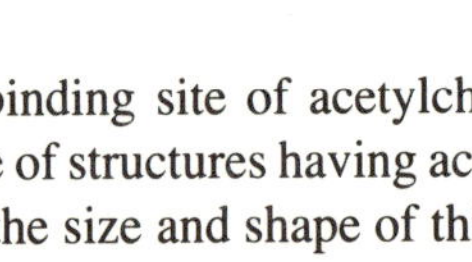

14

Diphenylacetic Acid

Little is known about the area surrounding the binding site of acetylcholine. Ariens[12] has suggested that it is hydrophobic in nature. The type of structures having activity as cholinergic antagonists reinforce his hypothesis.[11] However, the size and shape of this hydrophobic area are not known.

The limitation on the size of substituents which may be attached to dialkylaminoalkyl esters of substituted acetic acid suggests that there are size limitations to this area.

A suitable model which accounts for the previously discussed structure-activity relationships of cholinergic antagonists is shown in Figure 1. It is a more developed map of the muscarinic receptor than that presented in the preceding review.[13]

For the convenience of our discussion, this receptor map is divided into areas. Areas A, B, D, and E are essentially in the same plane, while area C is at right angles to the plane and represents the wall of a depressed receptor.

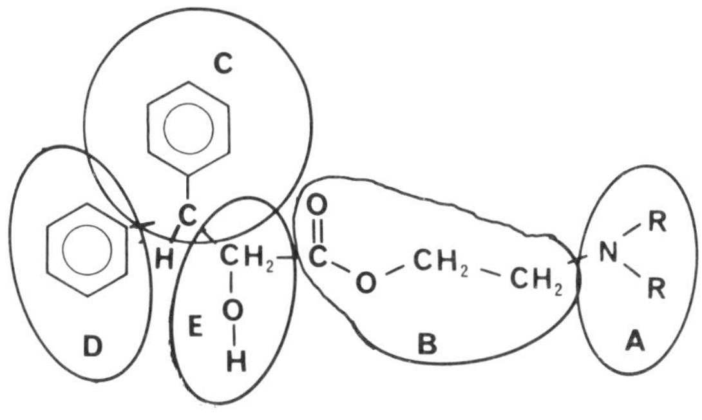

FIGURE 1. Model of muscarinic antagonist receptor binding.

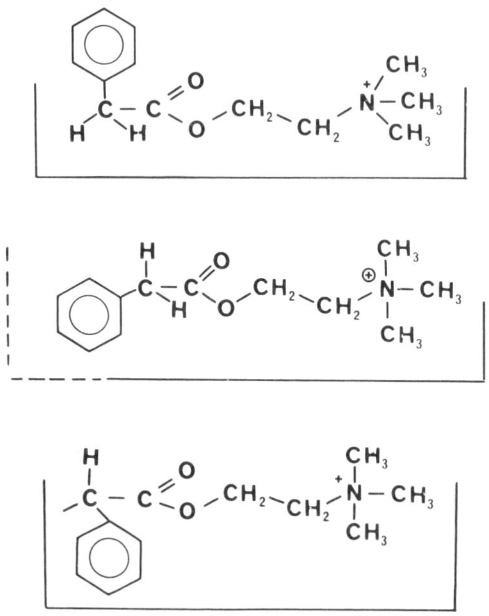

FIGURE 2. Binding of a partial agonist to muscarinic receptor.

Area A is the anionic site which contains at least one negative charge to attract the drug to the receptor. A great deal of this area is hydrophobic and is free to alter its conformation to accept groups from methyl to propyl.

Area B as illustrated binds the ethylene bridge, which connects the substituted amino group and the hydrogen-bonding group. Areas D and E are in the plane of the receptor, are hydrophobic, but differ in size. Area D can accommodate a six-membered aromatic ring while area E is restricted in size and cannot accommodate more than a methyl group. Area C is at right angles to areas D and E and can accommodate a six-membered ring system. This fits experimental evidence which shows that dimethylaminoethyl esters of 2,2-diphenylpropionic acid (Structure 13) are more potent than the corresponding similarly substituted butyric and valeric acids and accounts for the inactivity of esters of triphenyl acetic acid.

An examination of the model as depicted in Figure 2 explains Belleau's perturbation theory and explains why dialkylaminoethyl esters of phenylacetic acid are partial agonists. Figure 2 assumes that compounds can bind at site C with the receptor in its active conformation, but that binding at sites D and E can occur only with an alteration of the active conformation of the receptor.

The solid lines represent the active conformation of the muscarinic receptor. The dotted lines represent the conformational change which must take place to accommodate cholinergic antagonist binding.

Binding as depicted in A_1 would occur to provide an active conformation of the receptor with an agonist response. With free rotation of the molecule, binding might occur as in B_1. This would require a conformational change in the receptor which would inhibit its agonist response. Binding as depicted in C_1 could not occur because the hydrophobic region E (Figure 1) is too small to accommodate the phenyl group.

In A_1 a cholinergic stimulation would occur, but in B_1 the receptor would be inactivated and blocked for interaction with acetylcholine. Added together, this would produce a partial agonist response.

It should be noted that the dimethylaminoethyl ester of 2-phenylpropionic acid (Structure 15) could not bind to the receptor, if it assumes the conformation with phenyl group in the position shown in C_1. However, binding in conformation could only occur with the creation of a receptor conformation unfavorable to receptor activity.

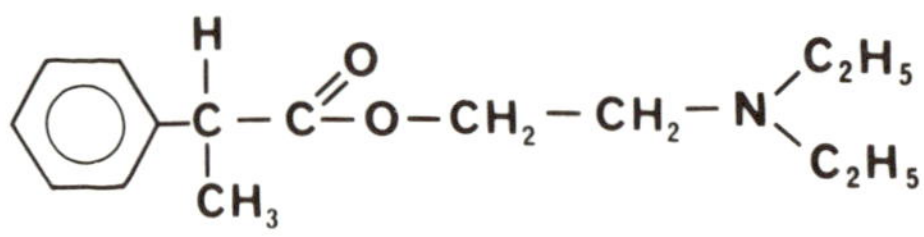

15

2-Phenylpropionic Acid

Cholinergic antagonists in which the nitrogen is quaternized often possess greater antisecretion than antispasmodic properties. Such compounds owe their activity to a favorable transport to the ganglia. There they compete with acetylcholine for these receptors.

Ganglionic Blocking Agents

These compounds are reversible antagonists of acetylcholine acting at the ganglia and as such block stimulation of both the adrenergic and cholinergic nervous system.[14,15] The prototype of such compounds is tetraethylammonium chloride (TEA) (Structure 16). The ability of this compound to reduce blood pressure led to an investigation of its mode of action.

$$(CH_3{-}CH_2)_4\overset{+}{N}\ Cl^-$$

16

Tetraethylammonium Chloride (TEA)

The poor absorption of TEA after oral administration reduced its clinical value and led to attempts to increase the hydrophobic nature of the compounds to aid in absorption and distribution.

The synthesis and biological evaluation of *bis*-quaternary ammonium compounds provided hexamethonium chloride (Structure 17). In a study of analogous *bis*-compounds, it was found that for optimum activity at the ganglia, quaternary ammonium groups should be separate by a penta- or hexamethalene chain. Like TEA, hexamethonium chloride has a short duration of action and its absorption from the GI tract is not entirely predictable.

$$(CH_3)_3\overset{+}{N}-(CH_2)_6-\overset{+}{N}(CH_3)_3\ 2\,Cl^-$$

17

Hexamethonium Chloride

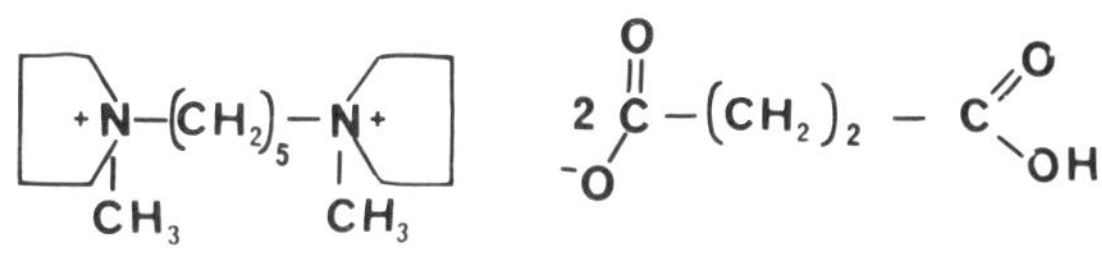

18

Pentolinium Tartrate

A compound with improved characteristics for oral administration is pentolinium tartrate (Structure 18).

While the more hydrophobic quaternary ammonium compounds with ganglionic blocking action may be administered orally, their absorption is poor and not always predictable. A secondary amino, mecamylamine, replaced most of these in therapy and is included in the monographs.

Antagonists at the Nicotinic Receptor

The muscle relaxant activity of curare (Structure 19) provided a chemical structure and a clue for the development of compounds acting at the nicotinic receptor. Curare and compounds of similar structure are used to relax skeletal muscles prior to surgery.

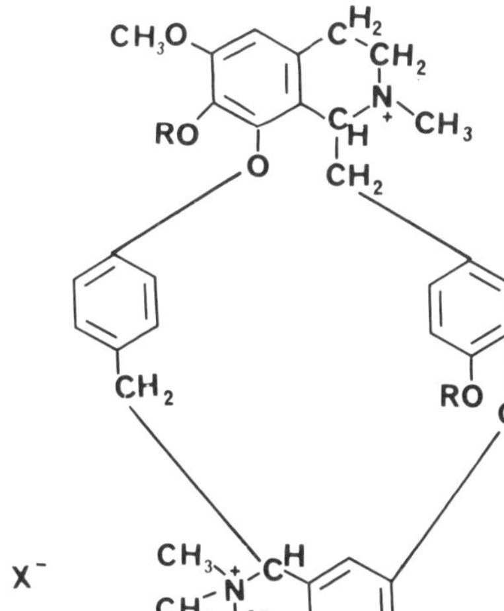

19

Curare

In the structure of d-tubocurarine or dimethyltubocurarine (Structure 19), two ammonium groups are separated by 10 to 12 atoms. This observation led to the development of a variety of *bis*-quaternary ammonium compounds which were capable of acting as antagonists of acetylcholine at the nicotinic receptor. While these resemble the previously discussed ganglionic blocking agents, the optimum distance between the quaternary ammonium groups for muscle relaxant activity is 10 to 12 carbon atoms, while that for ganglionic blocking action is 5 to 6 carbon atoms.

Among other skeletal muscle relaxants acting by a mechanism similar to curare are hexafluorenium bromide, succinylcholine chloride, and gallamine triethyliodide (see Table 2 and the monographs). While a molecule containing a single quaternary ammonium group may possess curare activity, those with two cationic groups separated by the proper length spaces are more potent.[16,17]

Table 2
IN VITRO MODELS FOR ASSESSING CHOLINERGIC ANTAGONISM

Tissue or organ preparation	Animal	Type of activity	Ref.
Vas deferens	Guinea pig	Spasmolytic (antispasmodic)	25—28
Rectus abdominus	Frog	Spasmolytic	
Ileum	Guinea pig, rat	Spasmolytic	29—31
	Mouse, rabbit	Spasmolytic	
Phrenic nerve diaphragm	Rat	Spasmolytic	

CHOLINERGIC ANTAGONISTS: ASSAY PROCEDURES

Blockade or inhibition of responses to parasympathetic stimulation or cholinergic agonists is effected by a group of drugs referred to as parasympatholytics, anticholinergics, or parasympathetic blocking agents. These drugs are antagonists rather than agonists; i.e., they have affinity for, but lack intrinsic activity on, cholinergic receptors.

In Vitro Test Systems

Some of the in vitro techniques described for the testing of cholinergic (parasympathetic) agonists can also be used in the assay of cholinergic antagonistic activity. The difference is that the latter procedure requires measuring first a standard cholinergic response to an agonist such as acetylcholine, carbamethacholine, or methacholine, and then in the presence of an antagonist such as atropine. The preparations in Table 2 have been used for the determination of such activity.

In Vivo Test Systems

Several assay procedures for the determination of anticholinergic activity have already been discussed in the previous review; consequently, only a cursory treatment will be given here of those already mentioned.

Measurement of Mydriatic Activity

This has been described previously. The procedure involves measuring the pupil diameter of the test animal before and after administration of the drug, e.g., atropine. The two measurements are compared for any change. An increase is indicative of mydriasis.

Antisialagogue Activity

Blockade of salivation in the rabbit with graded responses to parasympathetic stimulation was reported by Issekutz.[18] Cohen and Tvede[19] showed that there was a linear relationship when percent inhibition of pilocarpine-induced salivation in rabbits was plotted against the logarithm of the doses of atropine. Brown and Quinton[20] developed a quantitative method using rabbits for assessing inhibition of salivation. This involved the use of albino rabbits, about 1.5 to 2.0 kg in weight, anesthetized with 500 to 600 mg/kg of urethane administered orally to effect sedation. Initially, the dose of 5 mg/kg of pilocarpine nitrate is administered s.c. 20 to 30 min after the urethane. Next, an anticholinergic drug is administered about 60 min before pilocarpine is given. Each rabbit is positioned in such a manner that its head is over a funnel to aid in collection of the animal's saliva. If a rabbit in a preliminary experiment does not show a 50% reduction in salivation when a dose of 25 μg/kg atropine is given 15 min before pilocarpine, that animal should not be used for the assay procedure. The lack of sensitivity may be due to large amounts of the enzyme atropinase being present in the animal's plasma. This blockade may also be studied in rats.

Assessment of Antispasmodic Activity

Prevention of intestinal spasm experiments may be carried out in an anesthetized dog, cat, or rat as reported by Luduena and Lands.[21] The animal is fasted for 24 hr prior to use, allowing water freely. It is anesthetized with 30 mg/kg pentobarbital sodium i.v. (i.p. for the rat). The trachea and jugular vein are cannulated and laparatomy is performed. A motile loop of ileum is lifted out from a 2-cm incision made in the abdominal cavity. A 4-mm incision is made in the ileum and a thin rubber balloon (about 6 cm long and 1 cm wide) connected by means of a polyethylene tubing (PE 60 or PE 100) to a pressure transducer (Statham P23 DB) is inserted into the intestine at the incision of the ileum. The balloon is filled with saline (preferably air) and pressure recordings are made on an electronic recorder (polygraph). After placing and securing the balloon with purse string and sutures, the intestine is replaced in the abdominal cavity and the peritoneum and abdominal wall are carefully closed. Contractions are evoked by i.v. administration of 1 μg/kg of carbamethacholine (carbachol) at 10-min intervals. After constant responses are obtained, the antagonist is administered 5 to 10 min before the injection of the next dose of carbachol and the effects are studied for about 60 min. Contractions obtained after administration of the antagonist are expressed as percentages of the mean of two control values and plotted graphically. The mean percentage inhibition for 60 plotted minutes may then be calculated. The maximum inhibition and the duration of spasmolytic activity can be estimated from the time contraction curve.

Assessment of Antigastric Motility

The effect of compounds on GI charcoal propulsion in mice has been used as a method for determining anticholinergic activity.[22] The technique entails a charcoal meal fed to mice or rats as described. The test compound (anticholinergic agent) may be given (s.c.) or orally by gavage in a range of doses 15 or 30 min before administration of 1 mℓ of a 5% charcoal suspension in 5% gum acacia. The animals (whether mice or rats) should be fasted from food for 12 to 18 hr, allowing water freely. The animals are killed and the small intestine is removed 20 min after administration of the charcoal suspension. The extent of charcoal propulsion (i.e., distance traversed by the meal) in the small intestine is measured. By comparing the values obtained in treated and untreated animals, a reduction in measurements in treated animals is indicative of blockade at the parasympathetic neuroeffector junction (i.e., cholinergic antagonism). The inhibition of the test compound on intestinal transport may be determined graphically as the ED_{50} (i.e., the dose causing a 50% inhibition of propulsion in comparison to control animals).

Assessment of Antilacrimal Activity

The use of "bloody tears" (chromodacyorrhea) in rats as a method for detecting cholinergic antagonism has been described. Essentially, the technique entails the determinations of the dose of the antagonist which will protect against methacholine-induced lacrimation in 50% of the rats. Groups of 6 to 12 rats were used at each of four logarithmically spaced doses. A method for quantifying the lacrimination and chromodacryorrhea, using a piece of Whatman No. 1 filter paper to determine the amount of tear shed, has been described.

Cardiovascular Responses as Index of Anticholinergic Activity

Various animal species (dogs, cats, rabbits, guinea pigs, and rats) have been used to assess cholinergic antagonistic activity by measuring cardiovascular parameters such as arterial blood pressure and heart rate. The technique involves the determination of the extent to which the cardiovascular responses to parasympathetic stimulation of the vagus nerve or injected cholinergic agents such as methacholine, are inhibited by the use of an anticholinergic agent such as atropine. The details of the experimental procedure have been described.[23,24] Compounds with cholinergic blocking activity will partially or totally inhibit the decrease in heart rate and arterial blood pressure responses to parasympathetic stimulation or cholinergic agonists. In short, there may be a reversal (i.e., an increase) or no change in these parameters when cholinergic agonists are administered in the presence of an antagonist.

THERAPEUTIC USES OF CHOLINERGIC ANTAGONISTS

The antisecretory and antispasmodic activities of the cholinergic antagonists have led to their use in the treatment of peptic ulcer and in cases of hypermotility or spasms of smooth muscle in the GI or genitourinary tracts. Since the secretion of the exocrine glands is also controlled by histamine, the antisecretory action of these drugs is not as effective as desired and for this reason H_2 blockers have proven more effective.

Some cholinergic antagonists are available in dosage forms with phenobarbital and/or antacids. Their ability to dilate the pupil of the eye makes these drugs useful in ophthalmology as mydriatics and cycloplegics, and they are also useful in measuring refractive errors. Because of the long duration of action of atropine, a number of shorter-acting synthetic compounds, such as homatropine and cyclopentolate, have been developed.

Atropine and scopolamine are used preoperatively before general anesthesia to reduce excessive salivation and bradycardia. Scopolamine also induces sedation.

A number of cholinergic antagonists are used to reduce the symptoms of Parkinson's disease.[18] Since skeletal muscles are regulated from the CNS by both dopamine and acetylcholine, cholinergic antagonists aid in counteracting the reduced levels of dopamine resulting from this disease. It is reported that most are more effective in relieving muscle rigidity than in reducing the tremors associated with Parkinsonism.

PHARMACOKINETICS

While a number of cholinergic antagonists are acid derivatives, they are in general resistant to enzymatic hydrolysis and are excreted unchanged in the liver or bile. Exceptions will be noted in the monographs and tables of individual drugs.

With the exception of the compounds with quaternary ammonium groups, they are well absorbed when administered orally and exert their response in 1 to 2 hr. Most have a duration of action of 4 to 5 hr. Those with longer or shorter onset or duration of action will be noted in the monographs or tables.

The quaternary ammonium compounds, while poorly absorbed from the GI tract, are still effective when administered orally. The unabsorbed drug is usually excreted unchanged in the feces.

TOXICITY OF CHOLINERGIC ANTAGONISTS

The toxicity of drugs in this category, exemplified by atropine, is the opposite of those seen with cholinergic agonists. The usual therapeutic dose of atropine sulfate is 0.25 to 1 mg. It is assumed that doses higher than this are dangerous; however, the lethal dose of atropine is very high in many mammals. The rat has been shown to tolerate about 700 mg/kg; the cat can tolerate 2 mg/kg, or about 150 times the therapeutic dose in man.

In acute atropine poisoning the following symptoms are characteristic: in the GI tract, the effects include dry mouth, loss of taste, dysphagia, difficulty in swallowing, and hoarseness. Nausea and vomiting may occur and swallowing and talking become difficult (aphasia). Ocular effects include blurred vision, photophobia, cyclophlegia, mydriasis, and increased ocular tension. Dermatologic involvement includes the skin, which is hot, flushed, and dry, and evidence of hypersensitivity reactions such as skin rash. There is fever and elevated body temperature as high as 43°C. Increased body temperature is partly due to the central actions of atropine and partly to the inhibition of sweating. As a result of increased body temperature, respiration is rapid. Restlessness, excitement, mental confusion, hallucinations, mania, delirium, and convulsions may occur. Autonomic involvement includes tachycardia (increased heart rate) and urinary retention.

After the excitement subsides, severe depression and coma may ensue. Death may occur if treatment if not instituted.

The treatment of atropine poisoning calls for gastric-lavage of unabsorbed poison, administration of neostigmine or pilocarpine to overcome antimuscarinic effects, and sedation to reduce excitement and delirium. Other cholinergic antagonists may have similar side effects.

Cholinergic antagonists are contraindicated in patients with coronary disease and glaucoma. Drug interactions are reported in Table 3.

ATROPINE SULFATE

Uses and Dosage

The alkaloid is used as an antispasmodic and antisecretory agent. Besides its use in ulcer therapy and GI spasm, it has been incorporated into cold preparations to reduce nasal secretions. It is also used to reduce the symptoms of Parkinson's disease and as an antidote for overdosage of inhibitors or acetylcholinesterase.

The drug may be administered orally, s.c., i.m., or i.v. in doses of 400 to 600 μg every 4 to 6 hr. The oral dose is 250 μg three times daily. It is available in 300-, 400-, 500-, and 600-μg and 1.2-mg tablets. It is also available for injection in a solution of 400 μg/mℓ, and can be obtained in combination with meperidine hydrochloride and morphine sulfate.

Pharmacokinetics

Complete absorption from the GI tract occurs in 2 hr. Approximately 50% exists free in the blood with the remainder protein bound. Its mydriatic effect can last for up to 5 days. It is excreted chiefly unchanged in the urine, and less than 5% is hydrolyzed to tropanol and tropic acid.

Physical-Chemical Data

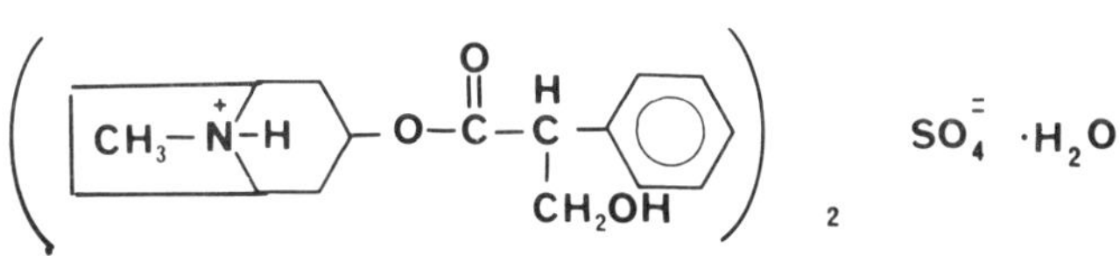

20

Atropine Sulfate

Atropine base $C_{17}H_{23}NO_3$ — mol wt 289.4; atropine sulfate $C_{34}H_{48}N_2SO_{10}{\cdot}H_2O$ — mol wt 694.82; pK_a 9.7.

Atropine sulfate is a white crystalline powder with a mp of 191 to 195°C (anhyd). The free base melts at 115 to 118°C. In terms of solubility, 1 g of the sulfate salt dissolves in 1 mℓ of water, 4 mℓ of alcohol, and is insoluble in chloroform and ether. Its chemical name is (±) tropyl tropinate; another name is Valpin®.

Table 3
DRUG INTERACTIONS: CHOLINERGIC ANTAGONISTS

Drug	May react with	To produce
Anticholinergics	Antihistamine, phenothiazine tranquilizers, meperidine-like analgesics	Additive anticholinergic effect
Anticholinergics	Anticholinesterases	Antagonism of antiglaucoma action of the cholinesterase inhibitors
Anticholinergics	Sympathomimetics	Enhanced mydriasis and bronchial relaxation
Atropine	Meperidine	Additive anticholinergic effect
Trihexylphenidyl	Diphenhydramine plus imipramine	Dry mouth

PROPANTHELINE BROMIDE

Mechanism of Action

Propantheline bromide competes with acetylcholine at the muscarinic receptor; at higher doses it acts at the receptor in the ganglia.

Pharmacokinetics

Its antisecretory effect can be noted in approximately 45 min after the usual oral dose and persists for 4 to 6 hr.

Uses and Dosage

Propantheline bromide is used in the treatment of peptic ulcer, gastritis, and hypermotility of the small intestine. It may be administered orally or by i.m. or i.v. injection. The adult dose is 15 mg three times daily and 30 mg at bedtime; 30 mg of the drug may be administered by injection. It is available in a 30-mg vial for injection, extended-release form (30 mg), and in 7.5- and 15-mg tablets. Since it slowly hydrolyzes in water solutions for injection, it should not be stored for long periods of time.

Physical-Chemical Data

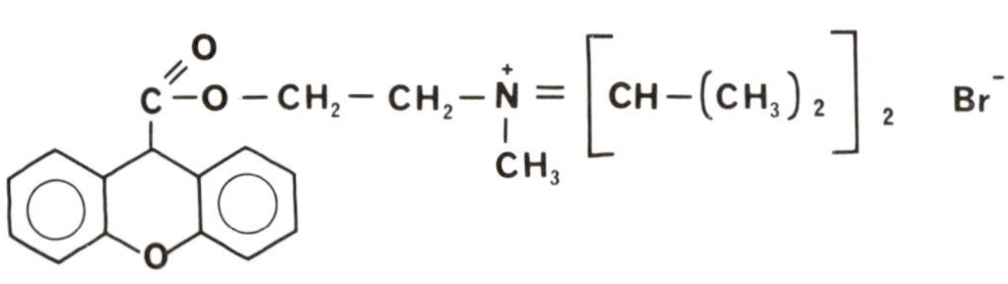

21

Propantheline Bromide

It is a white solid with a mp of 158 to 160°C. The compound is very soluble in water, alcohol, and chloroform, but is insoluble in ether. Its chemical name is 2-di-isopropylaminoethylxanthene-9-carboxylate methobromide; another name is Probanthine®. Its formula is $C_{23}H_{30}BrNO_3$. It has a mol wt of 448.4.

GLYCOPYRROLATE

Mechanism of Action

Glycopyrrolate competes with acetylcholine at the muscarinic receptor. It probably blocks the action of acetylcholine also at the ganglia.

Uses and Dosage

It is employed for its antispasmodic and antisecretory action in peptic ulcer, and in gastric conditions is characterized by excessive gastric motility. It is usually given orally in doses of 1 to 2 mg three times daily. When administered by injection the dose is 100 to 200 μg at 1-hr intervals. It is available as an injection (200 μg/mℓ) and in 1- and 2-mg tablets.

Physical-Chemical Data

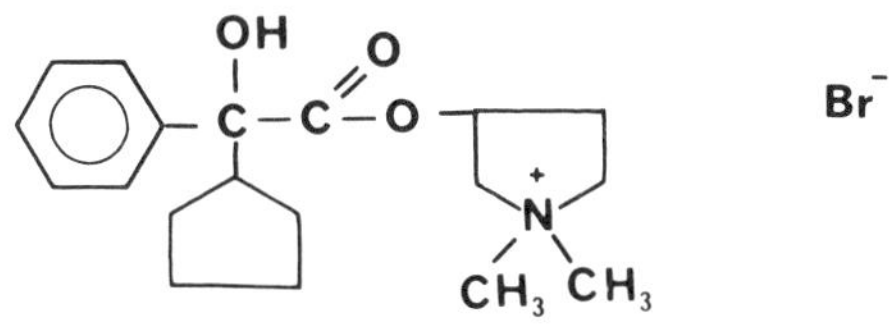

22

Glycopyrrolate

The chemical name is 2-(α-cyclopentyl)-α-phenylglycolloyloxy)-1,1-dimethylpyrrolidinium bromide. Other names are Robinul®, clycopyrronium bromide, Nodupton®, and Tarodyn®. The chemical formula is $C_{19}H_{28}BrNO_3$. The mol wt is 398.3. The compound is soluble in water.

EUCATROPINE HYDROCHLORIDE

Mechanism of Action

The drug is a competitive inhibitor of acetylcholine.

Pharmacokinetics

When applied to the eye there is little systemic absorption. The mydriatic effect occurs within 30 min and persists only for about 4 hr.

Uses and Dosage

It is applied to the eye as a mydriatic for ophthalmic examination. It produces no loss of accommodation or cycloplegia. Preparations available are as a 2 to 4 and 5% solution. Usual application requires two drops of the solution instilled in the eye 5 min apart.

Physical-Chemical Data

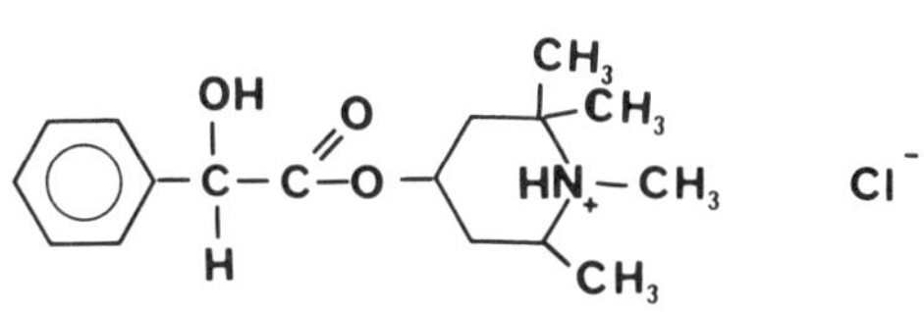

23

Eucatropine Hydrochloride

The chemical name is 1,2,2,6-tetramethyl-4-piperidyl mandelate; other names are Euphthalmine® (when combined with phenephrine), Eucadryl®, and Mydriatin®. The chemical formula is $C_{17}H_{26}ClNO_3$. The mol wt is 327.9. The compound melts at 183 to 186°C and is soluble in water, ethanol, and chloroform, but is insoluble in ether.

PROCYCLIDINE HYDROCHLORIDE

Mechanism of Action

Procyclidine hydrochloride acts by a competitive antagonism of acetylcholine.

Uses and Dosage

The compound is usually used in the symptomatic treatment of Parkinson's Disease. The initial dose is usually 2.5 mg three times daily and may be increased as needed to 5 to 10 mg three times daily. It is available in 5-mg tablets.

Physical-Chemical Data

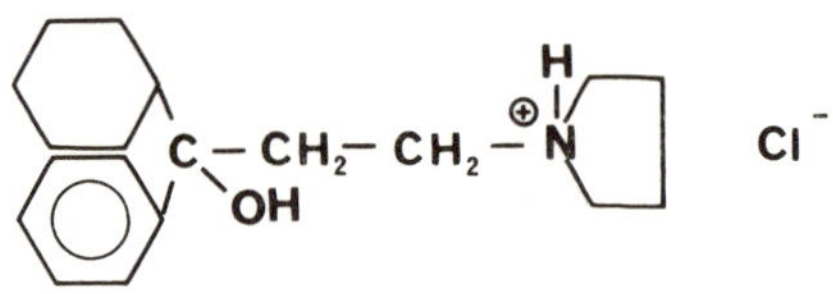

24

Procyclidine Hydrochloride

The chemical name is 1-cyclohexyl-1-phenyl-3′-[pyrrolidin-1-yl-1-propanol]. Another name is Kemodrin®. The chemical formula is $C_{19}H_{30}HOCl$. The mol wt is 323.9. It melts at 225 to 227°C. It is soluble in water (1 g in 40 mℓ) and in ethyl alcohol (1 g in 15 mℓ), but is insoluble in ether.

MECAMYLAMINE HYDROCHLORIDE

Mechanism of Action

Mecamylamine hydrochloride is a competitive inhibitor of acetylcholine at ganglia.

Pharmacokinetics

The drug is readily absorbed from the GI tract. Effects may be noted within 1 to 2 hr and persist for 10 to 12 hr. About 50% of the dose is slowly excreted unchanged in the urine. The remainder is metabolized to unidentified products which are believed to include the glucuronide and hydroxylated compounds.

Uses and Dosage

It is used in the treatment of moderate to severe hypertension. While the dosage must be modified for individual patients, the initial dose is 2.5 mg twice daily. It is given orally and is available in 2.5- and 10-mg tablets.

Physical-Chemical Data

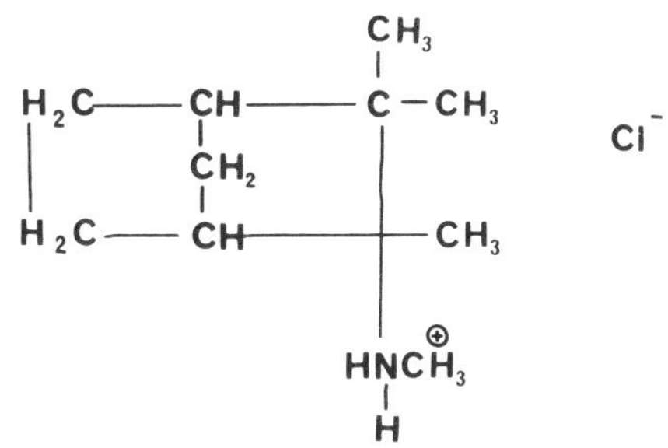

25

Mecamylamine Hydrochloride

Its chemical formula is $C_{11}H_{22}NCl$. Its mol wt is 203.8. Its chemical name is 3-methylaminoisocamphane; other names include Dimecamine and Inversine®. The base boils at 72°C. The hydrochloride melts at 245 to 249°C (dec). It has a solubility of 1 g in 5 mℓ of water and 12 mℓ of alcohol. It is insoluble in ether. Its pK_a is 11.2.

TUBOCURARINE CHLORIDE

Mechanism of Action

Tubocurarine chloride has a direct action on skeletal muscle.

Pharmacokinetics

It is a rapid-acting skeletal muscle relaxant. When administered i.v., muscle paralysis may occur within 5 min. About 30% of a dose may be recovered unchanged in the urine with the remainder being metabolized to a number of products by the liver. Appreciable amounts are excreted in the bile. Paralysis from a single dose starts to subside in 20 to 25 min.

Uses and Dosage

The drug is used to provide skeletal muscle relaxation during surgery. Usually, 6 to 10 mg of the compound is injected i.v. It is available as an injection.

Physical-Chemical Data

Its chemical formula (see Structure 20) is $C_{38}H_{44}Cl_2N_2O_6 \cdot 5\ H_2O$; its mol wt is 785.8. Its chemical name is (+)-7,12′-dihydroxy-6,6′-dimethoxy-2,2′,2′-trimethyltubocuraranium chloride. Another name is d-tubocuraranium chloride. Its mp is 270°C (dec). In terms of solubility, 1 g is soluble in 20 mℓ of water and 25 mℓ of alcohol; it is insoluble in ether and chloroform. Its pK_a is 7.4.

Table 4
OTHER CHOLINERGIC BLOCKING AGENTS

Drug	Use	Other data
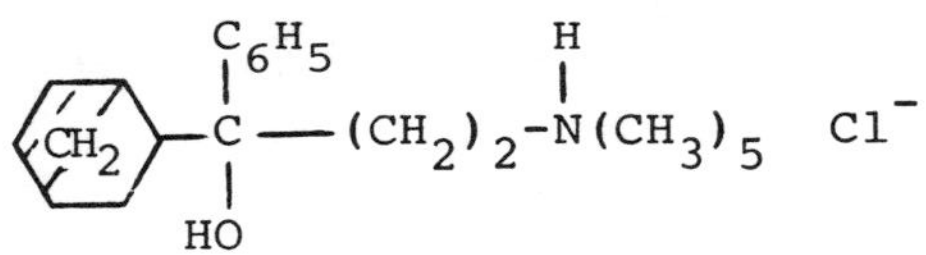 Biperiden Akineton®	Parkinson's Disease, extrapyramidal drug reaction	Available as the lactate salt in injectables
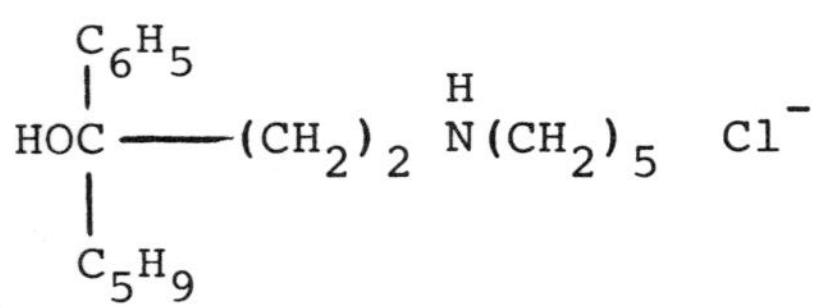 Cycrimine hydrochloride Pagitane® hydrochloride	Parkinson's Disease	
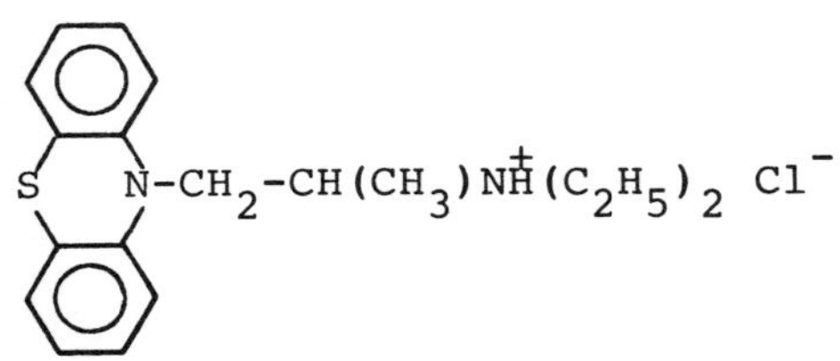 Ethopropazine hydrochloride Parsidol® hydrochloride, Profunamine hydrochloride	Parkinson's Disease, extrapyrimidal drug reaction	Relieves both tremor and rigidity, CNS depression

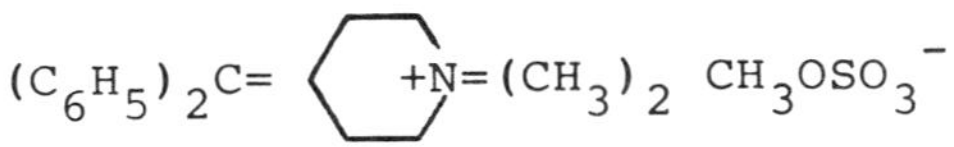 Diphemanil methyl sulfate	Antispasmodic, antisecretory, peptic ulcer	
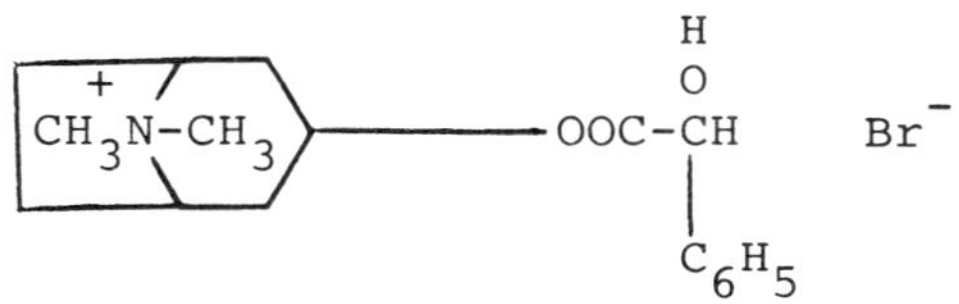 Homatropine hydrobromide	Antispasmodic, antisecretory, mydriatic, cycloplegic	Duration — 24 hr, less CNS stimulation than atropine
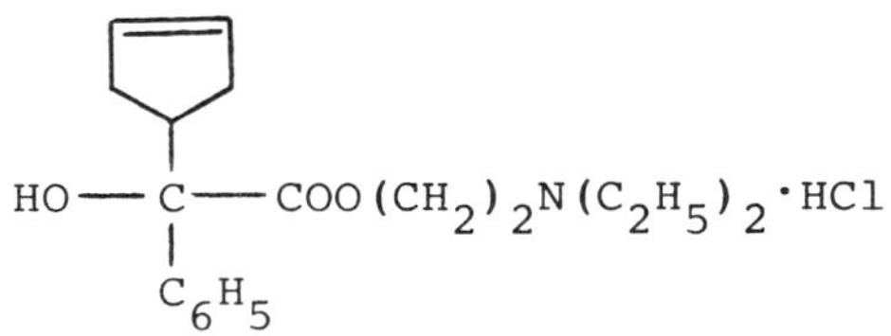 Cyclopentolate hydrochloride Cyclogyl® hydrochloride	Mydriatic	Onset — 30 to 50 min, duration — 24 hr, slight hystaminic absorption
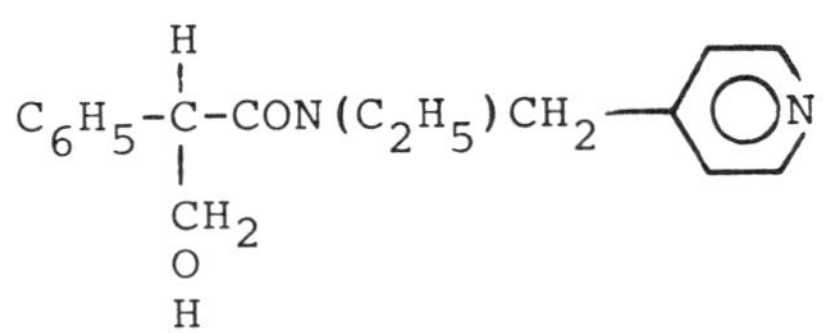	Mydriatic	Onset — 25 min, duration — 20 min

Table 4 (continued)
OTHER CHOLINERGIC BLOCKING AGENTS

Drug	Use	Other data
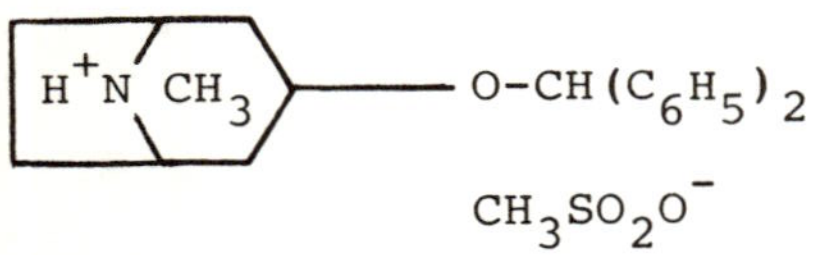 Benzotropine methanesulfonate Benzotropine mesulate Cogetin®, methanesulfonate	Parkinson's Disease, paralysis agitans, extrapyria	Potency approximately equivalent to atropine
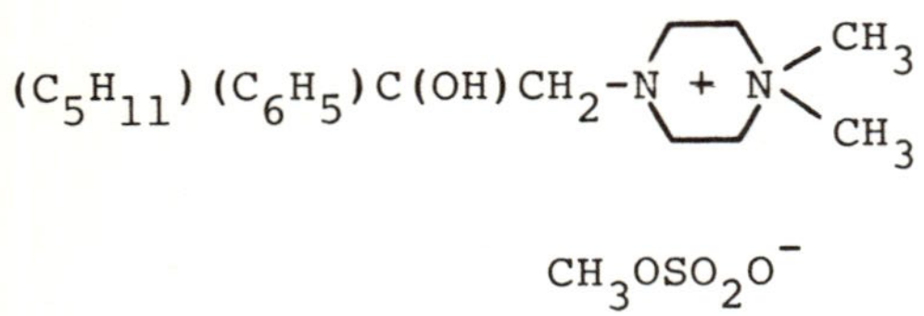 Hexocyclium methylsulfate Tral®	Antisecretory	High doses block ganglionic transmission
$NH_2OC\text{-}C(C_6H_5)_2\text{-}(CH_2)_2\text{-}\overset{+}{N}(CH_3)[CH(CH_3)_2]_2\ I^-$	Long acting, antisecretory, antispasmodic	Duration of action approximately 12 hr
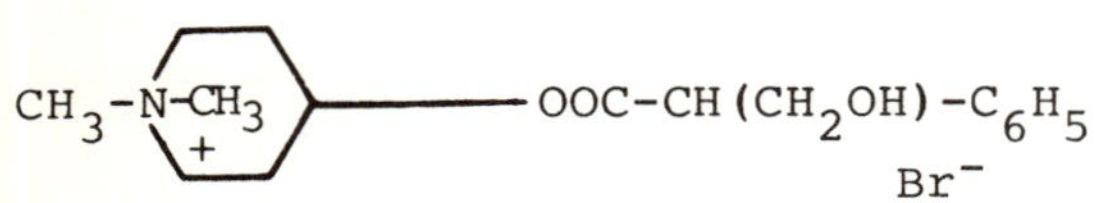 Methscopolamine bromide Hyoscine methylbromide Pumine® bromide Scopolamine methylbromide	Antispasmodic, antisecretory	Duration — 4 to 6 hr, excreted in urine and bile

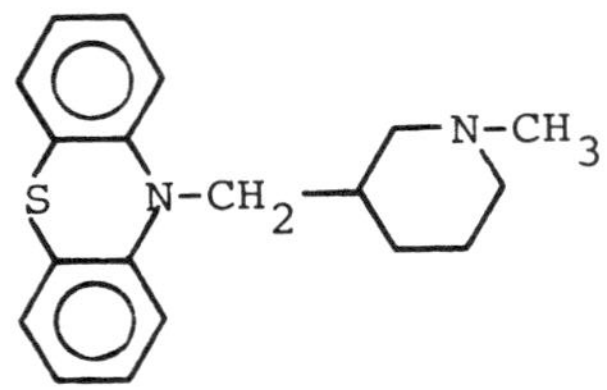 Methixine hydrochloride Trest®	Antispasmodic	Antihistaminic activity, metabolized to sulfoxide and also demethylated or both
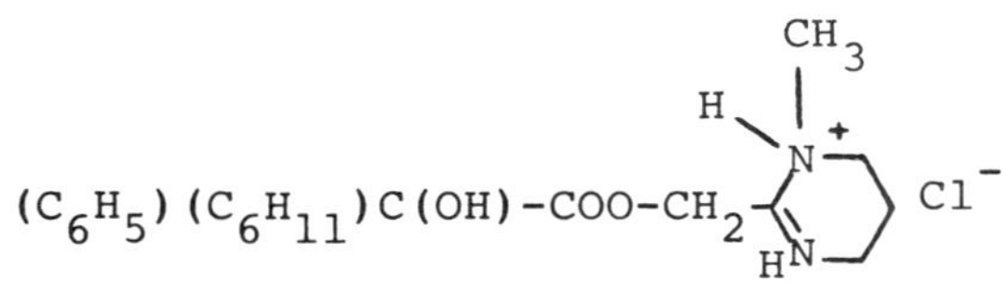 Oxyphencyclimine hydrochloride Daricon® Sutrol® Viothene®	Antispasmodic, antisecretory	Onset — 1 hr, duration 8—12 hr
$(C_6H_5)_2CH{-}O{-}(CH_2)_3{-}N(CH_3)_2$ Orphenedrine Disipal® Norflex®	Parkinson's Disease	Weak antihistaminic activity; 30 to 50% excreted unchanged in urine; remainder undergoes *N*-demethylation; available as citrate salt
$(C_6H_5)(C_6H_{11})C(OH)(CH_2)_2\overset{+}{N}(CH_3)[CH(CH_3)_2]_2\ Br^-$ Oxyphenonium bromide Artrenyl® bromide	Antispasmodic, antisecretory, peptic ulcer	

Table 4 (continued)
OTHER CHOLINERGIC BLOCKING AGENTS

Drug	Use	Other data
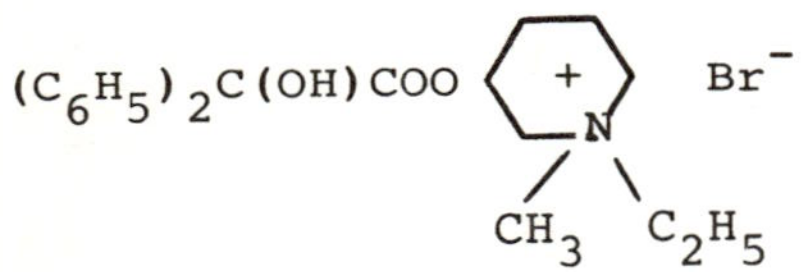 Pipenzolate bromide Piptol®	Antispasmodic, antisecretory, peptic ulcer	
$(C_6H_5)(C_6H_{11})\text{-}C(OH)\text{-}(CH_2)_2\overset{+}{N}(CH_3)_5\ Cl^-$ Trihexylphenidyl hydrochloride Artane® hydrochloride Pipanal® hydrochloride Tremin® hydrochloride	Antispasmodic, Parkinson's Disease	Duration 8—12 hr, weak mydriatic and antisialogogue activity
$(C_6H_5)(C_6H_{11})COH(CH_2)\underset{+}{\overset{H}{N}}(C_2H_5)_3\ Cl^-$ Tridihexethyl chloride Pathilon® chloride	Antispasmodic, antisecretory, peptic ulcer	

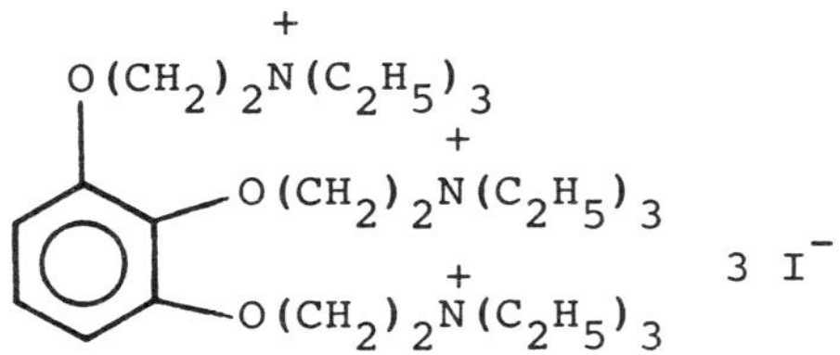 Gallamine triethiodide Floxedil®	Skeletal muscle relaxant	Solutions — stable to autoclaving, very fast acting
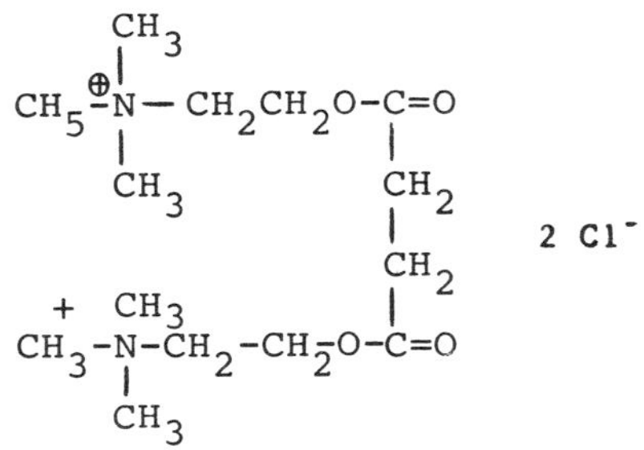 Succinyl choline chloride Anectine® chloride Quelicin® chloride Sucostrin® chloride	Skeletal muscle relaxant	Shorter in action than tubocurarine chloride

REFERENCES

1. **Wintersteinen, O. and Dutchern, J. D.,** Curare alkaloids from chondodendron tomentosum, *Science,* 97, 467, 1943.
2. **Casey, A. F.,** Stereochemical aspects of parasympathomimetics and their antagonists: recent developments, *Progress in Medicinal Chemistry,* North-Holland, Amsterdam, 1975, 11.
3. **Belleau, B. and Lacasse, G.,** Aspects of the chemical mechanism of the complex formation between acetylcholinesterase and acetylcholine related compounds, *J. Med. Chem.,* 7, 776, 1964.
4. **Ing, H. R.,** The structure action relationship of the choline group, *Science,* 109, 264, 1949.
5. **Belleau, B.,** Conformation perturbation in relation to regulation of enzyme and receptor behavior, *Adv. Drug Res.,* Academic Press, New York, 1965, 2.
6. **Beckett, A. H. and Albadr, A. A.,** A modified structure for the acetylcholinesterase receptor, *J. Pharm. Pharmacol.,* 27, 855, 1975.
7. **Landis, A. M., Hoppe, J. O., Lewis, I. R., and Ananenko, E.,** Pharmacologic action of basic esters of xenylacetic and *p*-xenylglyclic acid — new antispasmodic drugs, *J. Pharmacol. Exp. Ther.,* 100, 19, 1950.
8. **Long, J. P., Ludcenu, R. F., and Tullar, B. F.,** Stereochemical factors involved in cholinolitic activity, *J. Pharmacol. Exp. Ther.,* 117, 29, 1966.
9. **Deloomy, B. and Gearien, J. E.,** unpublished data.
10. **Wung, W. and Gearien, J. E.,** unpublished data.
11. **Babbington, A. and Brimblecomb, R. W.,** Muscarinic receptors in the peripheral and central nervous system, *Adv. Drug Res.,* 2, 143, 1965.
12. **Ariens, E. J.,** Receptor theory and structure-action relationship, *Adv. Drug Res.,* 3, 235, 1966.
13. **Huang, B. and Lu, M.,** unpublished data.
14. **Van Rossum, J. M.,** Classification and molecular pharmacology of ganglionic blocking agents, *Int. J. Neuropharmacol.,* 1, 97, 1962.
15. **Van Rossum, J. M.,** Classification and molecular pharmacology of ganglionic blocking agents. II, *Int. J. Neuropharmacol.,* 1, 403, 1962.
16. **Phillips, A. P. and Castillo, J. C.,** Synthetic curare substitutes from stilbazoline quaternary ammonium salt, *J. Am. Chem. Soc.,* 73, 5949, 1951.
17. **Cavallito, C. J., Soria, A. E., and Hoppe, J. O.,** Amino- and ammonium-alkylaminobenzoquinones as curarimimetic agents, *J. Am. Chem. Soc.,* 72, 2661, 1950.
18. **Issekutz, B. V.,** Uber die wirkung der Tropine und ihrer Quaternaren Ammoniumbasen, *Z. Exp. Pathol. Pharmacol.,* 19, 99, 1917.
19. **Cohen, R. L. and Tvede, K.,** Homatropine methylbromide: a pharmacological reevaluation, *J. Pharmacol. Exp. Ther.,* 105, 166, 1952.
20. **Brown, D. M. and Quinton, R. M.,** The assay of antiacetylcholine agents for antagonism of pilocarpine-induced salivation in rabbits, *Br. J. Pharmacol.,* 12, 53, 1957.
21. **Luduena, F. P. and Lands, A. M.,** An investigation of the pharmacological actions of three potent antispasmodic compounds and their corresponding metho-salts, *J. Pharmacol. Exp. Ther.,* 110, 282, 1954.
22. **Long, J. P. and Keasling, H. H.,** The comparative anticholinergic activity of a series of derivatives of 3-hydroxypiperidine, *J. Am. Pharm. Assoc.,* 43, 616, 1954.
23. **Domer, F. R. and Schueler, F. W.,** Pharmacologic investigations of a series of p-xylylene bis-ammonium congeners, *Arch. Int. Pharmacodyn.,* 114, 217, 1958.
24. **Yusef, S. M. and Riddle, O. D., Jr.,** A comparative study of the pharmacological activity of various anticholinergic drugs, *Arch. Int. Pharmacodyn.,* 152, 198, 1964.
25. **Bentley, G. A.,** Studies on sympathetic mechanisms in isolated intestinal and vas deferens preparations, *Br. J. Pharmacol.,* 19, 85, 1962.
26. **Hukovic, S.,** Responses of the isolated sympathetic nerve ductus deferens preparation of the guinea pig, *Br. J. Pharmacol.,* 16, 188, 1961.
27. **Birmingham, A. T. and Wilson, A. B.,** Preganglionic and postganglionic stimulation of the guinea pig isolated vas deferens preparation, *Br. J. Pharmacol.,* 21, 569, 1963.
28. **Bulbring, E., Crema, A., and Saxby, O. B.,** A method for recording peristalsis in isolated intestine, *Br. J. Pharmacol.,* 13, 440, 1958.
29. **Beleslin, D. B., Bogdanovic, S. B., and Rakic, M. M.,** Mechanism and site of action of methadone on the peristaltic reflex of the isolated guinea pig ileum, *Arch. Intern. Pharamcodyn.,* 164, 345, 1966.
30. **Bulbring, E. and Lin, R. C. Y.,** The effect of intraluminal application of 5-hydroxytryptamine and 5-hydroxytryptophan on peristalsis: the local production of 5-HT and its release in relation to intraluminal pressure and propulsive activity, *J. Physiol.,* 140, 381, 1958.
31. **Trandelenburg, P.,** Physiologische und pharmakologische versuche uber die dunndarnperistaltik, *Arch. Exp. Pathol. Pharmakol.,* 81, 55, 1917.

HISTAMINE, H_1 AND H_2 HISTAMINE RECEPTOR ANTAGONISTS, AND DISODIUM CROMOGLYCATE

Donald C. Schlegel and Lawrence W. Chakrin

HISTAMINE

Histamine, 2, is 2-(4-imidazolyl)ethylamine. It is synthesized by decarboxylation of the amino acid, histidine, 1, a reaction mediated by

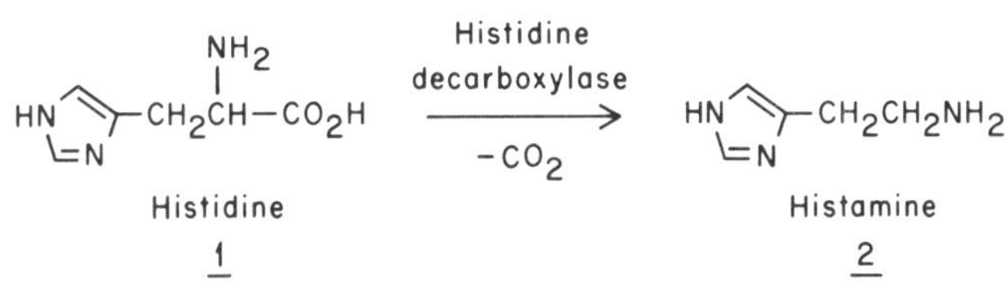

the enzyme, 1-histidine decarboxylase. Essentially every mammalian tissue contains histamine, stored as a preformed mediator, although the concentration of the amine is especially high in the skin, intestinal mucosa, and pulmonary tissue. The amine, especially that involved in immediate hypersensitivity reactions, is stored in discrete mast cell or basophil secretory granules, presumably as a complex. This stored amine undergoes rather slow turnover by comparison to the extra mast cell histamine of the GI mucosa and the central nervous system (CNS).

Although it remains the subject of some controversy, the calcium-dependent cellular release of histamine involves several processes, perhaps including the exocytosis of all or part of the subcellular cytoplasmic granules and the subsequent dissociation of histamine from the granule matrix. Once released, possibly through the ion exchange-like process, histamine presumably contributes to portions of the clinical syndrome of immediate hypersensitivity reactions, including extrinsic asthma and allergic rhinitis. Additionally, nonmast cell/basophil histamine has been thought to regulate the microcirculation, so as to provide for locally increased requirements for blood, contribute to anabolic processes, regulate secretion of gastric parietal cells, and presumably serve as a chemical transmitter in the CNS.

Subsequent to its yet poorly understood role in these processes, and apart from tissue reuptake, histamine apparently undergoes metabolism by one of two pathways: (1) through methylation of the ring structure by imidazole-*N*-methyltransferase, histamine yields methyl histamine, which is then oxidized through monoamine oxidase to methylimidazoleacetic acid and (2) histamine can undergo an oxidative deamination by histaminase or diaminooxidase to imidazoleacetic acid or its riboside. The products of this metabolism are excreted in the urine (see Scheme 1). For a more complete discussion the reader is referred to Reference 1.

It has been suggested recently[2] that the biological activity of histamine at H_1 and H_2 receptor sites might be explained by the various chemical forms of the amine. The N_1H form of histamine (Monocation, Scheme 1) may contribute significantly to the principal forms of histamine at both receptor sites. This form is the most prevalent in water at physiologic pH. However, other tautomeric forms and protonation states of histamine, such as the dicationic form at acidic stomach pH may also contribute to the interaction of the amine on these two different receptor types. The imidazole nitrogen *ortho* to the ammonium ethyl group apparently is of significance, and the imidazole ring may need to rotate and achieve coplanarity with the side chain of the molecule. With respect to the H_2 receptor,

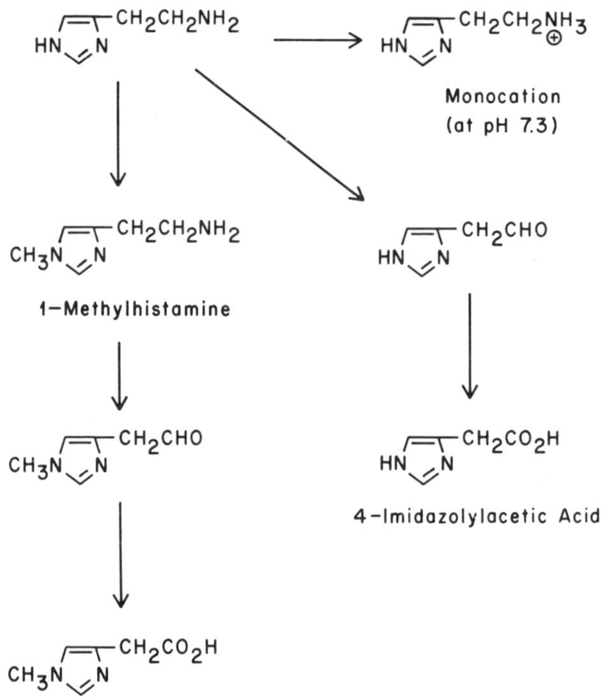

SCHEME 1.

the tautomeric properties of the imidazole ring may be of importance and histamine may be involved in multiple hydrogen bonding or may act as a proton transfer agent. These suggestions have developed from structure-activity studies of analogs of histamine in biologic models of the multireceptor pharmacology of histamine, i.e., contraction of guinea pig ileum (H_1), depression of blood pressure of the anesthetized cat (H_1), stimulating mammalian gastric acid secretion (H_2), and inhibiting the isolated rat uterus (H_2). The biologic activities of several analogs of histamine relative to histamine itself showed a separation in activities that may be explained by activation of each of these receptor types. The interested reader is referred to an indepth discussion by Ganellen and Durant.[2]

HISTAMINE ANTAGONISTS

General Discussion

The development of agents capable of antagonizing the actions of histamine has been a research effort for more than 4 decades. Such compounds, if pharmacologically selective, would serve both as valuable investigative tools and presumably as clinically useful therapeutic agents. For the most part, such compounds have been shown to be competitive, reversible antagonists of the amine at tissue sites.

Throughout the late 1930s and during the 2 decades that followed, a broad number of compounds, almost all containing a substituted ethylamine, were synthesized and tested as histamine antagonists. These compounds, later to be identified as histamine H_1, receptor blocking agents, were especially effective in reducing or abolishing several physiologic responses to histamine. However, the ineffectiveness of these agents in blocking histamine-stimulated gastric acid secretion, and other responses, such as increases in heart rate and inhibition of uterine contractions, prompted Black and colleagues through the late 1960s and the 1970s to search for what would eventually prove to be a series of H_2 receptor antagonists. The early H_2 receptor antagonists include compounds that are imidazole derivatives. They possess large structurally specific side chains and are less basic and more hydrophilic than histamine. The reader is referred to References 3 to 5.

HISTAMINE H_1 RECEPTOR ANTAGONISTS

Discussion

These compounds are strongly basic tertiary amines. They are capable of inhibiting the spasmogenic effects of histamine on tracheobronchial and GI smooth muscle. Moreover, these compounds inhibit the vasoconstrictor, and to some extent, vasodilator effects of histamine. They are especially effective in antagonizing histamine-induced increases in capillary permeability, thus preventing the edema and cutaneous wheel formation. With respect to the CNS, the H_1 receptor antagonists can occasionally stimulate, but more frequently depress, with sedation often the result. Those agents that produce CNS stimulation, e.g., the alkylamine antagonists, presumably do so by altering the neuronal storage and release of endogenous CNS amines.

In clinical practice, the H_1 antagonists are employed in the symptomatic treatment of some allergic responses where they presumably antagonize the actions of immunologically induced histamine release. In bronchial asthma, the antagonists are rather ineffective, but in seasonal allergic rhinitis, conjunctivitis, and allergic dermatitis, the therapy has been successful presumably due to the greater role of histamine in these diseases. Occasionally, the H_1 antagonists are employed in cough syrups where possible antiallergic activity and CNS sedation may be beneficial. However, the anticholinergic activity of these compounds may produce relatively viscous tracheobronchial secretions which are more difficult to expectorate.

The CNS effects of the H_1 antagonists have been utilized in the prophylaxis and treatment of motion sickness. While the antimuscarinic alkaloid scopolamine remains especially effective for this indication, the H_1 antagonists, diphenhydramine and promethazine, have been employed with success. The sedative effect of the ethanolamine H_1 receptor antagonists has been exploited as hypnotics. Additionally, the local anesthetic activity of H_1 receptor antagonists has been utilized to treat itching associated with several skin conditions, but topical application may be limited by sensitization.

Generally, in regular therapeutic dose, the H_1 receptor antagonists produce such side effects as sedation and to some extent the usual reports of dizziness, fatigue, blurred vision, tremors, and tinnitus. These CNS effects present significant concern in acute poisoning with the antagonists. Both the depressant and stimulant properties may be disclosed, which may include overt hallucinations, excitement, and convulsions. GI disturbances, including nausea, vomiting, and constipation or diarrhea, account for an additional group of effects, followed in frequency by a variety of miscellaneous difficulties, including dryness of the mouth and tracheobronchial tissues, headache, tingling, hypotension, palpitation, and tightness in the chest. Drug allergy can develop following the oral, and more frequently, the topical application of the compounds.

Drug Interactions

The H_1 antihistamines exhibit drug interactions with several other classes of compounds. Monoamine oxidase inhibitors prolong and intensify the anticholinergic effects of H_1 antihistamines. Moreover, the sedative activity of H_1 antihistamines may be additive to the sedative effects of CNS depressants and thus, such agents as barbiturates, alcohol, and narcotic analgesics should be used in reduced dosage or eliminated.

General Structural Requirements and Physical Properties of an H_1 Antihistamine

The specific chemical structures of the H_1 antihistamines vary widely, but all the more active compounds may be represented by the following general formula where X represents

Table 1
PARTITION COEFFICIENTS OF SOME HISTAMINE ANTAGONISTS

Antagonists	$P = C_{octanol}/C_{aq.\ buffer}$ [a]
Mepyramine	700
Diphenhydramine	2500
Triprolidine	8300
Histamine	0.2
Cimetidine	2.5
Metiamide	3.2

[a] Partition betweel l-octanol and aqueous buffer at 37°C.

an oxygen, nitrogen, or carbon atom connecting a cyclic aminoalkyl or disubstituted aminoalkyl side chain to an aromatic backbone, R.

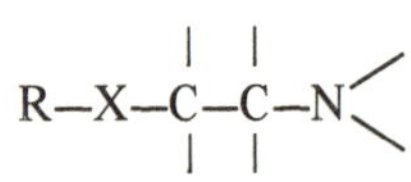

For maximum antihistaminic activity it has been found that the nitrogen atom of the side chain should be trisubstituted; two methyl radicals are often preferred as part of the nitrogen substitution pattern, although some cyclic aminoalkyl structures also show good activity. The aromatic backbone varies widely: (1) benzhydryl and various substituted analogs of it in the ethanolamine series, (2) benzyl, α-pyridyl, and various other heterocyclic systems in the ethylene diamine class, and (3) phenothiazine and other tricyclic ring systems in the tricyclic class.

As a class, the H_1 antihistamines are highly lipophilic molecules. This physical property can be measured by observing the partitioning of the compounds between octanol and aqueous buffer. A compound with a large P value is distributed mainly in the organic lipophilic phase. Representative partition coefficients for a few of the important antihistamines are listed in Table 1.[6]

This lipophilicity of the H_1 antihistamines is in sharp contrast to the hydrophilic nature of histamine itself and the H_2 histamine antagonists such as cimetidine and metiamide.

Another important characteristic of the H_1 antihistamines is that they all possess a strongly basic tertiary amine side chain that exists largely in a protonated form at physiological pH. The pK_a values for these amine side chains range between 8.5 and 10.

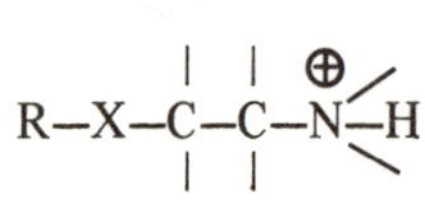

As free bases, the H_1 antihistamines are, in general, oils. Consequently, to increase their manageability and stability to air oxidation, they are formulated as acid salts. The two most commonly used acids for salt formation are hydrochloric and maleic.

The H_1 antihistamines have been divided into six major classes: ethylenediamine analogs, aminoalkyl ethers, dialkylaminopropyl derivatives, cyclizines, tricyclic compounds, and a miscellaneous class based largely on the type of X linkage and the type of aromatic backbone they possess. These classes, with their clinically most important members, will be considered in the following sections.

Classes of H_1 Histamine Antagonists*

Ethylenediamine Analogs

Structure-Activity Studies

Phenbenzamine, 3, of the ethylenediamine class, was the earliest clinically useful antihistamine. Replacement of the phenyl group of this compound by α-pyridyl gave a group of even more potent antihistamines — tripelennamine, 4a, being the base member.

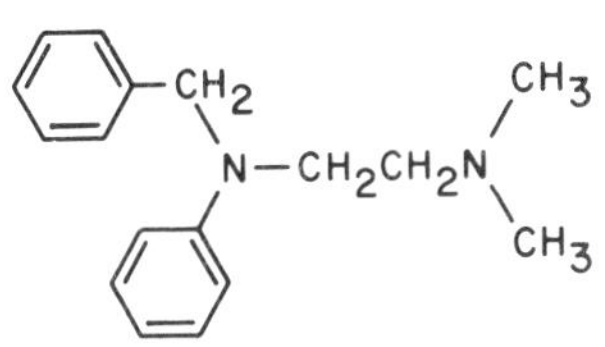

3

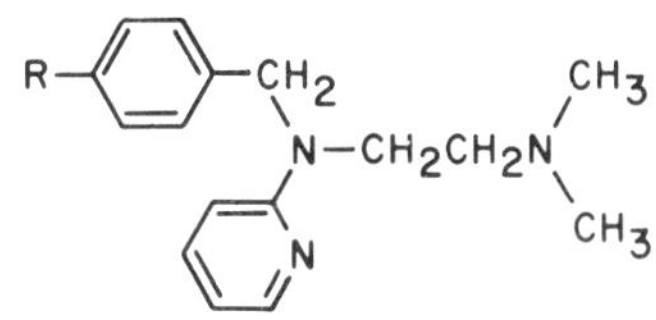

4a	R=H	Tripelennamine
4b	R=OCH_3	Pyrilamine = Mepyramine
4c	R=Cl	Chloropyramine
4d	R=Br	Bromtripelennamine

Placing substitutents in the *para* position of the benzyl group, methoxyl (pyrilamine=mepyramine), 4b, chlorine (chloropyramine), 4c, and bromine (bromtripelennamine), 4d, yields still more potent antihistamines.[7]

The benzyl group can also be replaced by a 2- or 3-thienylmethyl group to yield active compounds. Halogenation of the 2-thienylmethyl group in the 5-position, 5b, reportedly gave an equally active antihistamine of lower toxicity.[8,9]

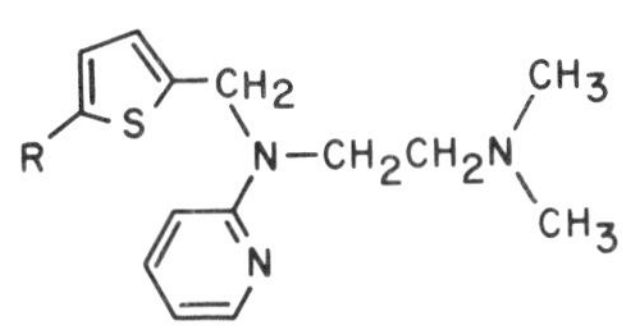

5a	R=H	Methapyrilene
5b	R=Cl	

Replacement of the α-pyridyl group by either the β- or γ-pyridyl isomer gave less active compounds.[10] The α-pyridyl group of mepyramine can be replaced by a 2-thiazoyl isostere (zolamine), 6a, or a pyrimidinyl moiety[11] (thonzylamine), 6b, to give compounds with antihistaminic activity similar to tripelennamine but with reportedly lesser toxicity and side effects.[8]

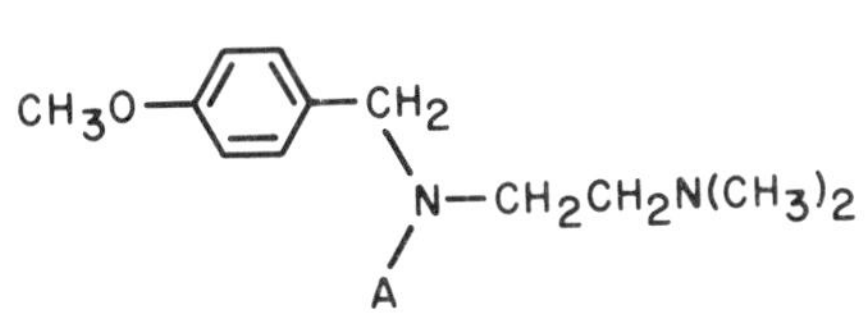

6a Zolamine A = 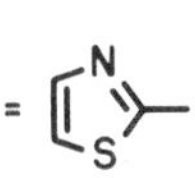

6b Thonzylamine A = 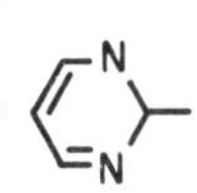

* As a consequence of its metabolism leading to the formation of a carcinogenic nitrosamine with nitrous acid, methapyrilene has been withdrawn from further use.

The high antihistaminic activity of this class seems to be associated with the dimethylaminoethyl group. Other side chains, such as the α-methyl, 7a, and the dimethylamino-*n*-propyl analog, 7b, of mepyramine are much less active.[12]

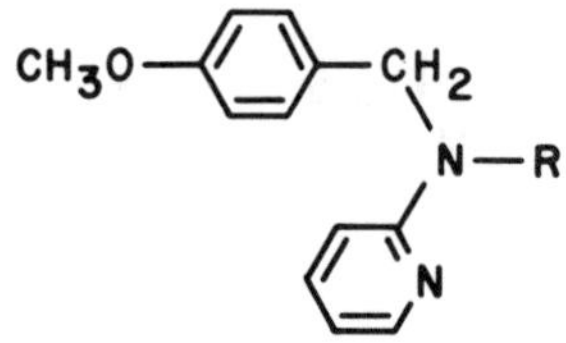

7a $R = CH_2\overset{CH_3}{C}HN(CH_3)_2$

7b $R = CH_2CH_2CH_2N(CH_3)_2$

Pharmacokinetics

The metabolism of tripelennamine, 8, is representative of the ethylene diamine class of antihistamines. In man, on oral administration, the major metabolite is the quaternary ammonium *N*-glucuronide, 9.[13] Other minor products are the N-oxide of tripelennamine, 10, and the glucuronides of 5-hydroxylated (in the pyridyl ring) tripelennamine, 11a, and its desmethyl analog, 11b. No metabolite with hydroxylation of the phenyl ring was observed in man.[13] On i.p. administration in the rat,

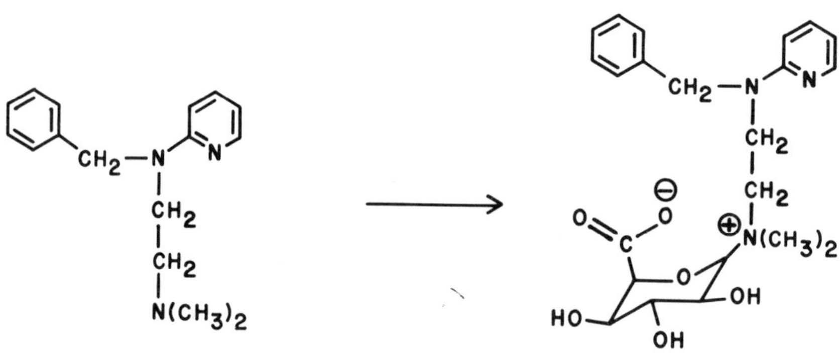

8

9

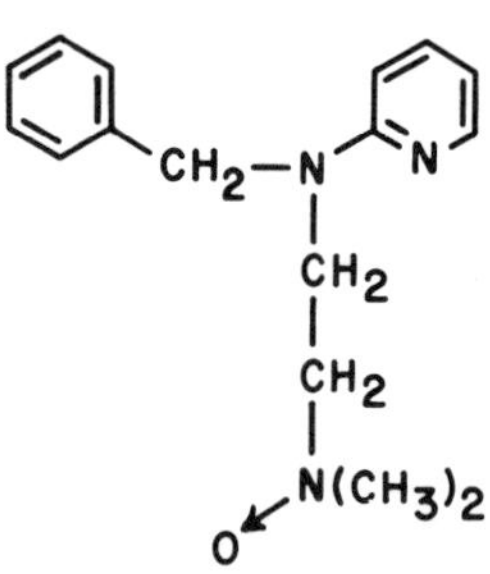

10

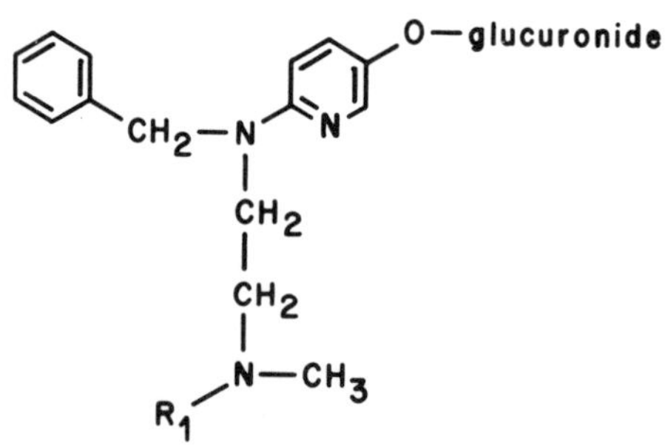

11a $R_1 = CH_3$

11b $R_1 = H$

however, extensive hydroxylation of the phenyl ring was observed and the major metabolite in the urine (representing 37.6% of the administered radioactivity) was the *p*-hydroxybenzyldesmethyl tripelennamine, 12, isolated as the glucuronide.[14]

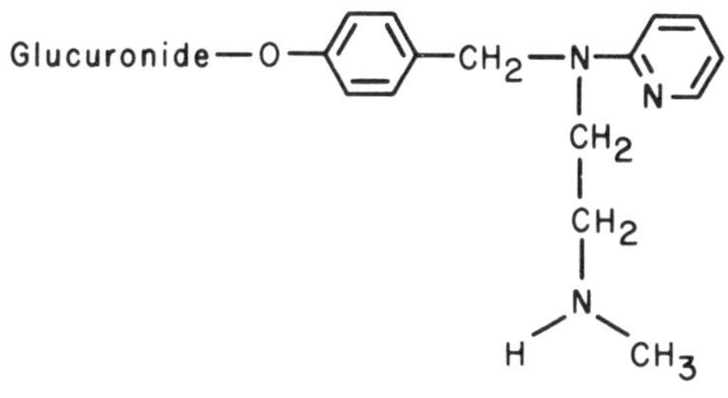

12

Other major glucuronide metabolites exhibited both mono- and dihydroxylation of the phenyl and pyridyl rings.[14] In neither the metabolic study in rats nor that in man was the didesmethyl tripelennamine or its hydroxylated analogs observed.[13,14]

Little has been reported on the pharmacokinetics of the various members of this class; however, one report on methapyrilene, 5a, states that the elimination of both i.v. and orally administered drug was rapid, with a plasma $t_{1/2}$ varying between 1.1 and 2.1 hr for a 20-mg i.v.-administered dose and 2.33 ± 0.78 hr for a 25-mg orally administered dose. Plasma clearance of drug on oral administration ranged between 0.013 to 0.048 ℓ/min/kg. The blood to plasma ratio was 0.99 ± 0.12. The bioavailability of the drug was low (4 to 46%), and showed large interindividual differences. Only 2% of the drug was recovered unchanged in the urine over 24 hr.[15]

Clinically Useful Members of the Class[16-20]

TRIPELENNAMINE HYDROCHLORIDE

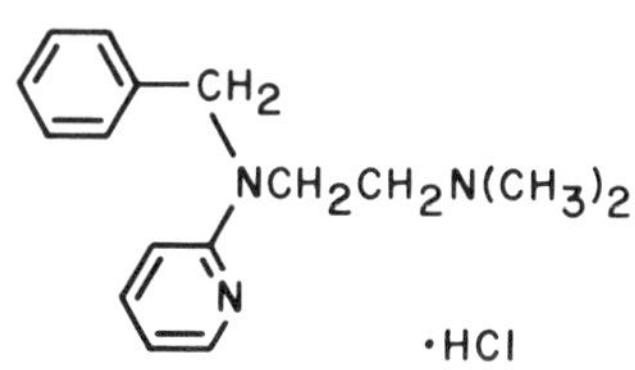

13

Uses

It is used for the treatment of seasonal and perennial allergic rhinitis and mild allergic skin reactions.

Dosage and Other Names

Alone it is administered in doses of 25 to 50 mg every 4 to 6 hr. This compound or the citrate salt equivalent of tripelennamine is used in combination with other drugs at doses of 20 to 30 mg four to six times a day. Other names are Pyribenzamine hydrochloride (Ciba-Geigy) and Tri-Tumine (Tutag).

Toxicity

It has the usual antihistamine side effects. The LD_{50} orally in mice is 210 mg/kg.

Physical-Chemical Data

The white, crystalline powder, mp 192.5 to 193.5°C, tastes bitter and slowly darkens on exposure to light. A 1-g sample dissolves in 0.77 mℓ of water, 6 mℓ of alcohol, and 6 mℓ of chloroform. The pH of an aqueous solution is about 6.5. The pK_a at 20°C is 8.95.

Chemical Name and Formula

The chemical name of tripelennamine hydrochloride is *N,N*-dimethyl-*N'*-(phenylmethyl)-*N'*-2-pyridinyl-1,2-ethanediamine monohydrochloride. The formula is $C_{16}H_{21}N_3 \cdot HCl$. The mol wt is 291.82.

PYRILAMINE MALEATE

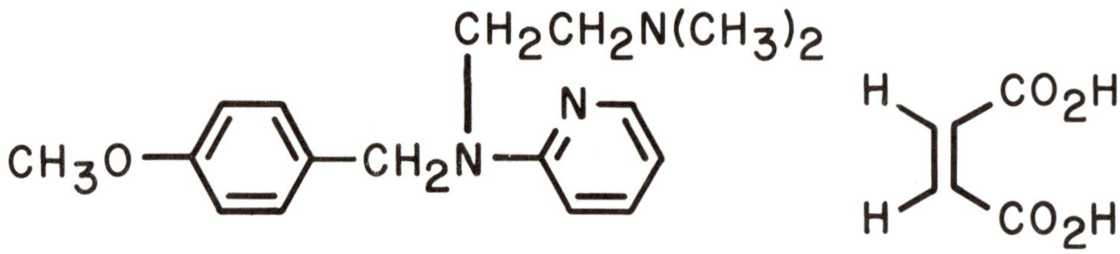

14

Uses

It is used in combination with other drugs in a wide variety of cough syrups, expectorants, and nasal decongestants for its general antihistaminic properties.

Dosage and Other Names

Many combination products contain about 8 mg of this drug per tablet. The tablets are administered q.i.d. Other names include Pyra-Maleate (Mallinckrodt), Neo-Antergan Maleate (Merck Sharp & Dohme), Dorantamin (Dorsey), Enrumay (Cooper), Histan (Cooper), Kriptin (Whitehall), Minihist (Ives), Pyramal Maleate (Philips Roxane), Paraminyl Maleate (Cooper), Pymafed (Hoechst-Roussel), Stamine (Tutag), Thylogen Maleate (Rorer), component of Emesert (Arnar-Stone), and component of Pyrdex (Savage).

Toxicity

It has the usual antihistamine side effects.

Physical-Chemical Data

The bitter-tasting crystals, mp 100 to 101°C, are stable in air. A 1-g sample is soluble in 0.4 mℓ of water and 15 mℓ of alcohol. The pH of a 10% solution is approximately 5.1, and the pK_a at 20°C is 8.85.

Chemical Name and Formula

Its chemical name is *N*-[(4-methoxyphenyl)methyl]-*N'*,*N'*-dimethyl-*N*-2-pyridinyl-1,2-

ethanediamine (Z)-2-butenedioate (1:1).* Its formula is $C_{17}H_{23}N_3O \cdot C_4H_4O_4$, and its mol wt is 401.36.

Aminoalkyl Ether Analogs

Structure-Activity Studies

Diphenhydramine, 15a, was the first medicinally useful member of the aminoalkyl ether class of antihistamines. Later it was found that *mono* substitution of one of the phenyl rings in the *para* position by methyl, methoxy, chloro, or bromo radicals, 15b, gave other useful antihistamines which may be advantageous because of fewer side effects.[23]

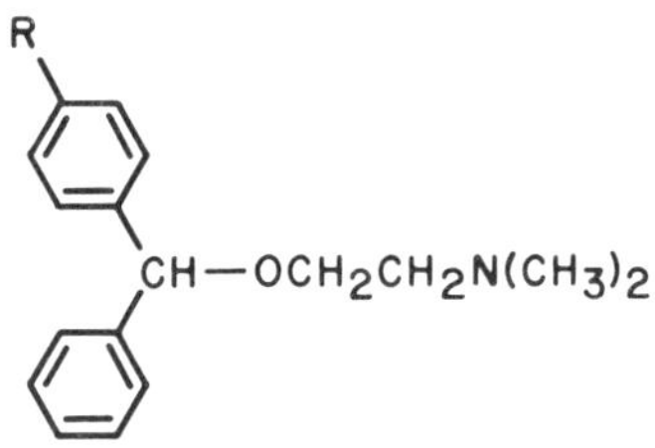

15a Diphenhydramine R=H

15b R=CH_3, OCH_3, Cl, Br

In an attempt to get rid of some of the usual side effects of drowsiness and nervousness exhibited by many individuals when taking diphenhydramine, other salts were prepared. One of these, the 8-chlorotheophyllinate salt (Dramamine), 16, has found great usefulness as an antimotion sickness medication.

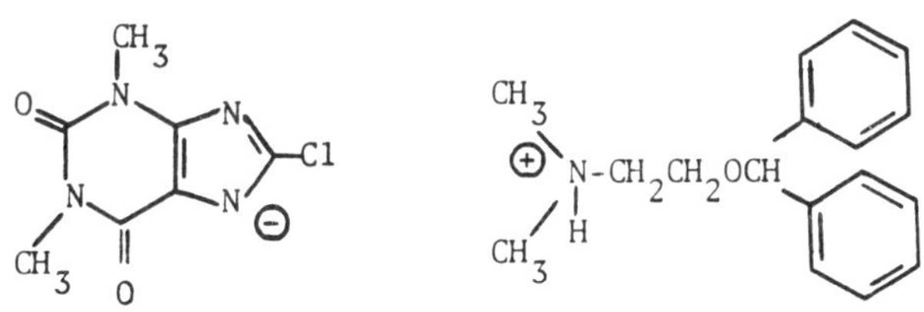

16 Dramamine

Replacement of one of the phenyl groups by a α-pyridyl group gave another series of compounds with even greater antihistaminic activity. Table 2 relates in more detail the antihistaminic activity observed in an in vivo study in guinea pigs of a variety of aminoalkyl

* What was earlier denoted as *cis-trans* isomerism about a double bond is now connotated by the symbols Z (*zusammen* = together) for *cis* (the two largest substituents, as described by the Cahn-Ingold-Prelog rules,[21] on the same side of the double bond) and E (*entgegen* = opposite) for *trans* (the two largest groups on opposite sides of the double bond).[22]

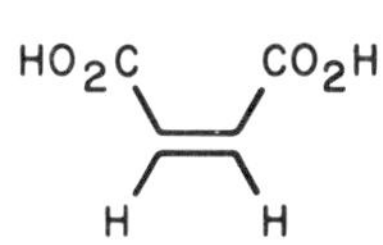

cis form (Z)
Maleic Acid
(Z)-2-Butenedioic Acid

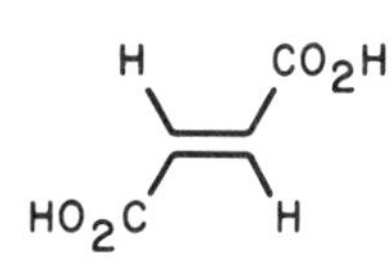

trans form (E)
Fumaric Acid
(E)-2-Butenedioic Acid

Table 2
RELATIVE ANTIHISTAMINIC ACTIVITY OF DIPHENHYDRAMINE AND ITS ANALOGS

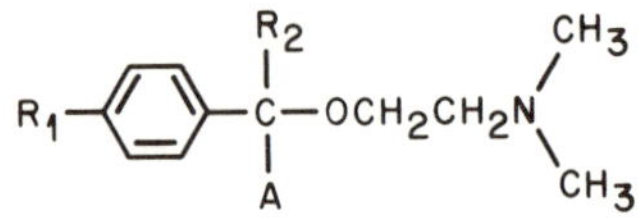

		A	R_1	R_2	Relative oral antihistaminic activity in guinea pig
17a	Diphenhydramine	Phenyl	H	H	1
17b	Experimental compound	2-Pyridyl	H	H	1.3
17c	Carbinoxamine	2-Pyridyl	Cl	H	39
17d	Doxylamine	2-Pyridyl	H	CH_3	2.4
17e	Experimental compound	2-Pyridyl	Cl	CH_3	2.8

ethers.[24] When the 2-pyridyl group was kept constant, a wide variety of substitutions were made for the phenyl group. Only the introduction of a *p*-chlorophenyl group (carbinoxamine), 17c, yielded a compound possessing significantly greater antihistaminic activity. Replacement of the hydrogen on the carbon α to the two aryl groups by a methyl group gave another usefully active compound, doxylamine, 17d.

Many of the aminoalkyl ether H_1 antihistamines (e.g., 17b to e) are optically active.* In general, the two possible enantiomeric forms are quite different in their antihistaminic activity, with the dextrorotatory form being the most active. Exemplary of this phenomenon are the (+) and (−) maleate salts of carbinoxamine. The (+) isomer was found at a dose 1/24 as large as a dose of the corresponding (−) isomer to be equiactive in preventing death when administered i.p. to guinea pigs that were challenged 20 min later with an i.v. lethal injection of histamine.[25] While histamine itself is not optically active, the receptor site for it must possess a degree of asymmetry to allow for this observed difference in activity of these two competitively inhibiting forms of carbinoxamine.

Pharmacokinetics

Drach and Howell[27] have reported that diphenhydramine, administered either orally or i.v. to rhesus monkeys, produced the metabolites shown in Scheme 2 and suggested the metabolic pathway to be as described. By far the major metabolite was the glutamine conjugate of (diphenylmethoxy) acetic acid, 24. A generally similar group of metabolites was observed in studies with dogs except that more of the N-oxide, 20, was formed and (diphenylmethoxy) acetic acid was conjugated with glycine instead of glutamine.[28] Studies in man with orphenadrine, 25, a close analog of diphenhydramine, showed a similar array of metabolites. The conjugate derivatives of *o*-methylbenzhydryloxyacetic acid, 26, and *o*-methylbenzhydrol, 27 with glucuronic acid (and/or ethereal sulfate), however, were the predominant products.[29]

* A molecule containing an asymmetric center, i.e., a carbon atom with four different groups attached to it, is termed "optically active". It has two enantiomeric forms, one of which rotates polarized light to the right and is denoted as the d (+) dextrorotatory form, and the other which rotates polarized light to the left and is denoted as the l (−) levorotatory form.[26]

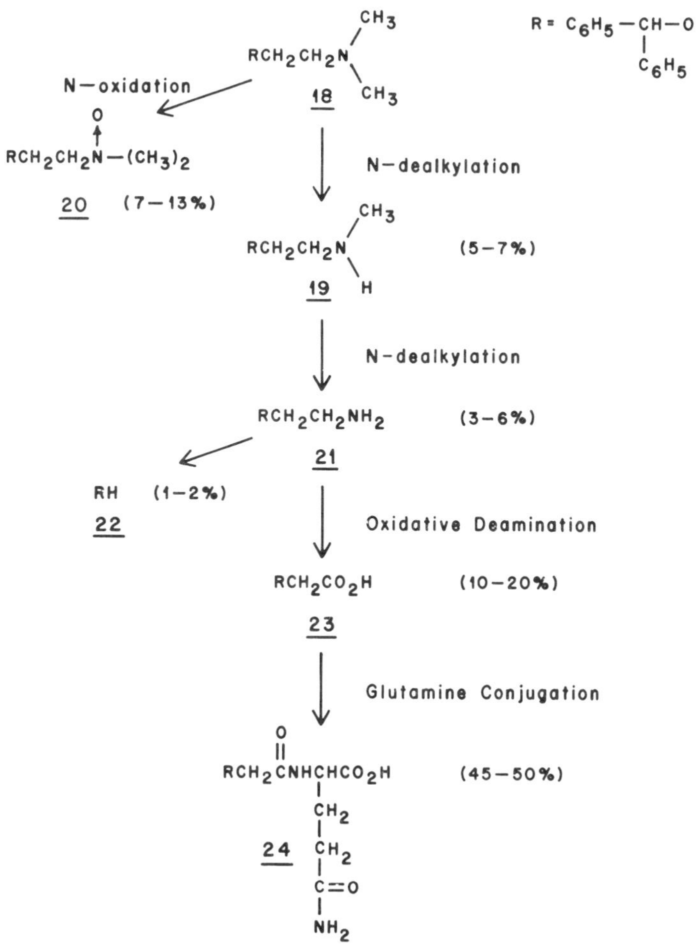

SCHEME 2.

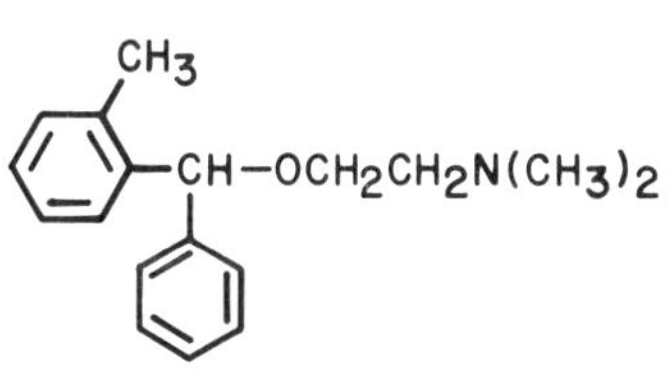

25

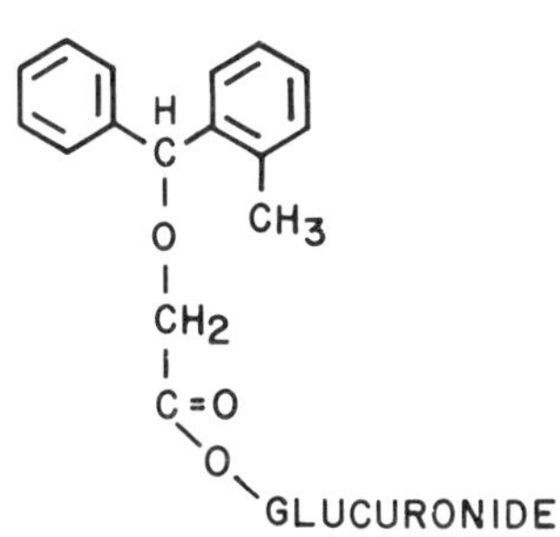

26

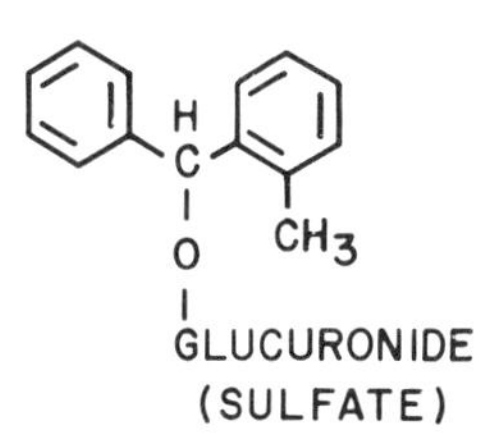

27

Diphenhydramine, the only compound of this class with some pharmacokinetic data reported on it, has a plasma elimination $t_{1/2}$ somewhere between 3[30] and an estimated 7 hr[31] on administration of a 5-mg oral dose. Conflict also arises on the bioavailability of the drug on oral dosing with one group reporting values of 26 to 60%[30] and another 83 to 100%.[31] Very little (4%) unchanged drug was isolated in the urine.

Clinically Useful Members of the Class[16-20]

DIPHENHYDRAMINE HYDROCHLORIDE

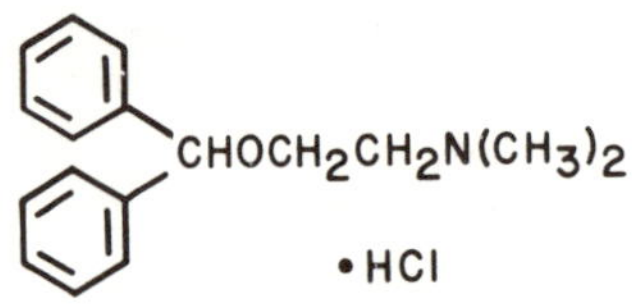

28

Uses

It is used in the treatment of urticaria, hay fever, vasomotor rhinitis, and some dermatoses. It has also been used in the treatment of mild cases of Parkinsonism and for the alleviation of motion sickness.

Dosage and Other Names

A 50-mg oral dose is administered three or four times daily. Other names include Bendylate (Tutag), Eldadryl (Elder), Benadryl (Parke-Davis), SK-Diphenhydramine (Smith Kline & French), component of Benacine (Parke-Davis), component of Bena-Fedrin (Parke-Davis), and component of Caladryl (Parke-Davis).

Toxicity

It has the usual antihistamine side effects, with quite strong sedative action. The oral LD_{50} in rats is 500 mg/kg.

A common side effect of most the H_1 antihistamines is reduction of salivation, causing a condition termed "dry mouth". This property may be related to the close structural similarity that these compounds have with anticholinergics, a class of compounds also known to produce an antisalivary effect. The anticholinergic, trasentine, may be used as an example to show the close structural similarity between this class and a typical H_1 antihistamine.[32]

```
C6H5 O                                C6H5
|    ||                               |
HC — COCH2CH2N(CH2CH3)2               HC—OCH2CH2N(CH3)2
|                                     |
C6H5                                  C6H5

      Trasentine                           Diphenhydramine
```

The only point of significant structural departure is the substitution of an ester group for the ether linkage in diphenhydramine. As a second example, some anticholinergics, such as centrine, lack the ester linkage and resemble compounds such as pheniramine of the propyl amine class of H_1 antihistamines.[33]

```
     O  C6H5                               C6H5
     ||  |                                  |
H2N—C—C—CH2CH(CH3)N(CH3)2              H—C—CH2CH2N(CH3)2
         |                                  |
        C6H5                            2—C5H4N

        Centrine                            Pheniramine
```

Physical-Chemical Data

It is a stable, bitter-tasting, white, crystalline powder, mp 166 to 170°C, which is very soluble in water (1:1) and alcohol (1:2) and slightly soluble in benzene and ether. The pH of a 1% aqueous solution is about 5.5. The pK_a at 20°C is 8.98.

Chemical Name and Formula

The chemical name is 2-(diphenylmethoxy)-*N*,*N*-dimethylethanamine hydrochloride. Its chemical formula is $C_{17}H_{21}NO \cdot HCl$, and the mol wt is 291.82.

DOXYLAMINE SUCCINATE

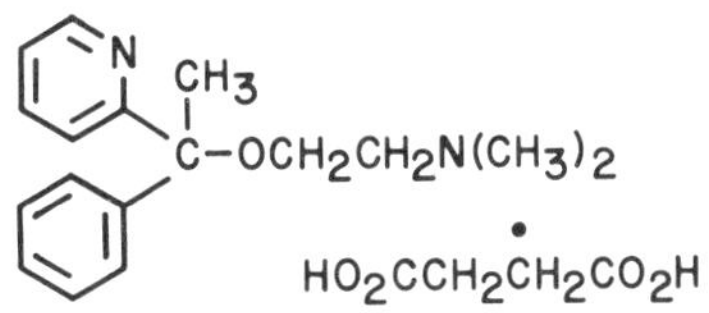

29

Uses

It is a highly effective antihistamine that is used in several night-time cold and cough preparations because of its pronounced sedative effects.

Dosage and Other Names

Several common combination products contain 7.5 mg of this antihistamine per tablespoon, with the adult dose being two tablespoons before bed. Other names are Decapryn succinate (Merrell-National) and Unison (Pfizer).

Toxicity*

It has the usual antihistamine side effects, with pronounced sedative properties. The oral LD_{50} in mice is 470 mg/kg.

Physical-Chemical Data

The crystalline product, with a mp of 100 to 104°C, is readily soluble in water (1:1) and alcohol (1:2). The pH of a 1% aqueous solution is 4.9 to 5.1.

Chemical Name and Formula

The chemical name is *N*,*N*-dimethyl-2-[1-phenyl-1-(2-pyridinyl)ethoxy]ethanamine butanedioate (1:1). The chemical formula is $C_{17}H_{22}N_2O \cdot C_4H_6O_4$. The mol wt is 388.46.

CARBINOXAMINE MALEATE

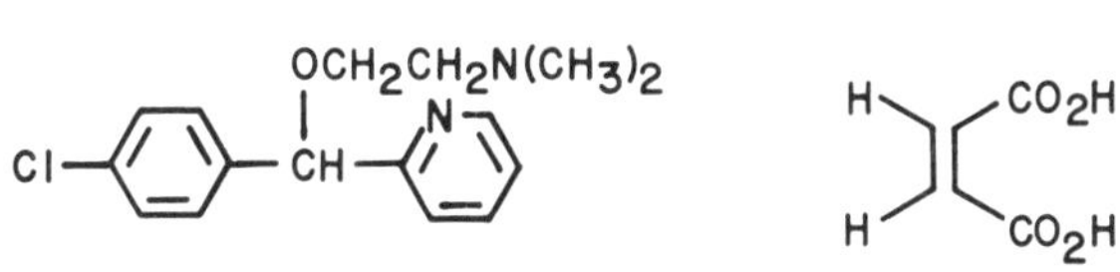

30

* The combination product, Bendectin, which contains doxylamine succinate, has been voluntarily withdrawn by its manufacturer because it has been occasionally, but inconclusively, implicated as a possible teratogen in test animals.

Uses

It is used in the treatment of seasonal and perennial rhinitis.

Dosage and Other Names

Rondec is a combination product from Ross which comes in tablet, syrup, or drop form and contains carbinoxamine maleate (CM) and pseudoephedrine hydrochloride (PH); tablets — 4 mg (CM) and 60 mg (PH); syrup — 4 mg (CM) and 60 mg (PH) in 5 mℓ; one tablet or 5 mℓ q.i.d. Other names include Clistin (McNEILAB), a component of Rondec DM (Ross), and Rondec (Ross).

Toxicity

It produces a low incidence of the usual antihistamine side effects.

Physical-Chemical Data

It is a bitter-tasting, white, crystalline powder, mp 117 to 119°C, which is very soluble in water, ethanol, and chloroform. The pH of a 1% aqueous solution is 4.6 to 5.1 and the pK_a at 20°C is 8.7.

Chemical Name and Formula

Its chemical name is 2-[(4-chlorophenyl)-2-pyridinylmethoxy]-*N*,*N*-dimethylethanamine (Z)-2-butenedioate (1:1). Its formula is $C_{16}H_{19}ClN_2O \cdot C_4H_4O_4$. It has a mol wt of 406.87.

CLEMASTINE FUMARATE

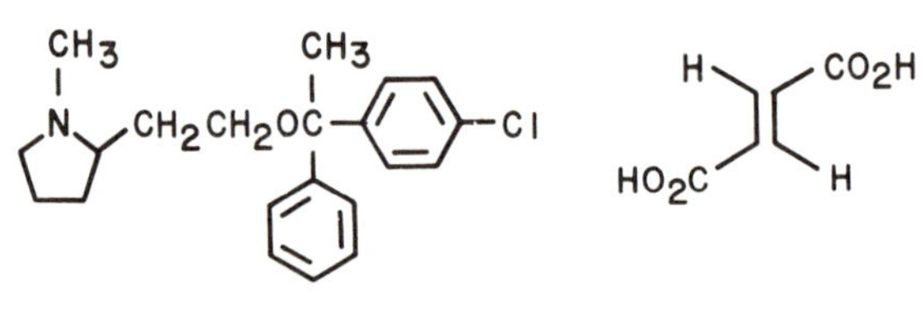

31

Uses

It is a long-acting antihistamine used in the relief of symptoms associated with allergic rhinitis such as sneezing, rhinorrhea, pruritus, and lacrimation. In larger doses it can also be used to treat mild allergic skin manifestions.

Dosage and Other Names

The dose is usually in the form of a 1.34-mg tablet, b.i.d. or a 2.68-mg tablet, one per day. It is also known as Tavist (Sandoz) and Tavegyl (Sandoz).

Toxicity

It has the usual antihistamine side effects. The oral LD_{50} in mice and rats is 730 and 3550 mg/kg, respectively. Intravenously, the LD_{50} is 43 and 82 mg/kg in mice and rats, respectively.

Physical-Chemical Data

It is a faintly yellow, crystalline powder, mp 176 to 181°C (dec) which is very slightly soluble in water, chloroform, and ether.

Chemical Name and Formula

Its chemical name is [R-(R*,R*)]-2-[2-[1-(4-chlorophenyl)-1-phenylethoxy]ethyl]-1-

methylpyrrolidine (E)-2-butenedioate (1:1).* Its chemical formula is $C_{21}H_{26}ClNO \cdot C_4H_4O_4$. Its mol wt is 459.97.

DIPHENYLPYRALINE HYDROCHLORIDE

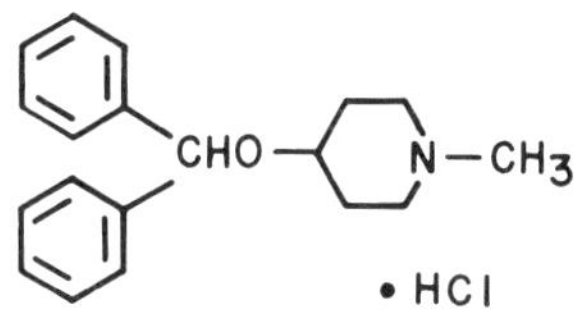

32

Uses

It is a long-acting antihistamine which is effective in the treatment of perennial and seasonal allergic rhinitis, vasomotor rhinitis, and various allergic skin reactions.

Dosage and Other Names

It is administered in a dose of one 5.0-mg tablet one or two times per day. It is also known as Diafen (Riker) and Hispril (Smith Kline & French).

Toxicity

It has the usual antihistamine side effects, with fairly pronounced anticholinergic and sedative properties.

Physical-Chemical Data

The crystalline compound, mp 206°C, is soluble in water and ethanol and practically insoluble in ether and benzene.

Chemical Name and Formula

Its chemical name is 4-(diphenylmethoxy)-1-methylpiperidine hydrochloride. Its chemical formula is $C_{19}H_{23}NO \cdot HCl$, and it has a mol wt of 317.86.

Propylamine Derivatives

Structure-Activity Studies

The propylamine class of antihistamines, which is related to the earlier classes by replacement of the nitrogen or oxygen side chain connecting atom by carbon, proved to be very potent and relatively nontoxic. In a study of oral antihistaminic activity in the guinea pig a large number of compounds with general structure 33 were tested.[24] The following conclusions can be drawn from this study (see Table 3):

1. Halogenation (chlorine, 33b, or bromine) of the *para* position of the phenyl ring greatly enhances the antihistaminic activity, but halogenation at other positions decreases or eliminates the activity.

* The R and S system devised by Cahn, Ingold, and Prelog is used to specify the arrangement of functional groups and consequently the absolute stereochemical configuration about an asymmetric center. If the groups, arranged according to a set of priority rules, are ordered in a fashion that allows counting from highest to lowest priority in a clockwise manner, the center is assigned an R (right = rectus) configuration. Conversely, in a counterclockwise arrangement of highest to lowest priority groups, the center is assigned an S (left = sinister) configuration. Please consult Reference 22 for group priority rules and further elaboration on the subject.

Table 3
RELATIVE ANTIHISTAMINIC ACTIVITY OF PHENIRAMINE AND ITS ANALOGS[24]

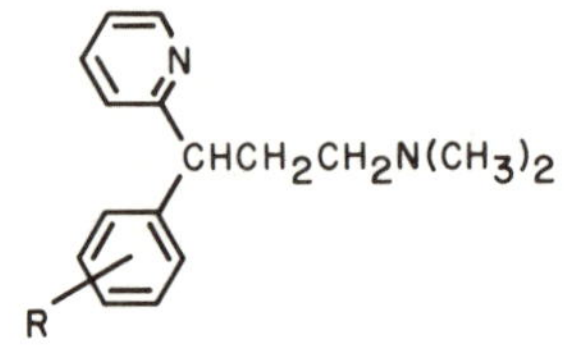

Compound	Name	R	Relative oral antihistaminic activity in guinea pigs
33a	Pheniramine	H	1
33b	Chlorpheniramine	4-Cl	19
33c	Experimental compound	4-CH_3	5
33d	Experimental compound	4-$(CH_3)N$	2

2. A *p*-methyl, 33c, or *p*-dimethylamino group, 33d, in the phenyl ring also increases activity.
3. In the base compound, pheniramine, where R=H (33a), replacement of the dimethylamino group with a diethylamino group removes all antihistaminic activity.[24]
4. Replacement of the α-pyridyl group or the phenyl group with other pyridyl isomers, with 2-thienyl, 2-thiazolyl, or α-naphthyl, gives compounds less active than pheniramine.[24]

While histamine itself does not contain an asymmetric center, the various propyl amine antihistamines, 34, do and their degree of activity is very much dependent upon their stereochemistry. As an example, the d isomer of chorophenirامine possesses almost all of the antihistaminic

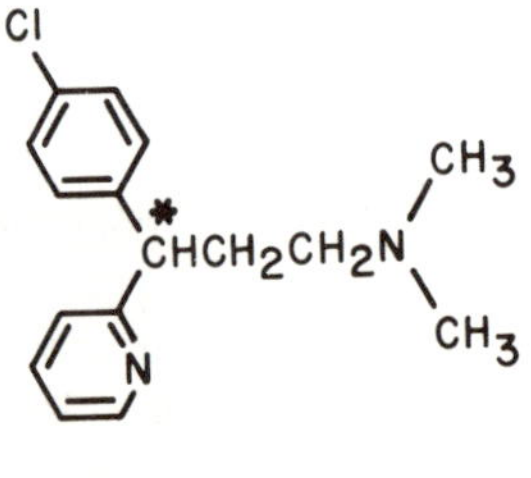

34

activity of the racemic dl mixture. In vivo in the guinea pig the effective protection of the d isomer was two to three times that of the dl mixture on i.v. histamine administration and slightly less than two times on aerosol administration. The l isomer was only 1% as effective as the d isomer in this in vivo i.v. challenge.[34]

This stereochemical feature also exhibits itself in other related structures such as pyrrobutamine, 35, which shows marked differences in the antihistaminic activity of its geometrical (*cis/trans*) isomers[35] and triprolidine, 36, which has almost all of its antihistaminic activity in the *trans* isomer.[36]

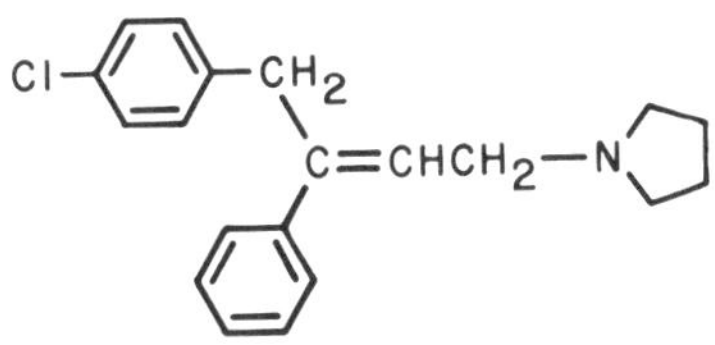

35

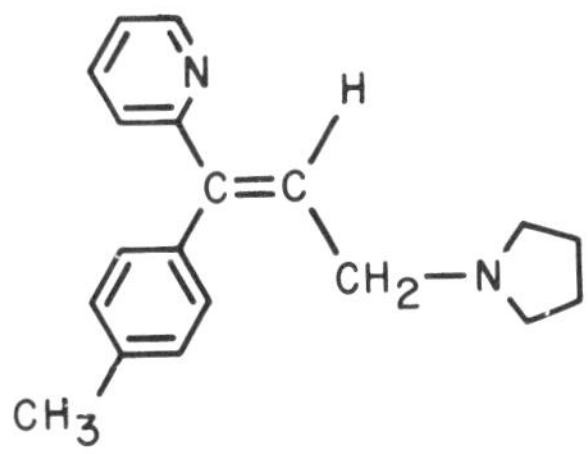

36

Pharmacokinetics

The metabolism of the diarylpropylamines is exemplified by studies of pheniramine and its halogenated analogs.[37,38] When chlor-, 37a, or brompheniramine, 37b, was administered either orally or i.v. to dogs and man, the mono-, 37c, and didesmethyl, 37d, derivatives were the major metabolites.[37,38]

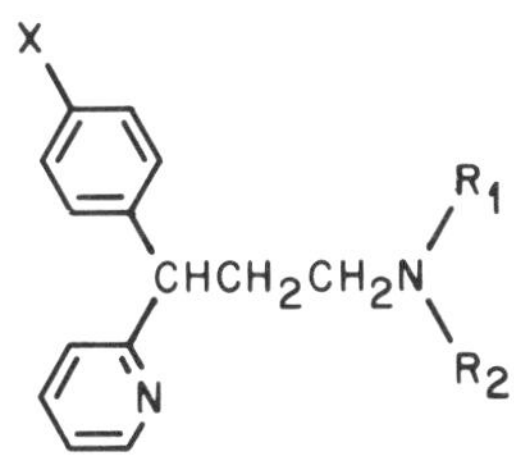

37a $R_1 = R_2 = CH_3$, X = Cl
37b $R_1 = R_2 = CH_3$, X = Br
37c $R_1 = H$, $R_2 = CH_3$, X = Cl or Br
37d $R_1 = R_2 = H$, X = Cl or Br

Metabolic studies of brompheniramine have shown the propionic acid derivative, 38a, and its glycine conjugate, 38b, to be formed in lesser amounts in both man and dogs. The N-oxide, 39, was only observed as a metabolite in dogs.[37]

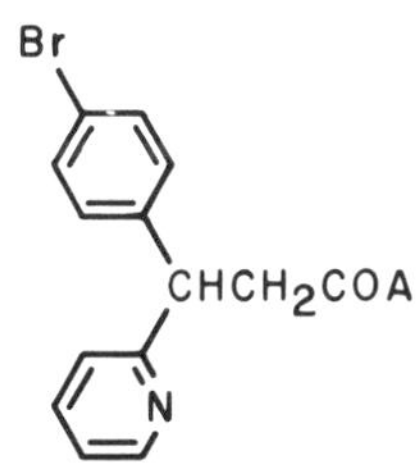

38a A = OH
38b A = $NHCH_2CO_2H$

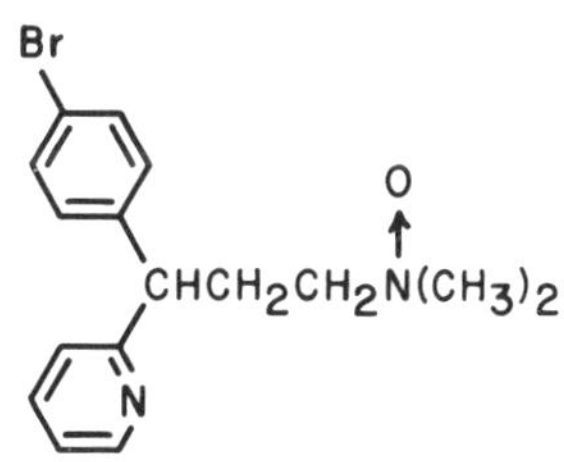

39

The pharmacokinetics and bioavailability of chlorpheniramine have been fairly extensively studied. In adults, the plasma $t_{1/2}$ on oral dosing has been reported to be 12 to 15 hr and on i.v. administration, 28 hr.[38] In children, the oral plasma $t_{1/2}$ has been reported to be 13.1 ± 6.3 hr.[39] The drug was reported to be 72% bound to plasma protein.[38] In children, histamine-induced wheal and flare responses were suppressed when chlorpheniramine serum levels were 4.1 to 10.0 ng/mℓ.[39] Oral bioavailability was reported to be low (47%) after administration of a solution and 35% after administration of a tablet.[40]

A long plasma $t_{1/2}$ (24.9 ± 9.3 hr) has also been reported for normal adults treated with brompheniramine.[41] The size of histamine-induced wheals was significantly suppressed at brompheniramine plasma levels of 8.2 ± 2 ng/mℓ.[41]

Clinically Useful Members of the Class[16-20]

PHENIRAMINE MALEATE

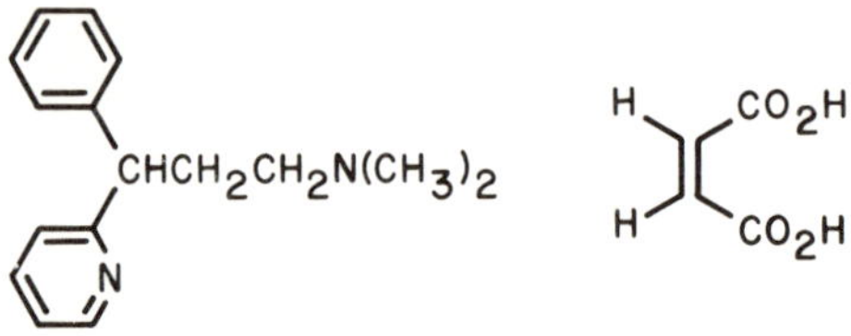

40

Uses

It is used in combination with several other drugs in many common cold preparations.

Dosage and Other Names

The combination products generally contain 6 to 12 mg per tablet or dose and are administered three or four times daily. It is also known as Inhiston (Schering-Plough) and Trimeton (Schering-Plough).

Toxicity

It has the usual antihistamine side effects. The LD_{100} s.c. in rats is 500 mg/kg.

Physical-Chemical Data

The white, crystalline solid, mp 107°C, is freely soluble in water and alcohol and slightly soluble in ether and benzene. The pH of a 1% aqueous solution is between 4.3 and 4.9. The pK_a at 20°C is 9.23.

Chemical Name and Formula

Its chemical name is *N,N*-dimethyl-γ-phenyl-2-pyridinepropanamine (Z)-2-butenedioate (1:1). Its chemical formula is $C_{16}H_{20}N_2 \cdot C_4H_4O_4$. Its mol wt is 356.34.

CHLORPHENIRAMINE MALEATE

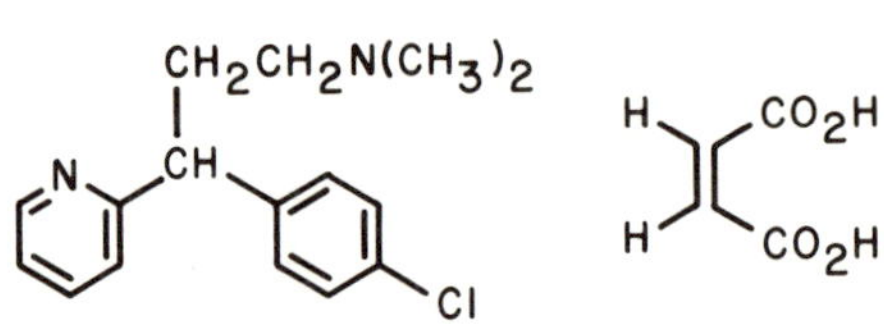

41

Uses

Chloropheniramine maleate, as a racemic mixture, is a long-acting potent antihistamine that is used alone or in combination for the relief of perennial and seasonal allergic rhinitis, allergic skin manifestations of urticaria and angioedema, and for adjunctive treatment of anaphylactic reactions after the acute manifestations have been controlled.

The dextro rotatary isomer, dexchlorpheniramine, is about twice as potent as racemic chlorpheniramine and has similar uses and liabilities.

Dosage and Other Names

It is the most widely used antihistamine and is administered alone in tablet (1 tablet — 4 mg every 4 to 6 hr) or syrup form or in a wide variety of combination cold preparations. It is also known as Alermine (Tutag), Chlor-Trimeton (Schering-Plough), Teldrin (Menley & James), Telodron (Norden), Histaspan (USV Pharm.), H-Stadur (Cooper), Nosocosed (Alcon), component of Coricidin (Schering-Plough), component of Cotylenol (McNEILAB), component of Demazin (Schering-Plough), component of Hycomine Compound (Endo), component of Intensin (McNEILAB), component of Triaminicin Chewables (Dorsey), component of Novahistine (Dow), component of Deconamine (Berlex), component of Coriforte (Schering-Plough), component of Corilin (Schering-Plough), component of Naldecon (Bristol), component of Diathal [Veterinary] (Schering-Plough), and component of Metrevet [Veterinary] (Schering-Plough).

Dexchlorpheniramine maleate is administered as 2-mg tablets four times per day. It is also known as Polaramine (Schering-Plough).

Toxicity

It has the usual antihistamine side effects.

Physical-Chemical Data

It is a white, crystalline solid, mp 130 to 135°C, which dissolves in 3.4 parts of water and 10 parts of alcohol. The pH of a 2% aqueous solution is 4 to 5. The pK_a at 20°C is 9.16.

Chemical Name and Formula

Its chemical name is γ-(4-chlorophenyl)-*N*,*N*-dimethyl-2-pyridinepropanamine (Z)-2-butenedioate (1:1). Its chemical formula is $C_{16}H_{19}ClN_2 \cdot C_4H_4O_4$. Its mol wt is 390.87.

BROMPHENIRAMINE MALEATE

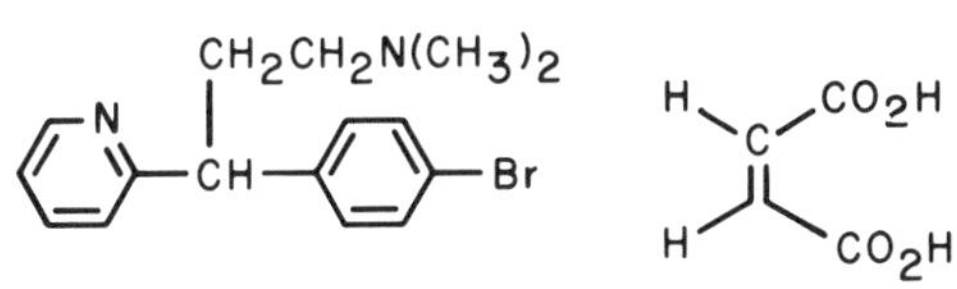

42

Uses

Brompheniramine maleate, as a racemic mixture, is used in combination with other drugs in a variety of preparations for treatment of the common cold and allergic rhinitis.

The dextrorotatory isomer, dexbrompheniramine, is about twice as potent as racemic brompheniramine and has similar uses and liabilities.

Dosage and Other Names

The dose in various combination products is 4 to 8 mg per tablet administrated three to four times per day. It is also known as Dimetane (Robins).

Dexbrompheniramine maleate (2 mg) is used in combination with pseudoephedrine sulfate and is administered three to four times per day. It is also known as Disophrol (Schering-Plough).

Toxicity

It has the usual antihistamine side effects.

Physical-Chemical Data

The white, crystalline material, mp 132 to 134°C, is rather soluble in water and somewhat less soluble in alcohol. The pH of a 2% aqueous solution is about 5. The pK_a at 20°C is 9.17.

Chemical Name and Formula

Its chemical name is γ-(4-bromophenyl)-*N*,*N*-dimethyl-2-pyridinepropanamine (Z)-2-butenedioate (1:1). Its chemical formula is $C_{16}H_{19}BrN_2 \cdot C_4H_4O_4$. Its mol wt is 435.32.

PYRROBUTAMINE PHOSPHATE

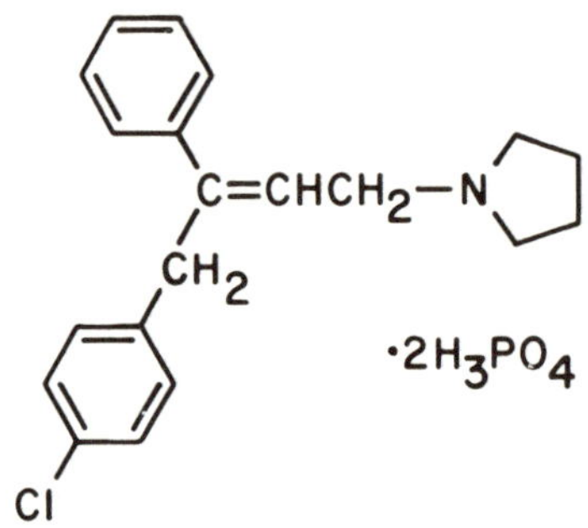

43

Uses

It is a long-acting antihistamine used for the treatment of seasonal and perennial allergic rhinitis and in vasomotor rhinitis.

Dosage and Other Names

It is administered in a 15-mg dose, two to three times daily. It is also known as Pyronil (Lilly).

Toxicity

It has the usual antihistamine side effects.

Physical-Chemical Data

The white crystalline solid, mp 129.5 to 130°C, is soluble to the extent of 10% in warm water and 5% in alcohol. It is practically insoluble in chloroform and ether.

Chemical Name and Formula

Its chemical name is 1-[4-(4-chlorophenyl)-3-phenyl-2-butenyl]pyrrolidine phosphate (1:2). Its chemical formula is $C_{20}H_{22}ClN \cdot 2H_3PO_4$. Its mol wt is 507.84.

TRIPROLIDINE HYDROCHLORIDE

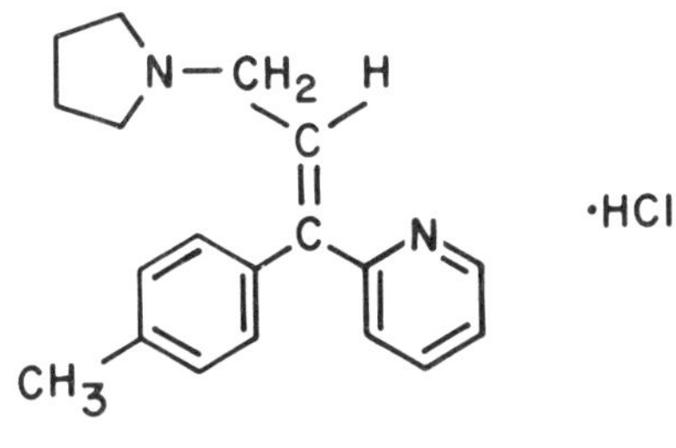

44

Uses

It is used for the treatment of seasonal and perennial allergic rhinitis and in vasomotor rhinitis.

Dosage and Other Names

It is administered in combination with other drugs at a dose of 2.5 mg per tablet given three to four times daily. It is also known as Actidil (Burroughs Wellcome).

Toxicity

It has the usual antihistamine side effects.

Physical-Chemical Data

The crystalline solid, mp 116 to 118°C, is moderately soluble in water, ethanol, and methanol.

Chemical Name and Formula

Its chemical name is (E)-2-[1-(4-methylphenyl)-3-(1-pyrrolidinyl)-1-propenyl]pyridine monohydrochloride monohydrate. Its chemical formula is $C_{19}H_{22}N_2 \cdot HCl \cdot H_2O$. Its mol wt is 332.87.

Cyclizines

Structure-Activity Studies

The cyclizines are a class of *N,N'*-disubstituted piperazines. Chlorcyclizine, 45a, is a potent antihistamine with a slow onset but prolonged duration of activity.[42] Its des-chloro analog, cyclizine, 45b, however, has only limited antihistaminic activity and is useful mainly for its antimotion sickness activity.[43] Two other analogs of chlorcyclizine, meclizine, 45c, and buclizine, 45d, are also potent, long-acting antihistamines, but have found more use as antinauseants.[42]

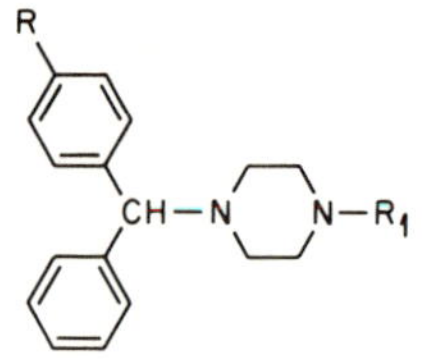

		R	R_1
45a	Chlorcyclizine	Cl	CH_3
45b	Cyclizine	H	CH_3
45c	Meclizine	Cl	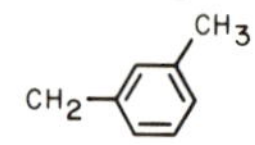
45d	Buclizine	Cl	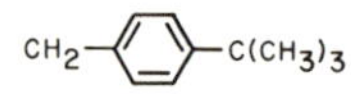

Pharmacokinetics

Several metabolic pathways have been observed for the general cyclizine class. Kuntzman et al.[44,45] have shown that, on oral administration, cyclizine, 46a, and chlorcyclizine, 46b, are demethylated in rats, dogs, and man to norcyclizine, 48a, and norchlorcyclizine, 48b. The hydroxymethyl derivative, 47, may be an intermediate in the formation of norchlorcyclizine.[44,46] Kuntzman has also shown that chlorcyclizine *N*-oxide[44] is a metabolite of 46b in rats and man, but this metabolite does not appear to be an intermediate in the conversion of chlorcyclizine, 46b, to norchlorcyclizine, 48b.[44] Interestingly, a significant sex difference is observed in the metabolism of chlorcyclizine in rats. On oral administration the male metabolizes 46b to 48b at a rate eight times greater than the female.[45] No significant antihistaminic activity is observed for either norchlorcyclizine or chlorcyclizine N-oxide.[45]

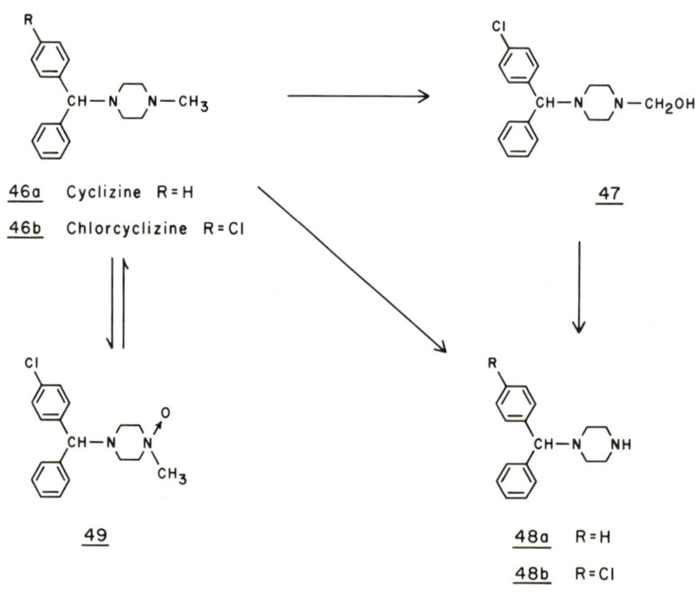

Clinically Useful Members of the Class[16-20]

CHLORCYCLIZINE HYDROCHLORIDE

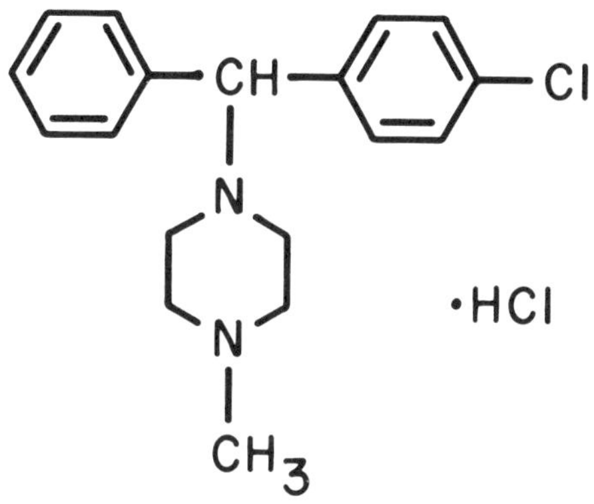

50

Uses

It is a long-acting antihistamine used in the treatment of allergic and vasomotor rhinitis, hay fever, and acute and subacute sinusitis.

Dosage and Other Names

The oral dosage is 25 mg three times per day. It is also known as Di-Paralene (Abbott), Perazil Histantin (Burroughs Wellcome), component of Fedrazil (Burroughs Wellcome), and component of Mantadil (Burroughs Wellcome).

Toxicity

It has the usual antihistamine side effects. The i.p. LD_{50} in mice is 137 mg/kg. It has been found on oral dosing to be teratogenic in animals.[77]

Physical-Chemical Data

It is a light sensitive, white, crystalline powder, mp 226 to 227°C, which is freely soluble in water (1:2) and soluble in alcohol (1:11). The pH of a 1% aqueous solution is between 4.8 and 5.5. The pK_a at 20°C is 8.15.

Chemical Name and Formula

Its chemical name is 1-(4-chlorophenyl)phenylmethyl-4-methylpiperazine monohydrochloride. Its chemical formula is $C_{18}H_{21}ClN_2 \cdot HCl$. Its mol wt is 337.29.

Tricyclic Structures

Structure-Activity Studies

The last broad class of H_1 antihistamines consists of compounds which possess one of a variety of tricyclic ring systems. These compounds can be thought of as analogs of the earlier-mentioned classes where the aromatic rings are now tied together at two junction points, A and B, by carbon, carbon and heteroatom, and heteroatom linkages, 51.

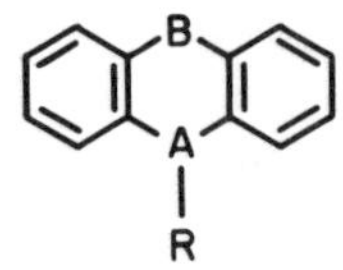

51

A—R = CH—OR, N—R, CH—R

B = Heteroatom, CH═CH, CH_2—CH_2,

$COCH_2$, CH_2—Heteroatom, etc.

Aminoalkyl side chains similar to those found active in the ethylene diamine, and alkylaminopropyl classes are again found active in this class.

Phenothiazine served as the core tricyclic nucleus in the earliest investigations. The most prominent member of the phenothiazine group is promethazine, 52.

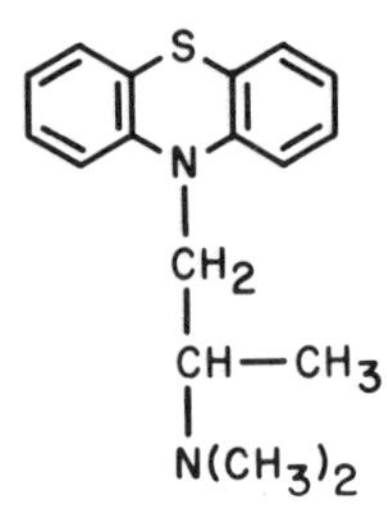

52

This compound is capable of counteracting 1500 lethal doses of histamine in laboratory animals.[23] In man, when administered orally, it proved to be about seven times more active than pyrilamine-mepyramine in inhibiting histamine-induced wheal formation and had a 3.75 times longer duration of action, which allowed for a once-a-day drug administration.[47] Removal of the 2-methyl group from promethazine gave 10-(2-dimethylamino-ethyl)phenothiazine which was 75 times less active than promethazine.[23] On resolution of promethazine into its (+) and (−) enantiomers, it was found that both forms had similar antihistaminic activity and toxicity.

Other side chains have been attached to the phenothiazine nucleus. Pyrathiazine, 53a, has a pyrrolidinylethyl group appended. This long-acting compound is less active than promethazine and about as potent as tripelennamine.[42,43] Trimeprazine, 53b, which has a 3-(dimethylamino)-2-methylpropyl group attached, is reported to have antihistaminic activity which is 1.5 to 5 times that of promethazine.[42] Methdilazine, 53c, possesses a (1-methyl-3-pyrrolidinyl)methyl side chain. In studies on the guinea pig intestine it is reported to have antihistaminic activity four times greater than promethazine.[43]

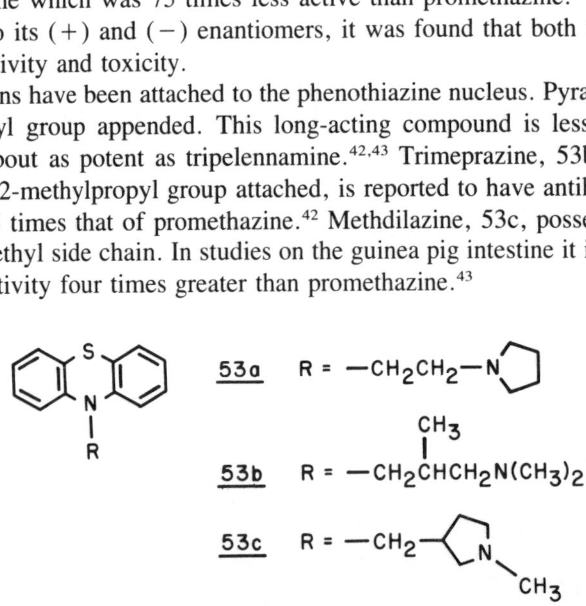

A second important general class of tricyclic structures is connected at point B by a vinyl group (cyproheptadine), 54a, an ethylene group, 54b, or a sulfur atom, 54c. These compounds possess potent H_1 antihistaminic and antiserotonin activity.

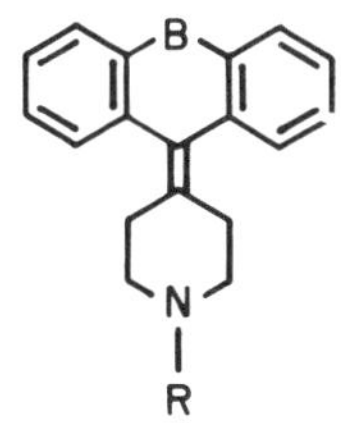

	B	R
54a	HC=CH	CH_3
54b	H_2C-CH_2	CH_3
54c	S	CH_3

Cyproheptadine, 54a, and the sulfur bioisoteric thioxanthene, 54c, have been shown to be equiactive.[48] Substitution of an oxygen atom at point B or alkyl groups other than methyl or ethyl for R, however, gave compounds possessing reduced antihistaminic activity.[48] Replacement of one of the aryl groups of 54b with a pyridyl moiety gave another interesting antihistaminic compound, azatadine, 55.

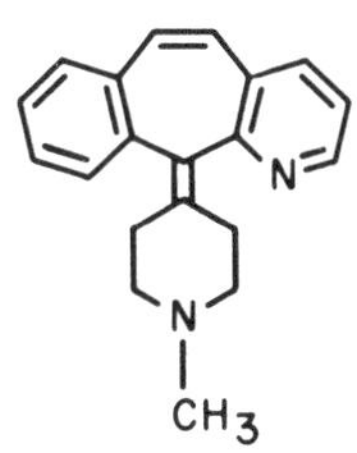

55

Oxidation of the ethylene group of 54b to a ketomethylene unit and substitution of a thiophene unit for one of the aryl rings yields ketotifen, 56. This compound lacks antiserotonin activity but is a very potent antihistamine with antianaphylactic properties.[49]

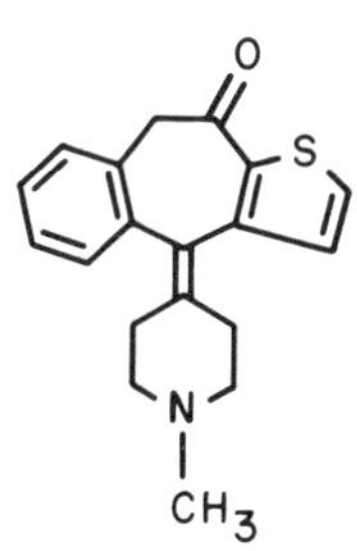

56

Pharmacokinetics

Tricyclic phenothiazine antihistamines such as promethazine, 57a, and methdilazine, 57b, are probably metabolized in a fashion similar to the psychotherapeutic agent, promazine, 57c.

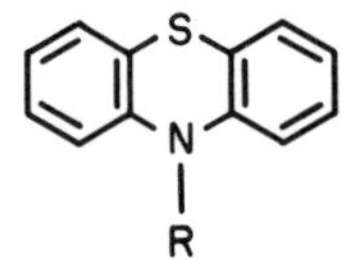

57a Promethazine $R = CH_2CH(CH_3)N(CH_3)_2$

57b Methdilazine

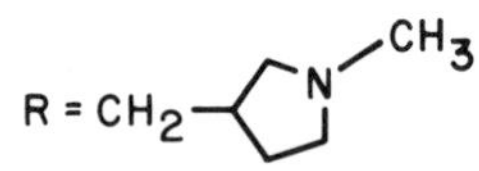

57c Promazine $R = CH_2CH_2CH_2N(CH_3)_2$

Promazine, 58a, on oral administration in dogs and man, undergoes mono, 58b, and di-demethylation, 58c, of the tertiary side chain.[46] These two metabolites, along with promazine itself, can then undergo oxidation of the

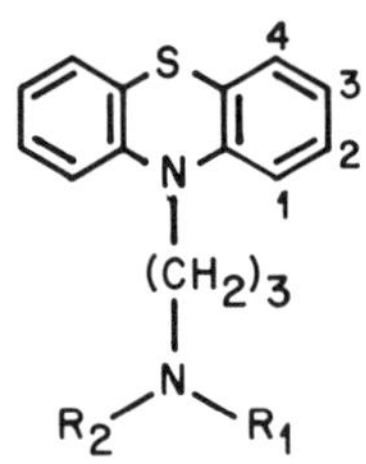

58a $R_1 = R_2 = CH_3$

58b $R_1 = H, R_2 = CH_3$

58c $R_1 = R_2 = H$

sulfur atom to give the corresponding sulfoxides (S = O).[50] Further aromatic hydroxylation at the 3-position of all of these compounds can also occur to give the phenols (–OH) which are isolated primarily as their glucuronides.[50] Two other metabolites, the tertiary amine oxide, 59,[50] and phenothiazine sulfoxide, 60,[51] (minus the side chain) have also been isolated.

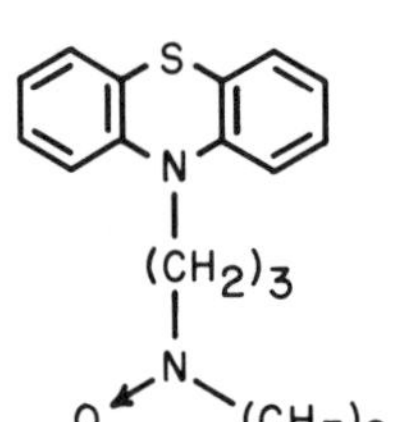

59

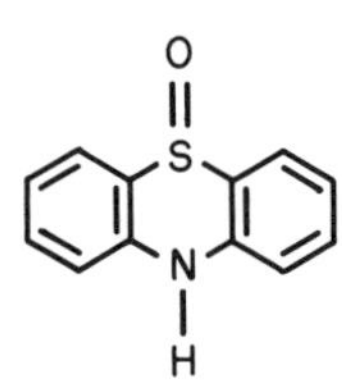

60

Another group of antihistaminic tricyclic compounds is represented by cyproheptadine, 61, azatadine, 62, and ketotifen, 63. The metabolism

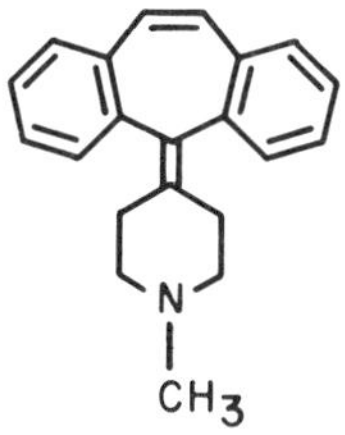

61

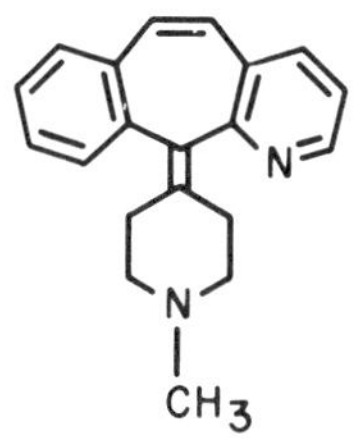

62

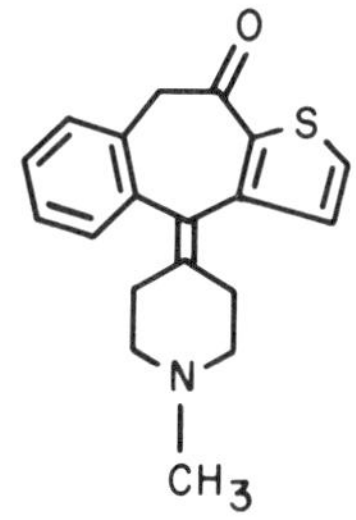

63

of cyproheptadine, representative of this group, has been actively studied in man and several other animal species. In man, on oral administration, the principal metabolite has been tentatively identified as a "quaternary ammonium glucuronide like conjugate of cyproheptadine", 64.[52] Other metabolic pathways produced products with aromatic ring hydroxylation and

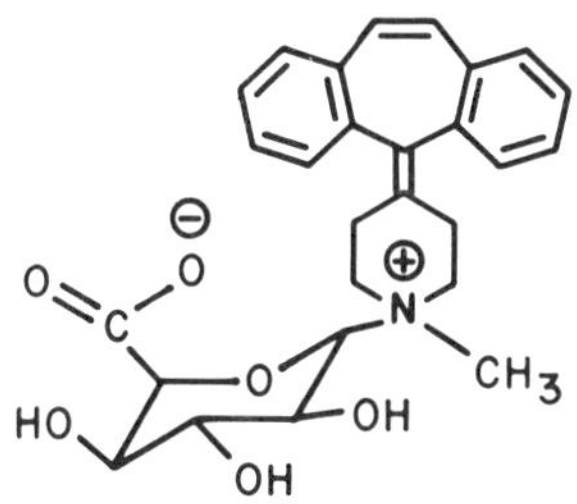

64

N-demethylation, 65, or oxidation of the piperidine ring, 66. These metabolites were isolated as the glucuronide conjugates.[52]

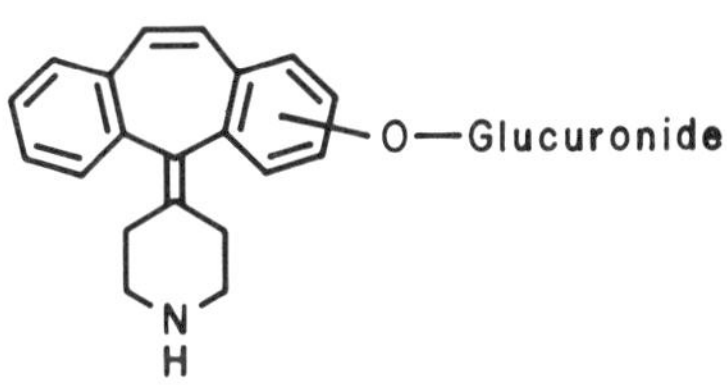

65

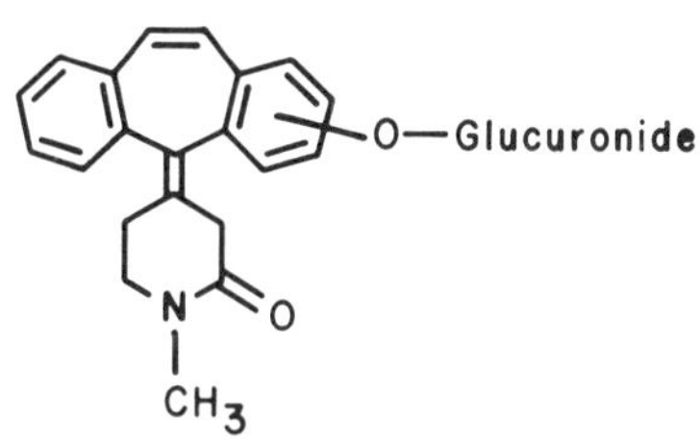

66

In animal species other than man, cyproheptadine is metabolized to the 10,11-epoxide analog, 67, and this is further converted to the *trans*, 68, and *cis*, 69, diols.

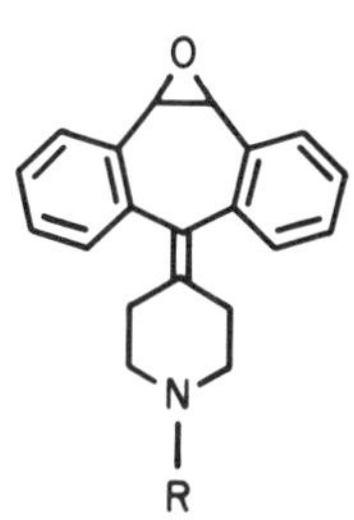

67a R = H

67b R = CH_3

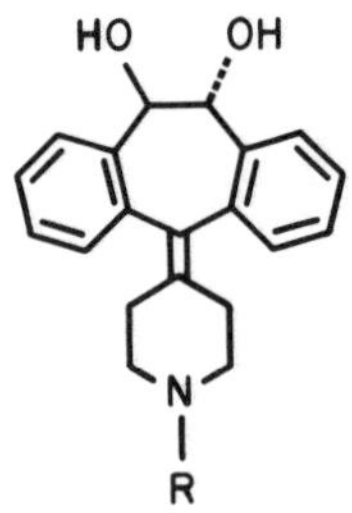

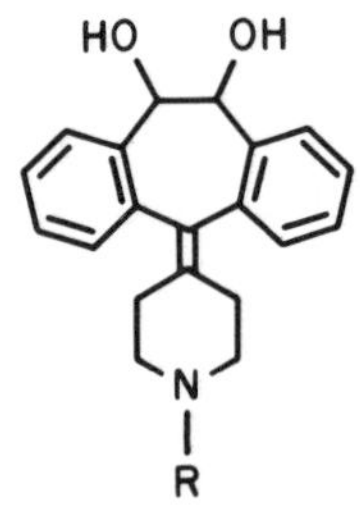

68 R = CH_3 69 R = CH_3

The epoxide, 67a, is a major metabolite in rats,[53-55] epoxide, 67b, and diol, 68, a major metabolite in dogs[53] and diol 69, important in cats.[53]

Clinically Useful Members of the Class[16-20]

PROMETHAZINE HYDROCHLORIDE

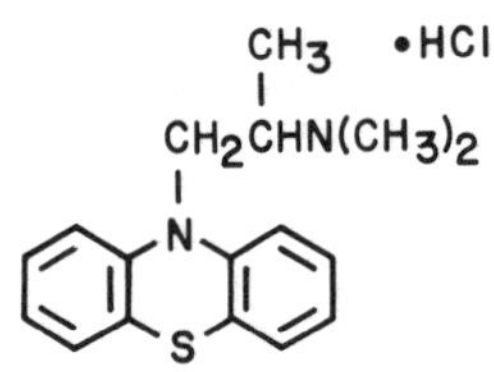

70

Uses

It has pronounced long-acting antihistaminic, antiemetic, and sedative activity and is used in the treatment of perennial and seasonal allergic rhinitis, vasomotor rhinitis, allergic skin reactions, prevention and control of nausea and vomiting, prevention of motion sickness, and for mild sedation.

Dosage and Other Names

The normal oral dose for allergy administration is one 25-mg tablet before bed. For motion sickness, one 25-mg tablet twice a day is often required. It is also known as Ganphen (Tutag), Fellozine (O'Neal, Jones & Feldman), Remsed (Endo), Phenergan (Wyeth), and component of Mepergan (Wyeth).

Toxicity

It has the usual side effects of antihistamines. The s.c. LD_{50} in rats is 400 mg/kg. The i.v. LD_{50} in mice is 55 mg/kg.

Physical-Chemical Data

It is a white, crystalline material, mp 230 to 232°C, that is very soluble in water and soluble in alcohol and chloroform. The pH of a 10% aqueous solution is 5.3. The pK_a at 20°C is 9.08.

Chemical Name and Formula

Its chemical name is *N,N*,α-trimethyl-10H-phenothiazine-10-ethanamine monohydrochloride. Its chemical formula is $C_{17}H_{20}N_2S \cdot HCl$. Its mol wt is 320.88.

METHDILAZINE HYDROCHLORIDE

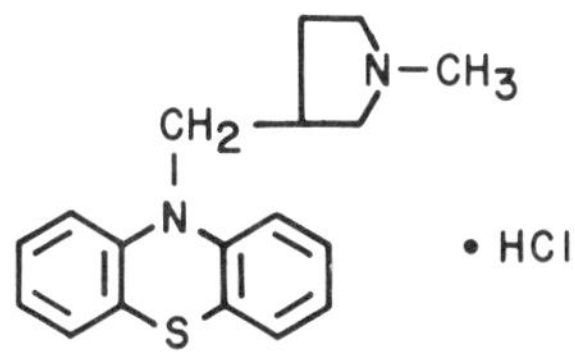

71

Uses

It is used for the symptomatic relief of pruritic symptoms in urticaria.

Dosage and Other Names

It is administered in an 8-mg tablet two to four times per day. It is also known as Tacaryl Hydrochloride (Westwood).

Toxicity

It has the usual side effects of antihistamines plus those of the phenothiazines.

Physical-Chemical Data

It is a crystalline compound, mp 187.5 to 189°C, which is freely soluble in water, alcohol, and chloroform. The pH of a 1% solution in water is 4.6 to 4.8.

Chemical Name and Formula

Its chemical name is [(1-methyl-3-pyrrolidinyl)methyl]-10H-phenothiazine monohydrochloride. Its chemical formula is $C_{18}H_{20}N_2S \cdot HCl$. Its mol wt is 332.89.

CYPROHEPTADINE HYDROCHLORIDE

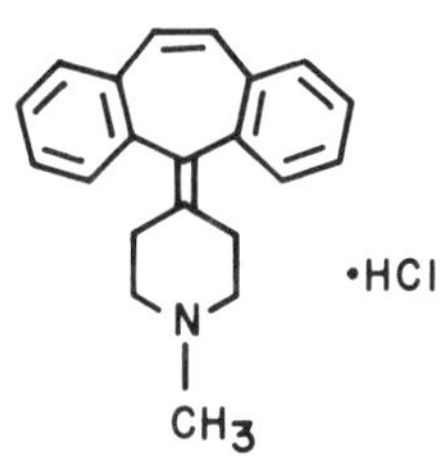

72

Uses

The compound has pronounced antiserotonin, as well as antihistaminic activity and is used for perennial and seasonal allergic rhinitis, vasomotor rhinitis, and allergic skin reactions.

Dosage and Other Names

A dose of one 4-mg tablet three to four times a day is administered. It is also known as Periactin (Merck Sharp & Dohme) and component of Dronactin (Merck Sharp & Dohme).

Toxicity

It has the usual side effects of antihistamines, with sedation being a significant problem on initiation of the drug. Cyproheptadine has presumably caused weight gain and increased growth in children, but the mechanism for this action is not well understood.

Physical-Chemical Data

It is a white to slightly yellow crystalline solid, mp 252.6 to 253.6°C (dec) and is slightly soluble in water.

Chemical Name and Formula

Its chemical name is 4-(5H-dibenzo[a,d]cyclohepten-5-ylidene)-1-methylpiperidine hydrochloride sesquihydrate. Its chemical formula is $C_{21}H_{21}N \cdot HCl \cdot 11/2\ H_2O$. Its mol wt is 350.89.

AZATADINE MALEATE

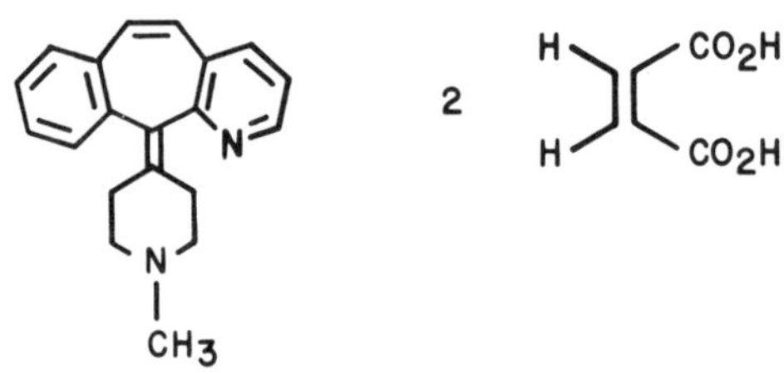

73

Uses

It has antihistamine, antiserotonin, and anticholinergic activity and is used in the treatment of perennial and seasonal allergic rhinitis and chronic urticaria.

Dosage and Other Names

The oral adult dose is one or two 1-mg tablets twice a day. It is also known as Optimine (Schering-Plough) and Idulian (Unilabo, Fr.).

Toxicity

It has the usual side effects of antihistamines.

Physical-Chemical Data

The crystalline solid, mp 145 to 153°C, is soluble in water, methanol, and chloroform and insoluble in ether.

Chemical Name and Formula

Its chemical name is 6,11-dihydro-11-(1-methyl-4-piperidinylidene)-5H-benzo[5,6]cyclohepta[1,2-b]pyridine (Z)-2-butenedioate (1:2). Its chemical formula is $C_{20}H_{22}N_2 \cdot 2(C_4H_4O_4)$. Its mol wt is 522.55.

KETOTIFEN FUMARATE

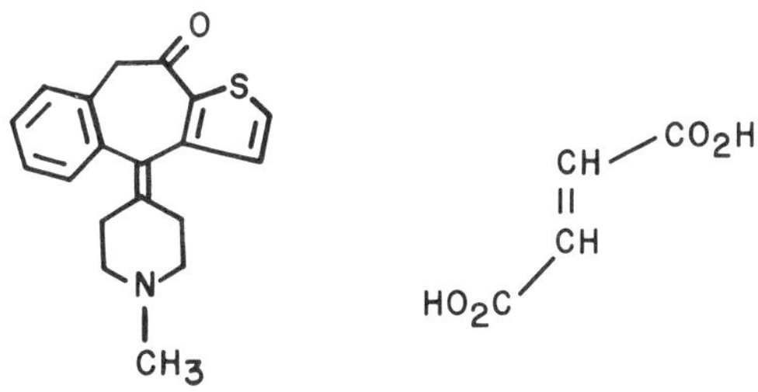

74

Uses

It is used in the prophylactic treatment of bronchial asthma, urticaria, and other dermatoses.

Dosage and Other Names

It is supplied in 1-mg capsules and syrup containing 0.2 mg/mℓ. The adult oral dose is one capsule twice daily. It is also known as Zaditen (Sandoz).

Toxicity

It has the usual antihistamine side effects.

Physical-Chemical Data

It is a crystalline material with a mp of 192°C (dec).

Chemical Name and Formula

Its chemical name is 4,9-dihydro-4-(1-methyl-4-piperidylidene)-10H-benzo[4,5]cyclohepta-[1,2-b]thiophen-10-one (E)-2-butenedioate (1:1). Its chemical formula is $C_{19}H_{19}NOS \cdot C_4H_4O_4$. Its mol wt is 425.50.

Miscellaneous Compounds

Clinically Useful Members of the Class[16-20]

DIMETHINDENE MALEATE

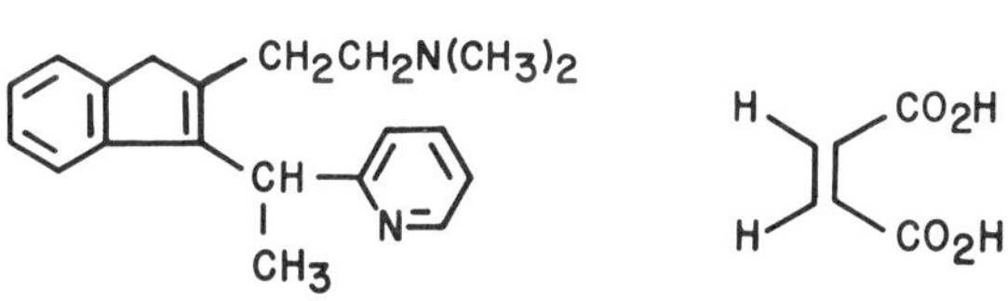

75

Uses

It is a long-acting antihistamine which is effective in the treatment of perennial and seasonal allergic rhinitis, vasomotor rhinitis, and various allergic skin reactions.

Dosage and Other Names

It is administered in a dose of one 2.5-mg tablet one or two times per day. It is also known as Forhistal Maleate (Ciba Geigy).

Pharmacokinetics

The elimination half-life of a 3.0-mg dose of the drug has been reported to be 5.89 hr.[56]

Toxicity

It has the usual antihistamine side effects. The i.v. LD_{50} in rats is 26.8 mg/kg and the oral LD_{50} is 618.2 mg/kg.

Physical-Chemical Data

The white, light-sensitive crystalline powder, mp 159 to 161°C, is free of the unpleasant, bitter taste of many antihistamines. It is freely soluble in methanol, soluble in chloroform, and sparingly soluble in water.

Chemical Name and Formula

Its chemical name is *N,N*-dimethyl-3-[1-(2-pyridinyl)ethyl]-1H-indene-2-ethanamine (Z)-2-butenedioate (1:1). Its chemical formula is $C_{20}H_{24}N_2 \cdot C_4H_4O_4$. Its mol wt is 408.50.

PHENINDAMINE TARTRATE

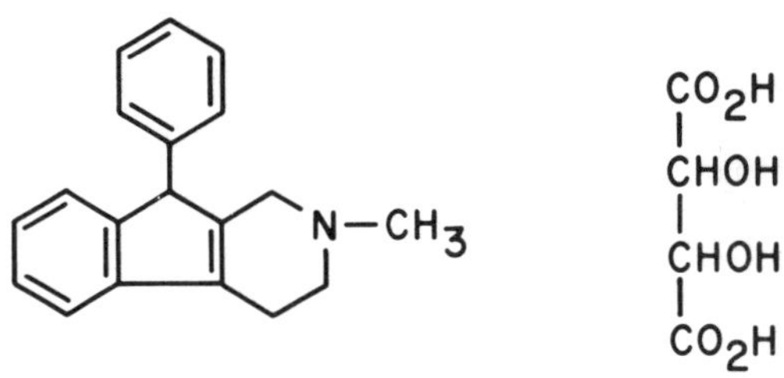

76

Uses

It is used in combination with other drugs in a variety of cold preparations and exhibits typical antihistamine effects. However, it has a mild stimulatory effect on some patients.

Dosage and Other Names

In a variety of combination products the adult oral dose is about 24 mg administered two to three times per day. It is also known as Nolamine (Carnrick) and P-V-Tussin (Reid-Provident).

Toxicity

It has the usual antihistamine side effects. The oral LD_{50} in rats is 280 mg/kg.

Physical-Chemical Data

The white, crystalline powder, mp 165 to 167°C, has a solubility of 2.5% in water. The pH of a 2% aqueous solution is about 3.5. The pK_a at 20°C is 8.98.

Chemical Name and Formula

Its chemical name is 2,3,4,9-tetrahydro-2-methyl-9-phenyl-1H-indeno[2,1-c]pyridine [R-(R*,R*)]-2,3-dihydroxybutanedioate (1:1). Its chemical formula is $C_{19}H_{19}N \cdot C_4H_6O_6$. Its mol wt is 411.35.

HISTAMINE H_2 RECEPTOR ANTAGONISTS

General Discussion of H_2 Histamine Receptors and Antagonists

More than 15 years have passed since Ash and Schild first postulated the existence of at least two types of histamine receptors. During this period, extraordinary interest has been shown in the second receptor type, termed the H_2 receptor, which, among other physiologic functions, presumably mediates histamine-induced secretion of gastric acid from parietal cells, increases heart rate, and inhibits uterine contractions. This interest resulted almost exclusively from the significant work of Black and associates, who provided pharmacologic characterization to, and therefore substantiated the existence of, histamine H_2 receptors.

With the discovery of the initial series of imidazole derivative H_2 receptor antagonists (prototype — burimamide; clinical candidate — metiamide, and lastly cimetidine) substantial research tools were provided to gastroenterologists. For some years earlier, opinions differed as to the role histamine played in the normal physiological control of gastric acid secretion. With these compounds it was demonstrated that the secretory response to gastrin and other secretogogues, such as choline esters, 2-deoxyglucose, insulin, food, etc., could be inhibited. Thus, recently, some evidence for the concept that a common mediator, histamine, as an obligatory physiologic step between the receptor interactions of gastrin and acetylcholine and the production of gastric acid has been accumulated. While the exact relationship between the secretogogues remains to be established, the successful introduction of cimetidine therapy into clinical medicine has provided strong support for the importance of the endogenous amine in the regulation of gastric secretion.

Cimetidine is a competitive antagonist of histamine at H_2 receptors and is virtually without effect at H_1 receptors. Administered either i.v. or orally, the compound inhibits histamine-produced gastric acid secretion as a function of its dose, as it does gastrin or pentagastrin, and to an extent acetylcholine-induced secretion. Cimetidine inhibits both the volume and the hydrogen ion concentration of gastric secretion and thus is of proven value in duodenal ulcer disease.

The compound has been shown to be effective as determined endoscopically in the treatment of active duodenal ulcer, and, at reduced dosage, in the prevention of recurrent ulcer. Thus, both therapeutic and prophylactic indications have responded to cimetidine, as have pathological hypersecretory conditions, including Zollergic-Ellison Syndrome, systemic mastocytasis, and multiple endocrine adenomas. In the management of active duodenal ulcer, antacids provided concomitantly offer early relief of associated pain.

While there are no outstanding contraindications to the use of cimetidine, some precautions have been noted, including possible effects from its weak antiandrogenic effect. Some reports of confusion have been reported in elderly or severely ill patients. In general, however, few adverse reactions to the drug itself have been reported and those that have are of a mild and transient nature and include diarrhea, muscle pain, dizziness, drug rash, and mild gynecomastia.

The success of cimetidine has prompted the search for other H_2 receptor antagonists and has yielded ranitidine (Glaxo) and etintidine (Bristol). Second-generation clinical candidates, presumably providing significantly improved potency or duration of activity, are presently being investigated. The reader is referred to References 3 to 5.

Drug Interactions

The H_2 receptor antagonist, cimetidine, reduces the hepatic metabolism of warfarin-type anticoagulants and such other drugs as propranolol, chlordiazepoxide, diazepam, and theophylline presumably through effects on hepatic microsomal enzyme systems. This interaction may delay the elimination and increase the blood levels of these drugs.[57] To date, ranitidine has not been shown to have much effect on the hepatic microsomal enzyme system.

Clinically Useful H_2 Receptor Antagonists

Structure-Activity Studies

The major H_2 antagonist on the market today is cimetidine, 77. This compound was initially developed on the premise that an antagonist of histamine

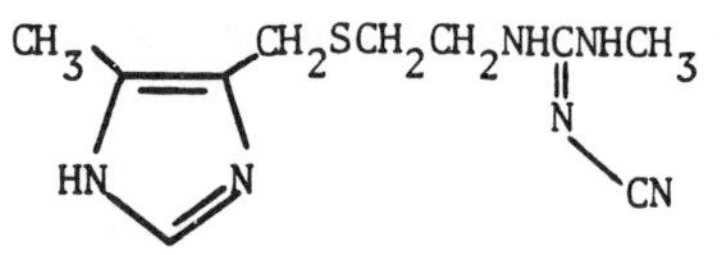

77

should look at least somewhat like it. The earliest synthesized analogs which exhibited limited H_2 antagonism contained a 4-imidazolyl group attached to a two- or three-carbon side chain which was in turn appended to an isothiourea, 78a, or guanidine, 78b, group.[58] Further extensive

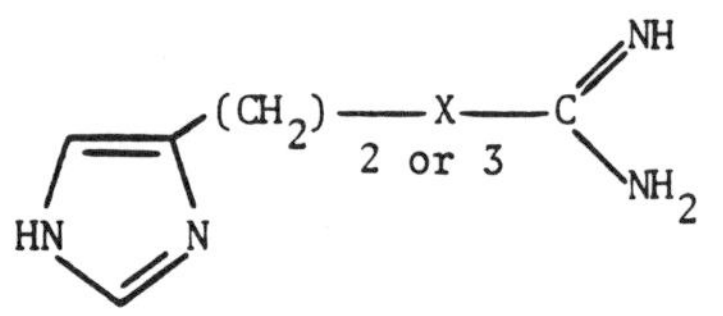

		X
78a	isothiourea	S
78b	guanidine	N

investigation at Smith Kline & French yielded burimamide, 79, the first H_2 antagonist to be investigated in man. In this compound the alkyl side chain had been lengthened to four carbons and attached to an *N*-methylthiourea group.[58]

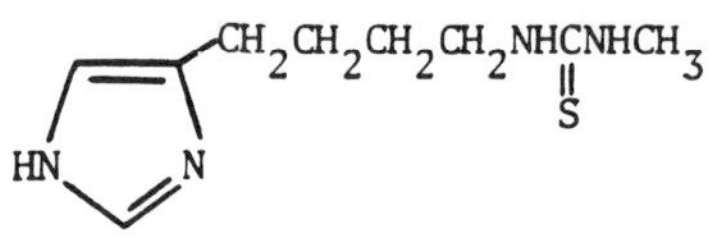

79

Burimamide, unfortunately, lacked adequate oral activity.

A more potent compound, metiamide, 80, was next taken to the clinic. This compound differed from burimamide in that it had a methyl group in the

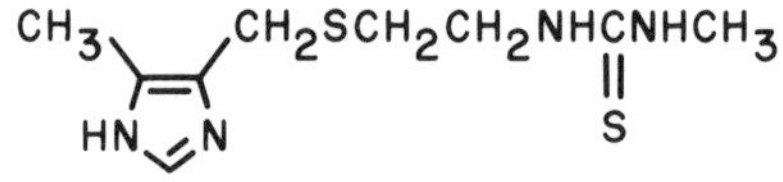

80

5-position of the imidazole ring and a sulfur atom substituted for one of the side chain carbon atoms. While effective in healing duodenal ulcers in man, it caused granulocytopenia in some patients.[58] Since it was believed possible that the thiourea moiety in metiamide caused the observed granulocytopenia, this group was replaced by the isosteric guanidine, 81a, and urea, 81b, radicals. These compounds proved to be 20 times less active than metiamide.[58] Replacement of the thiourea group by the nonbasic nitroguanidine, 81C, and cyanoguanidine moieties, 81d, however, yielded compounds equally active to metiamide, and the cyanoguanidine compound (cimetidine) was selected for advancement.[58]

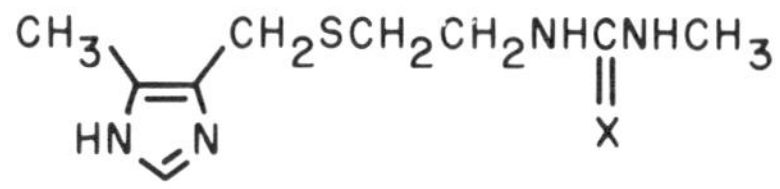

		X
81a	Guanidine	NH
81b	Urea	O
81c	Nitroguanidine	NNO_2
81d	Cyanoguanidine	NCN

Later work has shown H_2 antagonists can contain other heterocycles than imidazole. Ranitidine, 82, a recently marketed H_2 antagonist of Glaxo which has been reported to be four to ten times more effective than cimetidine in inhibiting histamine-induced gastric acid secretion in a variety of animal species and man,[59] possesses the furan group. Two other compounds presently under clinical study, tiotidine, 83a,[60] and BL-6341A, 83b,[61] incorporate the thiazole heterocyclic nucleus.

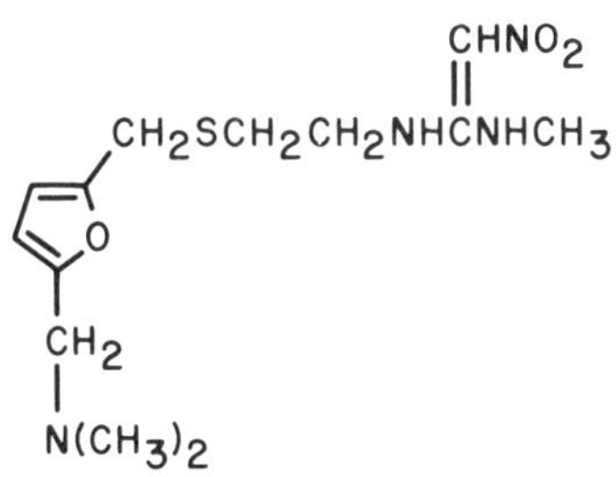

82

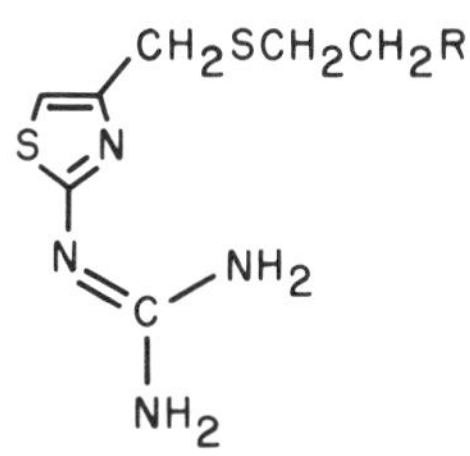

83a

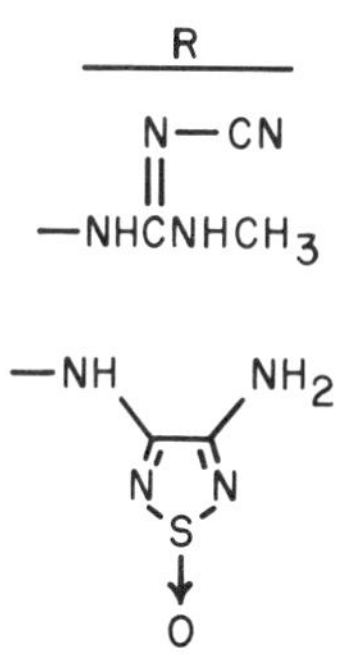

83b BL—6341A

Side chain groups other than cyanoguanidine are also useful as seen in ranitidine, BL-6341A and oxmetidine, 84.

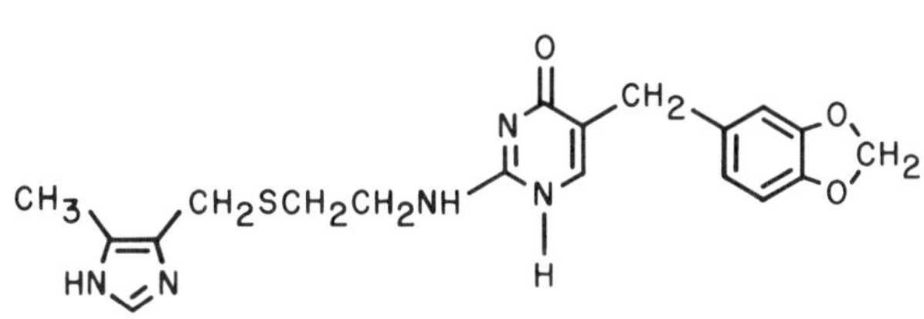

84

Metabolism

Cimetidine, 85a, administered orally or i.v., is largely excreted by rats, dogs, and man in the urine as unchanged drug. The major metabolite representing 12 to 30% of the administered drug and dependent upon the species and sex of the animals, is the sulfoxide, 85b.[62,63] The 5-hydroxymethyl compound, 85c, is also produced in small (3 to 8%) amounts.

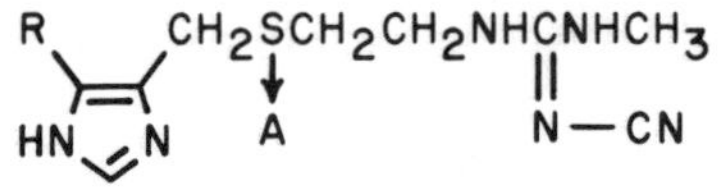

	R	A
85a	CH_3	--
85b	CH_3	O
85c	CH_2OH	--

Ranitidine, 86a, upon oral administration, is largely excreted unchanged in the urine of man, dog, rat, mouse, rabbit, and marmoset.[64] The major metabolite in man and dog is the N-oxide, 86b. In the other species the N-oxide is observed in near equal amount with two other metabolites, the sulfoxide, 86c, and the desmethyl analog, 86d.[67]

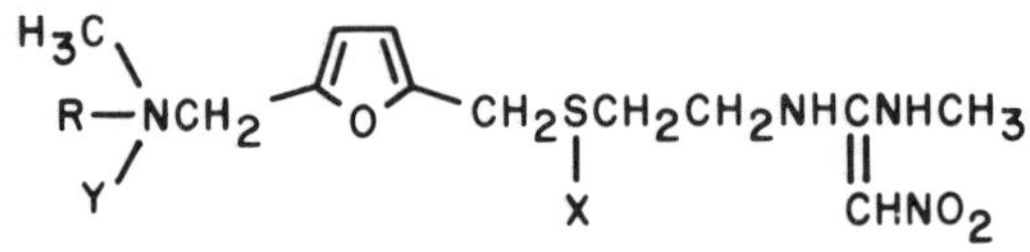

		R	Y	X
86a	Ranitidine	CH_3	--	--
86b	N-oxide	CH_3	O	--
86c	Sulfoxide	CH_3	--	O
86d	Desmethyl	H	--	--

***Clinically Useful H_2 Antihistamines*[16-20]**

CIMETIDINE

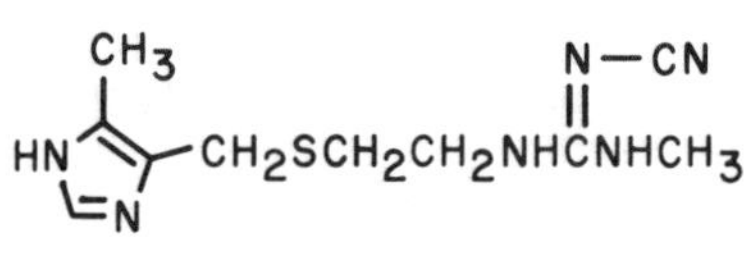

87

Uses

Cimetidine is used for the treatment of active and recurring duodenal ulcer and in the treatment of pathological hypersecretory conditions such as Zollergic-Ellison Syndrome.

Dosage and Other Names

In the treatment of an active duodenal ulcer, one 300-mg tablet is administered four times a day, while in prophylactic ulcer treatment two 200-mg tablets are taken before bed. It is also known as Tagamet (Smith Kline & French).

Pharmacokinetics[65-66]

Absorption of cimetidine after oral administration is rapid (45 to 90 min), and bioavailability ranges in the 60 to 70% area. The plasma half-life of cimetidine is approximately 2 hr. Plasma clearance is very age dependent, with the rate for patients less than 45 years old being more than double that for patients 53 to 64 years old. The principal route of elimination of the drug is the urine, and 24 hr after i.v. administration, 60 to 75% of the drug is recovered there in unchanged form. On oral administration the recovery is much more variable, with a range of 16.2 to 60.6%. Approximately 19% of the drug is protein bound.

Toxicity

In a small number of patients it has caused mild and transient diarrhea, muscular pain, dizziness, and rash. Mild gynecomastia has been reported in 4% of the patients under long-term treatment with cimetidine for pathological hypersecretory states. The oral LD_{50} in mice is 2600 mg/kg and in rats is 5000 mg/kg. The i.v. LD_{50} in mice and rats is 150 and 106 mg/kg, respectively.

Physical-Chemical Data

It is a crystalline material, mp 141 to 143°C, which is soluble in water to the extent of 1.14%. The pK_a is 6.80.

Chemical Name and Formula

Its chemical name is *N*-cyano-*N'*-methyl-*N''*-[2-[;(5-methyl-1H-imidazol-4-yl) methyl]thio]ethyl] guanidine. Its chemical formula is $C_{10}H_{16}N_6S$. Its mol wt is 252.34.

RANITIDINE HYDROCHLORIDE

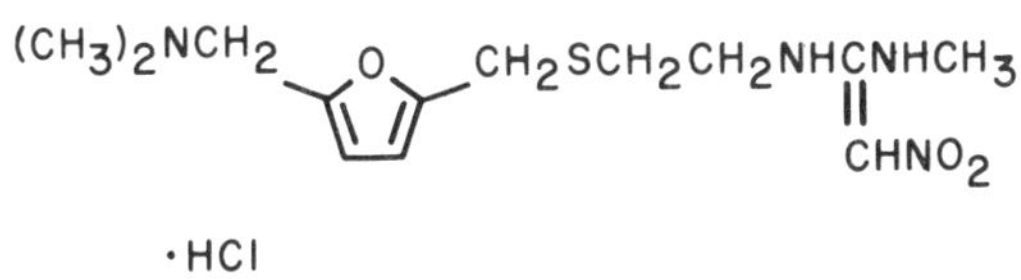

88

Uses

It is used for the treatment of active and recurring duodenal ulcer and in the treatment of pathological hypersecretory conditions. The compound inhibits histamine- and pentagastrin-induced gastric acid secretions.

Dosage and Other Names

In the treatment of active duodenal ulcer, one 150-mg tablet is administered two times a day.[59] It is also known as Ranitidine (Glaxo Holdings), Zantac (Glaxo Holdings).

Pharmacokinetics

Several studies have been reported on the pharmacokinetics of ranitidine.[67-69] Representative values are given in Table 4.

Physical-Chemical Data

The hydrochloride salt is an off-white solid, mp 133 to 134°C, which is soluble in water and methanol.[70]

Table 4
PHARMACOKINETIC PARAMETERS FOR RANITIDINE

Parameters	i.v. Dose (100 mg)	Oral dose	
		100 mg	150 mg
Plasma $t_{1/2}$ (min)	127 ± 7	179 ± 24	187 ± 37
Plasma clearance (mℓ/min)	568 ± 61		
Apparent vol. of distribution (ℓ/kg)	1.16 ± 0.09		
Bioavailability		56 ± 10	52 ± 13
Serum protein binding	14.8%		
Average hepatic clearance	30%	73%	

Chemical Name and Formula

Its chemical name is *N*-[2-[[[5-[(dimethylamino)methyl]-2-furanyl]methyl]thio]ethyl]-*N′*-methyl-2-nitro-1,1-ethenediamine monohydrochloride. Its chemical formula is $C_{13}H_{22}N_4O_3S \cdot HCl$. Its mol wt is 350.90.

INHIBITORS OF HISTAMINE RELEASE

DISODIUM CROMOGLYCATE

General Discussion

While neither an H_1 nor an H_2 receptor antagonist, cromolyn sodium has been classified with these agents, as significant portions of its pharmacology may be explained by its ability to inhibit the release of histamine and other mediators of anaphylaxis from pulmonary tissue.

Some 25 years ago, a Fisons Pharmaceutical Corporation[71,72] research team initiated a search for compounds having selective actions on smooth muscle. Research was focused on khellin, extracted from a Mediterranean plant, as the chemical prototype since the seeds had been used since ancient times as a smooth muscle relaxant. A series of chromone-2-carboxylic acids was developed, but the compounds were inactive as direct smooth muscle relaxants and as antagonists to several chemical spasmogens. However, they did partially inhibit, but not reverse, antigen-induced bronchospasm in a human asthmatic volunteer. The interest in chromones led to the bis-chromones and in 1965, one of these, cromolyn sodium (disodium cromoglycate), was synthesized and tested. Later, in 1968, it was introduced into therapy in the U.K.[72]

Asthma is a disease marked by recurrent, quickly developing, bronchial obstruction leading to paroxysmal dyspnea. The condition results from a spasm of hypersensitive bronchial smooth muscle, generally in the segmental and smaller bronchi. The spasm is often accompanied by a mucosal edema and respiratory mucus hypersecretion. The principal types of the disease are allergic or extrinsic asthma and ''nonallergic'' or intrinsic asthma — the latter of uncertain origin.

The symptomatology of immediate-type allergic diseases, including extrinsic asthma and allergic rhinitis, presumably results from the immunologically (antigen) induced release of various pharmacologically active substances from mast cells, and basophilic leukocytes previously sensitized with an antibody generally of the immunoglobulin E class,[71] and possibly from other cells as well. The substances thus released and referred to as primary mediators of anaphylaxis, i.e., those released as a direct consequence of antigen-antibody interaction, include histamine, slow-reacting substance, eosinophil chemotactic factor, and platelet-activating factor. There are probably other primary mediators as well as secondary mediators released indirectly, such as some of the prostaglandins and kinins involved in the modulation of allergic processes. The sequence of biochemical events associated with me-

Table 5
PCA ACTIVITY IN RATS OF DSCG AND ITS POSITIONAL ISOMERS

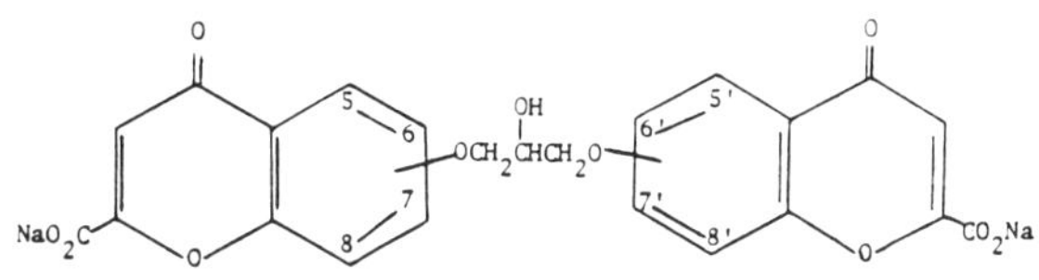

	Linkage positions	PCA ID_{50} (mg/kg)
89a	5,5′	0.7
89b	6,6′	0.3
89c	7.7′	0.5
89d	8,8′	Little activity at 10

diator release is not clearly understood and has been the subject of a substantial research effort, particularly during the past decade.

Presumably, the effectiveness of cromolyn sodium is based on its ability to inhibit the immunologically induced release of mediators from sensitized cells. The compound has no bronchodilator or anti-inflammatory activity, and it is not effective for the treatment of the acute asthmatic episode. Rather, it is a clinically effective, prophylactic, anti-allergic agent in a limited patient population consisting principally of children. The compound is poorly absorbed from the GI tract, so it is administered as a powder inhalation.

Although some inconvenience and a need for education are associated with its use, cromolyn sodium offers a rather substantial degree of safety, with only occasional cough or bronchospasm generally associated with its use. While enjoying broader use elsewhere, in the U.S. the compound is indicated as an adjunctive therapy in the management of patients with severe bronchial asthma. Cromolyn sodium is continued on a more chronic basis if it significantly reduces the severity of symptoms, allows for a reduction in or elimination of steroids, or allows for satisfactory management of individuals who have ''intolerable'' side effects to sympathomimetic drugs or the methylxanthines.

The clinical efficacy of cromolyn sodium as an antiasthmatic agent has prompted a significant world-wide research effort directed to the discovery of more potent, orally active anti-allergic drugs. This effort has not been especially successful to date.

Structure-Activity Studies

Structure-activity studies of disodium cromoglycate (DSCG), 89a, and related compounds have depended on the passive cutaneous anaphylaxis (PCA) reaction in rats to evaluate relative activities. From these studies it has been shown that the hydroxypropyldiether unit can join the two chromone rings at the 5-, 6-, and 7-positions and yield compounds of approximately similar activity (Table 5); linkage at the 8-position, however, gave a compound with little or no activity.[73]

The hydroxypropyldiether linkage can be replaced by an alkylene bridge of two to five carbons in length, unsubstituted by an hydroxyl group, 90, and still preserve activity; a one-carbon bridge (n = 1) or chain lengths longer than five carbons (n > 5), however, gave much less active compounds.[73]

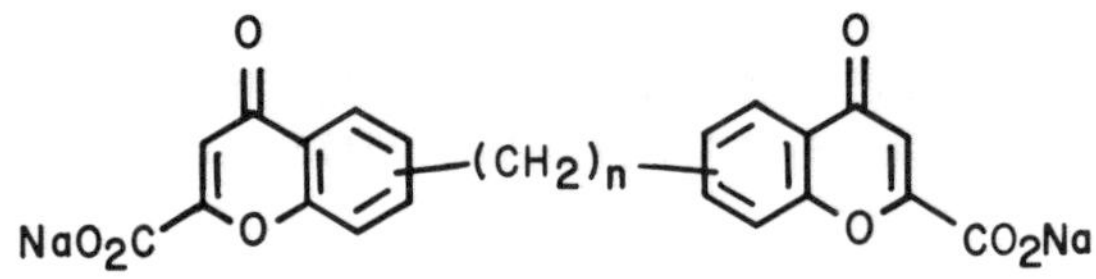

90

Direct linkage (n = 0) of the two chromone rings at the 6- or 7-positions also gave active compounds.

Pharmacokinetics

Disodium cromoglycate is very poorly absorbed orally and consequently is administered by inhalation using a Spinhaler. By this route in man about 7.5% of the administered dose reaches the lung, from which it is rapidly absorbed.[74] Whether administered orally, intravenously, or by inhalation, the compound is rapidly excreted in the bile and urine of man,[74,76] and nine other animal species which have been studied.[75] In the studied species the drug is always isolated in an unchanged form with no metabolites being observed.[74,76] Whether the compound is mainly excreted in the bile or urine depends upon the animal species.[75]

CROMOLYN SODIUM[16-20]

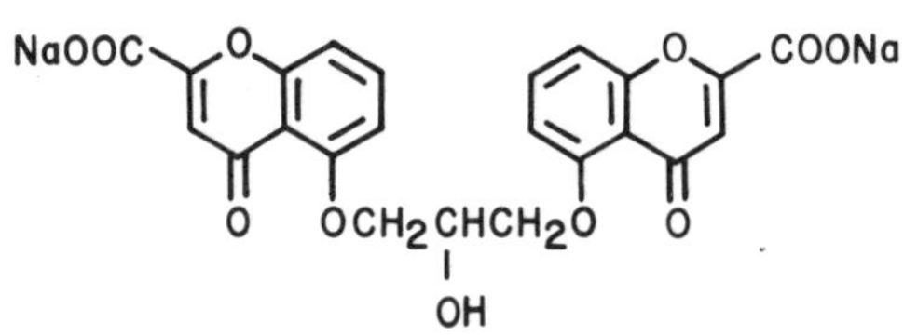

91

Uses

It is useful in the prophylactic treatment of severe bronchial asthma where the episodes are frequent and of predictable intensity.

Dosage and Other Names

It is administered by inhalation via a Spinhaler in 20-mg doses four times a day. It is also known as Disodium cromoglycate, FPL 670, Intal (Fisons).

Toxicity

The compound has few side effects although in a limited number of patients bronchospasm and cough have been observed.

Physical-Chemical Data

It is a white powder, mp 241 to 242°C, which is freely soluble in water (100 mg/mℓ at 20°C) and practically insoluble in chloroform and alcohol. The pK_a at 20°C is estimated to be 1.2 to 2.0.

Chemical Name and Formula

Its chemical name is 5,5′-[(2-hydroxy-1,3-propanediyl)*bis*-(oxy)]*bis*[4-oxo-4H-1-benzopyran-2-carboxylic acid, disodium salt]. Its chemical formula is $C_{23}H_{14}O_{11} \cdot Na_2$. Its mol wt is 512.34.

REFERENCES

1. **Douglas, W. W.,** *The Pharmacological Basis of Therapeutics,* 6th ed., Gilman, A. G., Goodman, L. S., and Gilman, A., Eds., Macmillan, New York, 1980, 609.
2. **Ganellin, C. R. and Durant, G. J.,** Histamine H_2-receptor agonists and antagonist, in *Burger's Medicinal Chemistry,* Part III, 4th ed., Wolff, M. E., Ed., Wiley-Interscience, New York, 1981.
3. **Wood, C. J. and Simkins, M. A., Eds.,** *International Symposium on Histamine H_2-Receptor Antagonists,* Published Deltakas (U.K.), London, 1973.
4. **Burland, W. L. and Simkins, M. A.,** *Cimetidine — Second International Symposium on Histamine H_2-Receptor Antagonists,* Excerpta Medica, Amsterdam, 1977.
5. **Torsoli, A., Lucchelli, P. E., and Brenblecombe, R. W., Eds.,** *H_2-Receptor Antagonists in Peptic Ulcer Disease and Progress in Histamine Research,* Excerpta Medica, Amsterdam, 1980.
6. **Ganellin, C. R. and Durant, G. J.,** Histamine H_2-receptor agonists and antagonist, in *Burger's Medicinal Chemistry,* Part III, 4th ed., Wolff, M. E., Ed., Wiley-Interscience, New York, 1981, 543.
7. **Barlow, R. B.,** *Introduction to Chemical Pharmacology,* 2nd ed., Methuen & Co., Bungay, 1968, 359.
8. **Witiak, D. T.,** Antiallergenic agents, in *Medicinal Chemistry,* Vol. 2, 3rd ed., Burger, A., Ed., Wiley-Interscience, New York, 1970, 1647.
9. **Clapp, R. C., Clark, J. H., Vaughan, J. R., English, J. P., and Anderson, G. W.,** Halogenated thiophene derivatives as antihistamine agent, *J. Am. Chem. Soc.,* 69, 1549, 1947.
10. **Barlow, R. B.,** *Introduction to Chemical Pharmacology,* 2nd ed., Methuen & Co., Bungay, 1968, 358.
11. **Feinstone, W. H., Williams, R. D., and Rubin, B.,** Antihistamine and antianaphylactic effect of hetramine, a new synthetic pyrimidine compound, *Proc. Soc. Exp. Biol. Med.,* 63, 158, 1946.
12. **Barlow, R. B.,** *Introduction to Chemical Pharmacology,* 2nd ed., Methuen & Co., Bungay, 1968, 360.
13. **Chaudhuri, N. K., Servando, O. A., Manniello, M. J., Luders, R. C., Chao, D. K., and Bartlett, M. F.,** Metabolism of tripelennamine in man, *Drug Metab. Dispos.,* 4, 372, 1976.
14. **Rao, G. S., Krishna, G., and Gillette, J. R.,** Metabolism, tissue distribution and covalent binding of tripelennamine and its N-nitroso derivative in the rat, *J. Pharmacol. Exp. Ther.,* 195, 433, 1975.
15. **Calandre, E. P., Alferez, N., Hassanein, K., and Azarnoff, D. L.,** Methapyrilene kinetics and dynamics, *Clin Pharmacol. Ther.,* 4, 527, 1981.
16. **Wade, A. and Reynolds, J. E. F., Eds.,** *Martindale the Extra Pharmacopoeia,* 27th ed., Pharmaceutical Press, London, 1977.
17. **Baker, C. E., Jr.,** *Physicians' Desk Reference,* Medical Economics Co., Oradell, N.J., 1982.
18. **Griffiths, M. C., Fleeger, C. A., and Miller, L. C., Eds.,** *USAN and the USP Dictionary of Drug Names,* The United States Pharmacopeial Convention, Rockville, Md., 1981.
19. **Windholz, M., Budavari, S., Stroumtsos, L. V., and Fertig, M. N., Eds.,** *The Merck Index,* 9th ed., Merck & Co., Rahway, N.J., 1976.
20. **Lewis, A. J., Gonzales, G. D., and Winek, C. L., Eds.,** *Modern Drug Encyclopedia and Therapeutic Index,* 16th ed., Technical Publ., New York, 1981.
21. **Alworth, W. L.,** *Stereochemistry and Its Application in Biochemistry,* Wiley-Interscience, New York, 1972, 69.
22. **Carey, F. A. and Sundberg, R. J.,** *Advanced Organic Chemistry, Part A Structure and Mechanism,* Plenum Press, New York, 1977, 52.
23. **Witiak, D. T.,** Antiallergenic agents, in *Medicinal Chemistry,* Vol. 2, 3rd ed., Burger, A., Ed., Wiley-Interscience, New York, 1970, 1643.
24. **Labelle, A. and Tislow, R.,** Studies on prophenpyridamine (trimeton) and chlorprophenpyridamine (chlor-trimeton), *J. Pharmacol. Exp. Ther.,* 113, 72, 1955.
25. **Roszkowski, A. P. and Govier, W. M.,** Comparison of racemic carbinoxamine with its optically active isomers, *Pharmacologist,* 1, 60, 1959.
26. **Alworth, W. L.,** *Stereochemistry and Its Application in Biochemistry,* Wiley-Interscience, New York, 1972, 55.
27. **Drach, J. C. and Howell, J. P.,** Identification of diphenhydramine urinary metabolites in the rhesus monkey, *Biochem. Pharmacol.,* 17, 2125, 1968.
28. **Drach, J. C., Howell, J. P., Borondy, P. E., and Glazko, A. J.,** Species differences in the metabolism of diphenhydramine (benadryl), *Proc. Soc. Exp. Biol. Med.,* 135, 849, 1970.
29. **Ellison, T., Snyder, A., Bolger, J., and Okun, R.,** Metabolism of orphenadrine citrate in man, *J. Pharmacol. Exp. Ther.,* 176, 284, 1971.
30. **Carruthers, S. G., Shoeman, D. W., Hignite, C. E., and Azarnoff, D. L.,** Correlation between plasma diphenhydramine level and sedative and antihistamine effects, *Clin. Pharmacol. Ther.,* 23, 375, 1978.
31. **Albert, K. S., Hallmark, M. R., Sakmar, E., Weidler, D. J., and Wagner, J. G.,** Pharmacokinetics of diphenhydramine in man, *J. Pharmacokinet. Biopharm.,* 3, 159, 1975.
32. **Ramma Sastry, B. V.,** Anticholinergics: antispasmodic and antiulcer drugs, in *Burger's Medicinal Chemistry,* Part III, 4th ed., Wolff, M. E., Ed., Wiley-Interscience, New York, 1981, 381.

33. **Ramma Sastry, B. V.,** Anticholinergics: antispasmodic and antiulcer drugs, in *Burger's Medicinal Chemistry,* Part III, 4th ed., Wolff, M. E., Ed., Wiley-Interscience, New York, 1981, 386.
34. **Roth, F. E. and Govier, W. M.,** Comparative pharmacology of chlorpheniramine (chlor-trimeton) and its optical isomers, *J. Pharmacol. Exp. Ther.,* 124, 347, 1958.
35. **Adamson, D. W., Barrett, P. A., Billinghurst, J. W., Green, A. F., and Jones, T. S. G.,** Geometrical isomers in a series of antihistamines, *Nature (London),* 168, 204, 1951.
36. **Green, A. F.,** The antagonism of histamine and the anaphylactic response by phenylpyridylallylamines, *Br. J. Pharmacol.,* 8, 171, 1953.
37. **Bruce, R. B., Turnbull, L. B., Newman, J. H., and Pitts, J. E.,** Metabolism of brompheniramine, *J. Med. Chem.,* 11, 1031, 1968.
38. **Peets, E. A., Jackson, M., and Symchowicz, S.,** Metabolism of chlorpheniramine maleate in man, *J. Pharmacol. Exp. Ther.,* 180, 464, 1972.
39. **Simons, F. E. R., Luciuk, G. H., and Simons, K. J.,** Pharmacokinetics and efficacy of chlorpheniramine in children, *J. Allergy Clin. Immunol.,* 69, 376, 1982.
40. **Huang, S. M., Huang, Y. C., and Chiou, W. L.,** Oral absorption and presystemic first-pass effect of chlorpheniramine in rabbits, *J. Pharmacokinet. Biopharm.,* 9, 725, 1981.
41. **Simons, F. E. R., Simons, K. J., and Frith, E. M.,** The pharmacokinetics and antihistamine effects of brompheniramine, *J. Allergy Clin. Immunol.,* 69, 117, 1982.
42. **Doerge, R. F.,** Histamine and antihistaminic agents, in *Textbook of Organic Medicinal and Pharmaceutical Chemistry,* 7th ed., Wilson, C. O., Gisvold, O., and Doerge, R. F., Eds., J. B. Lippincott, Philadelphia, 1972, 648.
43. **Barlow, R. B.,** *Introduction to Chemical Pharmacology,* 2nd ed., Methuen & Co., Bungay, 1968, 344.
44. **Kuntzman, R., Phillips, A., Tsai, I., Klutch, A., and Burns, J. J.,** N-oxide formation: a new route for inactivation of the antihistaminic chlorcyclizine, *J. Pharmacol. Exp. Ther.,* 155, 377, 1967.
45. **Kuntzman, R., Klutch, A., Tsai, I., and Burns, J. J.,** Physiological distribution and metabolic inactivation of chlorcyclizine and cyclizine, *J. Pharmacol. Exp. Ther.,* 149, 29, 1965.
46. **McMahon, R. E. and Sullivan, H. R.,** The oxidative demethylation of 1-propoxyphene and 1-propoxyphene N-oxide by rat liver microsomes, *Life Sci.,* 3, 1167, 1964.
47. **Bain, W. A., Broadbent, J. L., and Warin, R. P.,** Comparison of anthisan (mepyramine maleate) and phenergan as histamine antagonists, *Lancet,* 257, 47, 1949.
48. **Engelhardt, E. L., Zell, H. C., Saari, W. S., Christy, M. E., and Colton, C. D.,** Structure-activity relationships in the cyproheptadine series, *J. Med. Chem.,* 8, 829, 1965.
49. **Martin, U. and Romer, D.,** The pharmacological properties of a new, orally active, antianaphylactic compound: ketotifen, a benzocycloheptathiophene, *Arzneim. Forsch.,* 28, 770, 1978.
50. **Goldenberg, H., Fishman, V., Heaton, A., and Burnett, R.,** A detailed evaluation of promazine metabolism, *Proc. Soc. Exp. Biol. Med.,* 115, 1044, 1964.
51. **Fishman, V. and Goldenberg, H.,** Side-chain degradation and ring hydroxylation of phenothiazine tranquilizers, *J. Pharmacol. Exp. Ther.,* 150, 122, 1965.
52. **Porter, C. C. and Arison, B. H.,** Human metabolism of cyproheptadine, *Drug Metab. Dispos.,* 3, 189, 1975.
53. **Hucker, H. B., Balletto, A. J., Stauffer, S. C., Zacchei, A. G., and Arison, B. H.,** Physiological disposition and urinary metabolites of cyroheptadine in the dog, rat, and cat, *Drug Metab. Dispos.,* 2, 406, 1974.
54. **Frigerio, A., Cavo-Briones, M., and Belvedere, G.,** Formation of stable epoxides in the metabolism of tricyclic drugs, *Drug Metab. Rev.,* 5, 197, 1976.
55. **Hintze, K. L., Wold, J. S., and Fischer, L. J.,** Disposition of cyproheptadine in rats, mice, and humans and identification of a stable epoxide metabolite, *Drug Metab. Dispos.,* 3, 1, 1975.
56. **Wermeille, M. M. and Huber, G. A.,** Gas-liquid chromatographic determination of free dimethindene in human serum and urine at low concentrations, *J. Chromatogr.,* 228, 187, 1982.
57. **Bauman, J. H. and Kimelblatt,** Cimetidine as an inhibitor of drug metabolism: therapeutic implication and review of the literature, *Drug Intell. Clin. Pharm.,* 16, 380, 1982.
58. **Ganellin, R.,** Medicinal chemistry and dynamic structure-activity analysis in the discovery of drugs acting at histamine H_2 receptors, *J. Med. Chem.,* 24, 913, 1981.
59. **Koch, H., Cluxton, R. J., Jr., and Mesquita, C.,** Ranitidine: histamine H_2-receptor antagonist, *Pharm. Intern.,* 2, 74, 1981.
60. **Winer, N., Greene, L. S., Boccagno, J. A., and Thyrum, P. T.,** The effect of increasing doses of tiotidine on twenty-four hour intragastric acidity, nocturnal and food-stimulated acid secretion, and serum gastrin, *Clin Pharmacol. Ther.,* 29, 289, 1981.
61. **Algieri, A. A., Luke, G. M., Standridge, R. T., Brown, M., Partyka, R. A., and Crenshaw, R. R.,** A new class of potent and long-acting H_2-receptor antagonists; BL-6341A, in Proc. spring A.C.S. meeting, Med. Chem. Sec. 1981, #42.

62. **Taylor, D. C. and Cresswell, P. R.,** The metabolism of cimetidine in the rat, dog and man, *Biochem. Soc. Trans.*, 3, 884, 1975.
63. **Taylor, D. C., Cresswell, P. R., and Bartlett, D. C.,** The metabolism and elimination of cimetidine, a histamine H_2-receptor antagonist, in the rat, dog and man, *Drug Metab. Dispos.*, 6, 21, 1978.
64. **Bell, J. A., Dallas, F. A. A., Jenner, W. N., and Martin, L. E.,** The metabolism of ranitidine in animals and man, *Biochem. Soc. Trans.*, 8, 93, 1980.
65. **Somogyi, A., Rohner, H. G., and Gugler, R.,** Pharmacokinetics and bioavailability of cimetidine in gastric and duodenal ulcer patients, *Clin. Pharmacokinet.*, 5, 84, 1980.
66. **Kastrup, E. K. and Boyd, J. R., Eds.,** *Facts and Comparisons,* J. B. Lippincott, St. Louis, 1982, 307f.
67. **Bogues, K., Dixon, G. T., Fowler, P., Jenner, W. N., Maconochie, J. G., Martin, L. E., and Willoughby, B. A.,** Pharmacokinetics and bioavailability of ranitidine in humans, *Br. J. Pharmacol.*, 73, 275P, 1981.
68. **Garg, D. C., Weidler, D. J., Baltodano, N., and Eshelman, F. N.,** Pharmacokinetics of ranitidine, a new histamine H_2-receptor blocker, *Clin Pharmacol. Ther.*, 29, 247, 1981.
69. **Garg, D. C., Weidler, D. J., and Eshelman, F. N.,** Clinical pharmacokinetics of ranitidine a new histamine H_2-receptor blocker following oral administration in increasing doses and following multiple dose administration, *J. Clin. Pharmacol.*, 22(No. 4), 8A, 1982.
70. U.S. Patent 4,128,658.
71. **Chakrin, L. W. and Krell, R. D.,** Pathophysiology and pharmacotherapy of asthma: an overview, *J. Pharm. Sci.*, 69, 236, 1980.
72. Intal-Cromolyn Sodium, (monograph), Fisons Corp., Bedford, Mass., 1973.
73. **Cairns, H., Fitzmaurice, C., Hunter, D., Johnson, P. B., King, J., Lee, T. B., Lord, G. H., Minshull, R., and Cox, J. S. G.,** Synthesis and structure-activity relationships of disodium cromoglycate and some related compounds, *J. Med. Chem.*, 15, 583, 1972.
74. **Moss, G. F., Jones, K. M., Ritchie, J. T., and Cox, J. S. G.,** Plasma levels and urinary excretion of disodium cromoglycate after inhalation by human volunteers, *Toxicol. Appl. Pharmacol.*, 20, 147, 1971.
75. **Ashton, M. J., Clark, B., Jones, K. M., Moss, G. F., Neale, M. G., and Ritchie, J. T.,** The absorption, metabolism and excretion of disodium cromoglycate in nine animal species, *Toxicol. Appl. Pharmacol.*, 26, 319, 1973.
76. **Walker, S. R., Evans, M. E., Richards, A. J., and Paterson, J. W.,** The fate of ^{14}C disodium cromoglycate in man, *J. Pharm. Pharmacol.*, 24, 525, 1972.
77. **Auerbach, C. and Barrow, M. V.,** Urogenital abnormalities produced in rat fetuses with chlorcyclizine, *Teratology*, 5, 23, 1972.

PROSTANOIDS AND LEUKOTRIENES

Michael P. L. Caton

INTRODUCTION

Prostanoids are a group of natural products which occur widely in animal tissues. Their potential as a source of new drugs was recognized in the early 1960s following the characterization of the first-discovered members of the series — the E and F prostaglandins — and the realization that they possess a variety of biological actions which might be harnessed to afford novel methods of treatment in a variety of clinical areas. It was soon established that prostaglandins are biosynthesized from the 20-carbon essential fatty acids and it is now known that these give rise to a large number of related physiologically active products. They include the original ''classical'' prostaglandins, the thromboxanes and prostacyclins, and the leukotrienes, which were characterized as recently as 1979 to 1980.

The massive investment in research in this area over the past 2 decades has resulted in a very extensive literature, which can only be summarized briefly here with particular reference to the overall objective of the handbook; several books, reviews, and publications based on symposia[1-18a] are therefore recommended to the reader for more detailed information. Two specialist journals, *Prostaglandins*[19] and *Prostaglandins, Leukotrienes* and *Medicine*,[20] are devoted entirely to prostanoid and leukotriene research and contain many of the original research papers.

Structure and Nomenclature

The entire group of substances derived from the essential fatty acids is now frequently referred to as the ''Eicosanoids'', which are broadly classified as follows:

- Eicosanoids
  - Leukotrienes
  - Prostanoids
    - Prostaglandins
    - Thromboxanes

The chemical structures of the prostaglandins and thromboxanes are illustrated in Figure 1. The prostaglandins (PG) are characterized by a cyclopentane ring with two adjacent side chains and may be regarded as derivatives of the hypothetical ''prostanoic acid''.[21]

They are divided into nine groups — designated by the letters A to I — which are generally distinguished by the nature and distribution of double bonds and oxygen functions in the ring. The PGG and PGH series, however, have identical ring arrangements and are distinguished by the presence of a peroxy function in the lower side chain (R_2) in the former, in place of hydroxyl found in the other series. The PGIs, generally known as the prostacyclins, have a bicyclic structure as a result of the linking of the upper α-chain to the cyclopentane ring via an ether grouping. In the case of the F series, a subscript α is added to define the stereochemistry of the C-9 hydroxyl group. The side chains R_1 (α-chain) and R_2 (ω-chain) are of three types, characterized by the number of double bonds they contain which are designated by the subscripts 1, 2, and 3.* Except for the minor differences referred to above,

* Although in some cases all the possible compounds have not been isolated.

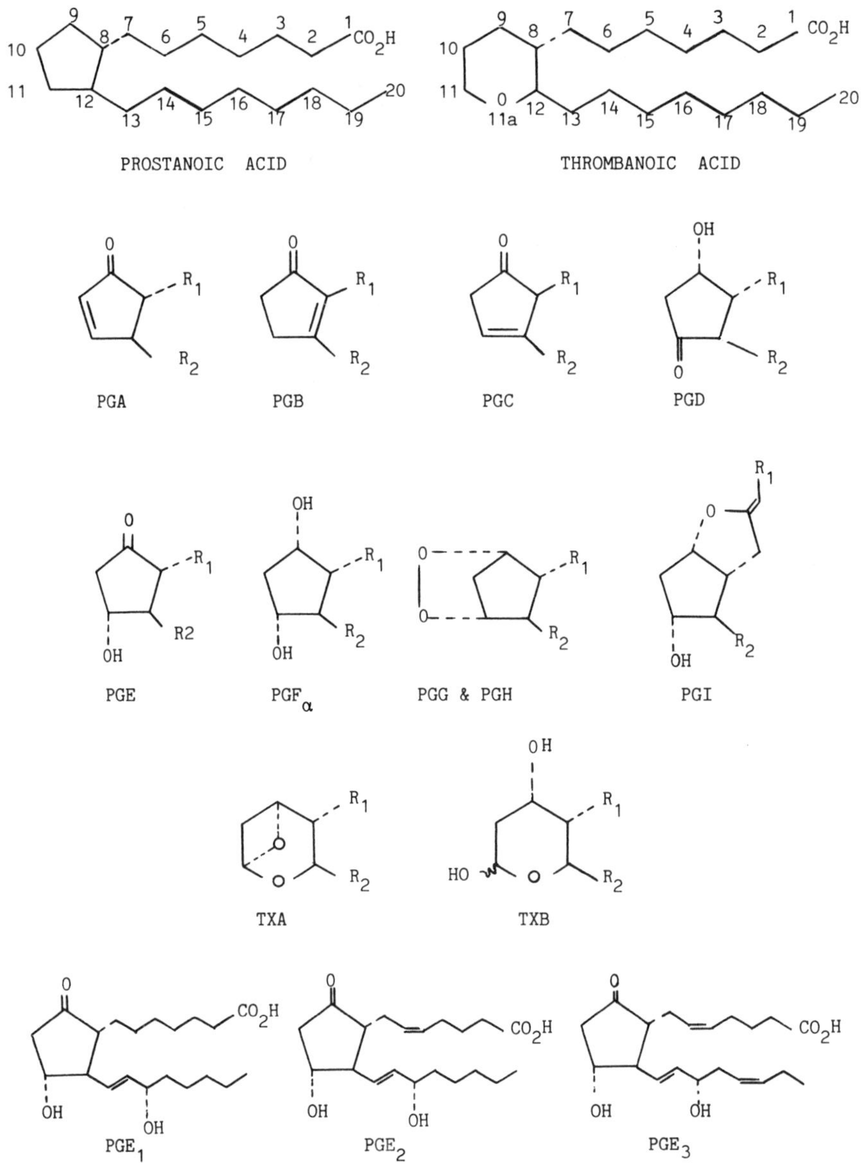

FIGURE 1. Prostaglandins and thromboxanes: structure and nomenclature.

the side chain arrangements are similar in all three groups; the arrangement illustrated in Figure 1 is for the PGE series only.

The thromboxanes (TX), which may be regarded as derivatives of thrombanoic acid,[22] are characterized by six-membered oxygen rings. They are of two types, TXA and TXB, and have the same side chains as the prostaglandins.

The leukotrienes (LT) are a group of straight chain fatty acids, some of which are attached to a peptide grouping (Figure 2). They are classified into five groups (A to E) which, as with the prostanoids, are further divided into three subgroups according to the number of double bonds present.*[23]

* The structure of the leukotriene B series has been modified since the publication of Reference 23 to that shown in Figure 2.

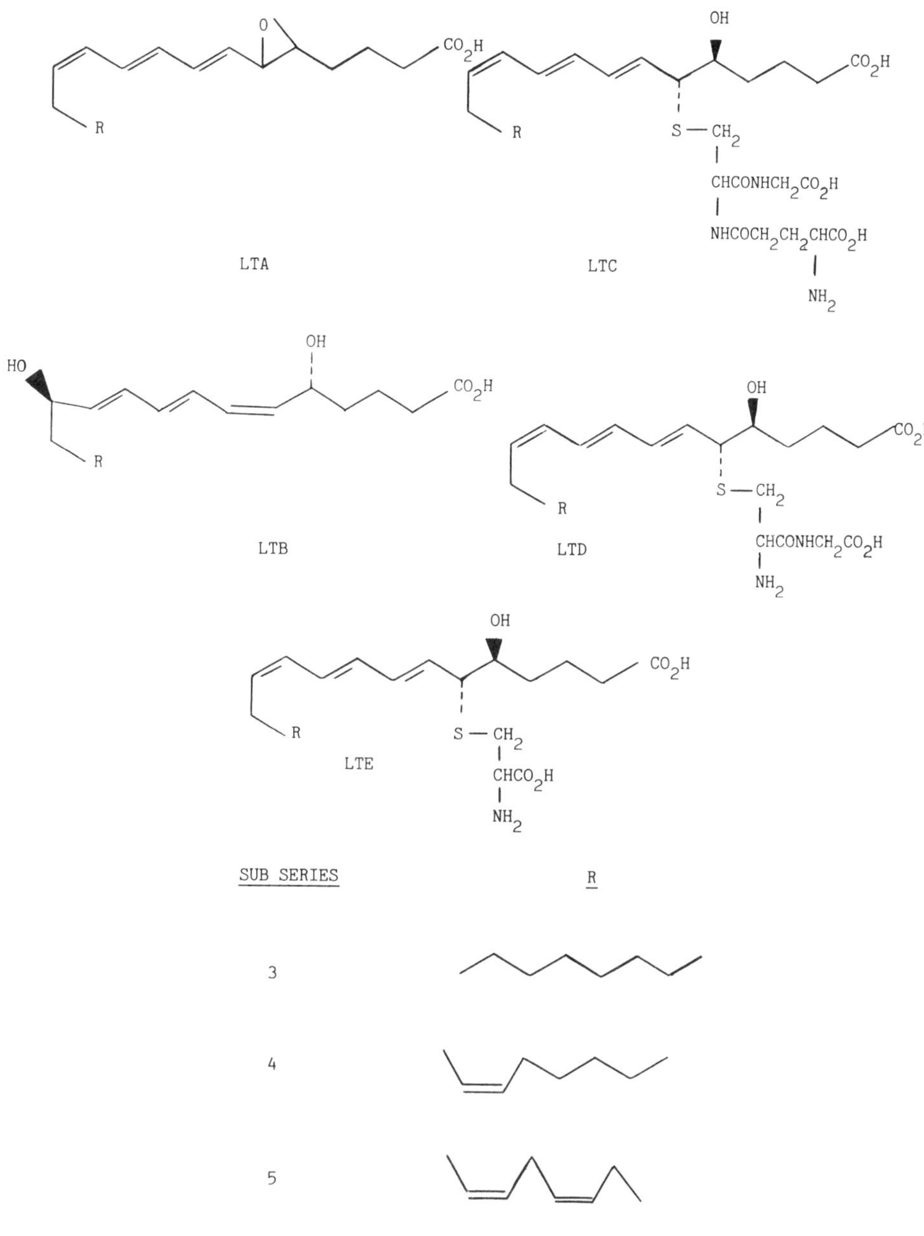

FIGURE 2. Leukotrienes: structure and nomenclature.

Occurrence and Natural Sources of Supply

Prostanoids occur in most mammalian tissues and are released on tissue stimulation or activation. The concentration and distribution of individual prostanoids vary considerably with the species and tissue, the highest concentrations being in the reproductive system. Some prostanoids, e.g., the E and F series, have a widespread distribution, whereas others, e.g., PGI and TXA, appear to be associated with a more limited number of tissues. Leukotrienes are found in the lung and various leukocytes, although further work on these recently characterized substances may reveal a more general distribution akin to that of the prostanoids. However, none of these natural sources affords sufficient quantities to be of practical value for providing significant supplies, although in the early 1960s laboratory-scale biosyntheses of the E and F prostaglandins were developed to afford gram quantities

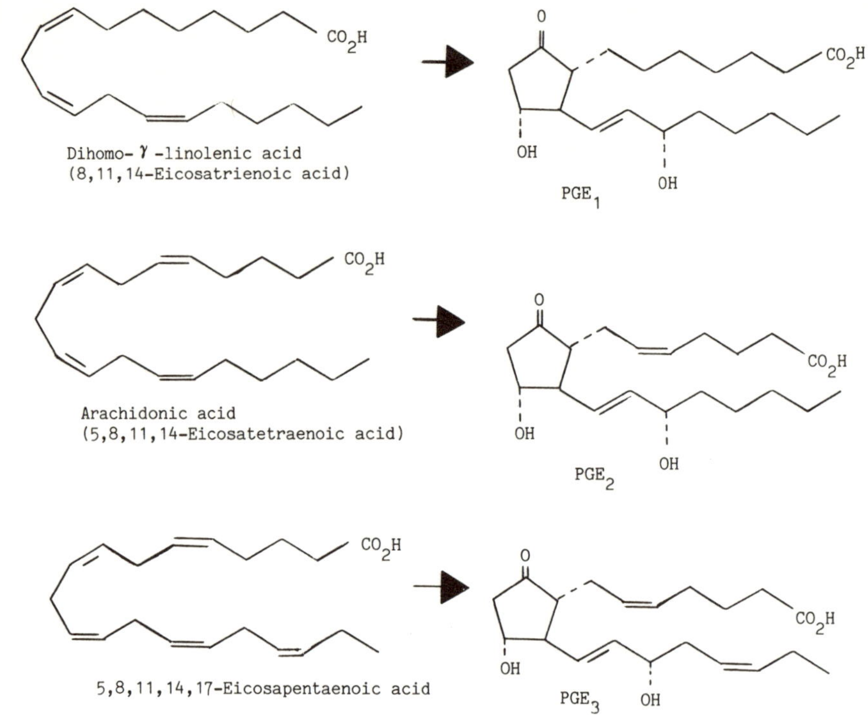

FIGURE 3. Biosynthetic origin of the 'E' prostaglandins.

of material, before total chemical syntheses were available. The unstable TXA_2, which has not been prepared by chemical synthesis, may be generated from platelets in vitro when required as a standard for biological testing. Details of tissue distribution of prostanoids are given in other reviews.[1a]

A member of the PGA series −15–R–PGA_2− is found in considerable quantity in a species of coral — the Gorgonian *Plexaura homomalla,* and this has been used as a starting material for the preparation of a number of PG analogs.[24] However, the scope for chemical modification is relatively limited unless multistage sequences are employed.

Chemical Synthesis

Most prostanoids are now conveniently obtained by total chemical synthesis. Although the syntheses are multistage, some of them have been developed to an overall level of efficiency suitable for the production of commercial quantities. The syntheses and their many variants are too complex and numerous to discuss here and the reader is referred to reviews and books devoted entirely to this subject.[12-15,25] Several syntheses of leukotrienes have recently been published and the subject has been reviewed.[26]

Biosynthesis

The three essential fatty acids (EFAs) — dihomo-γ-linolenic (8,11,14-eicosatrienoic), arachidonic (5,8,11,14-eicosatetraenoic), and 5,8,11,14,17-eicosapentaneoic acids — which differ from each other in the number of double bonds, give rise, respectively, to the 1, 2, and 3 subgroups of prostanoids and to the 3, 4, and 5 leukotriene series. This relationship is illustrated for the PGE series in Figure 3 and the total scheme of transformations from arachidonic acid, as now understood, is shown in Figures 4 to 6. The pathways from the other two EFAs are broadly similar, but in man the arachidonic acid-derived products are the most important.

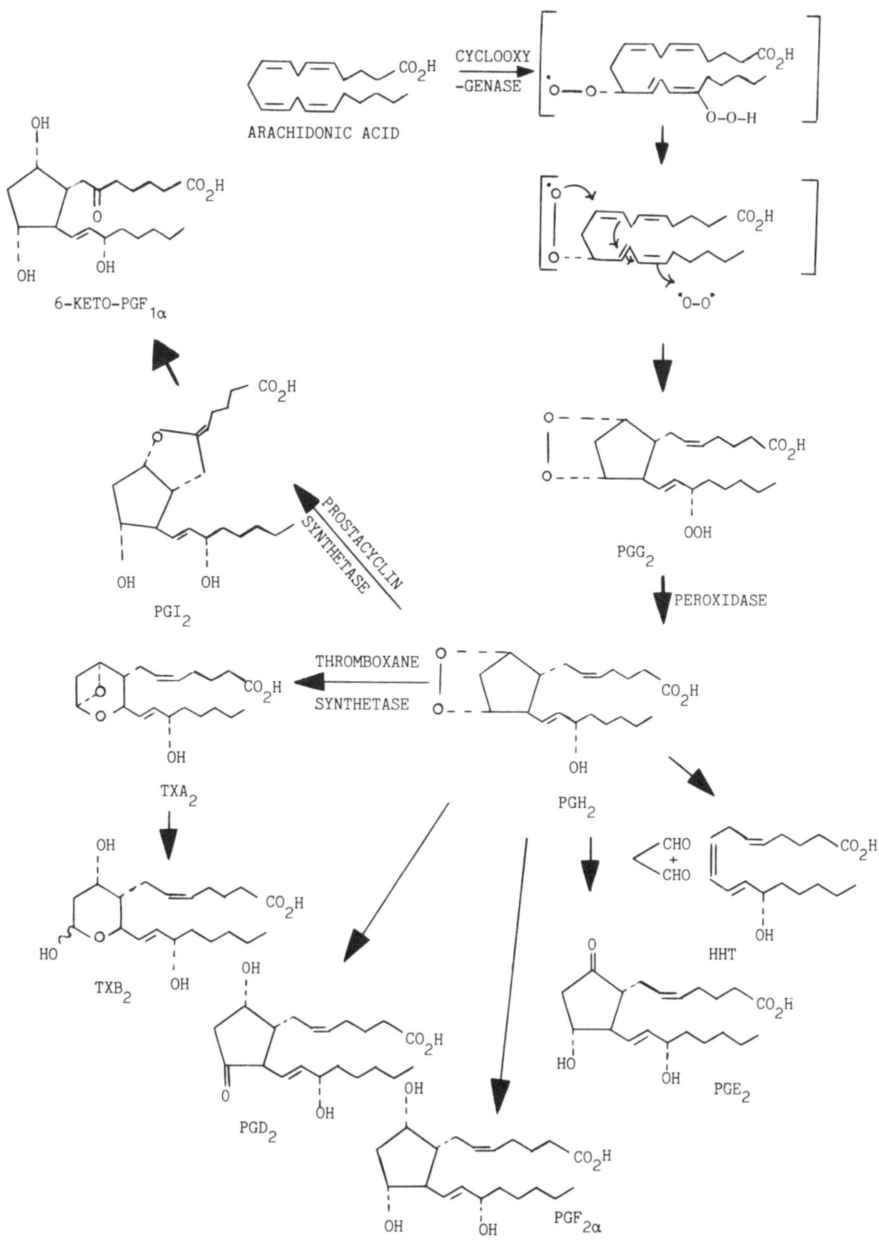

FIGURE 4. Biosynthesis of prostanoids from arachidonic acid.

The arachidonic acid is derived from dietary sources and esterified as a phospholipid, triglyceride, or cholesterol ester. The free acid is then released from the ester by the enzyme phospholipase A_2, a step which is believed to be rate limiting. Thus, the concentration of free acid is kept low and the subsequent steps, which are catalyzed by microsomal enzymes, then occur in rapid succession.

The first step in the transformation of the arachidonic acid to prostanoids (Figure 4) is a concerted process involving combination with molecular oxygen, and which is catalyzed by a cyclooxygenase using heme as a cofactor. This affords the endoperoxide PGG_2, which is then transformed into PGH_2 by an enzyme requiring tryptophan or a related substance as a

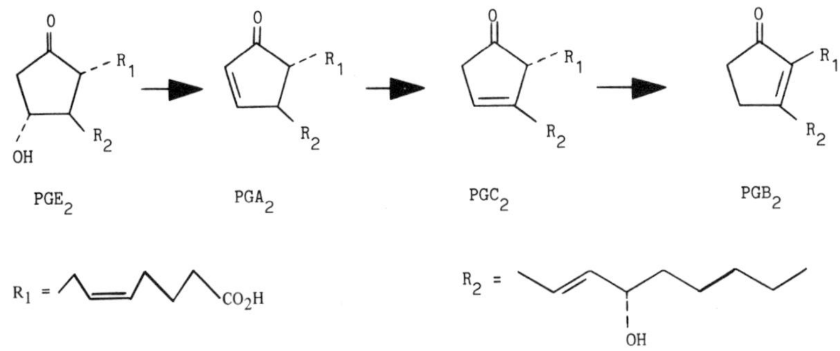

FIGURE 5. Formation of prostaglandins (PGs) A_2, C_2, and B_2.

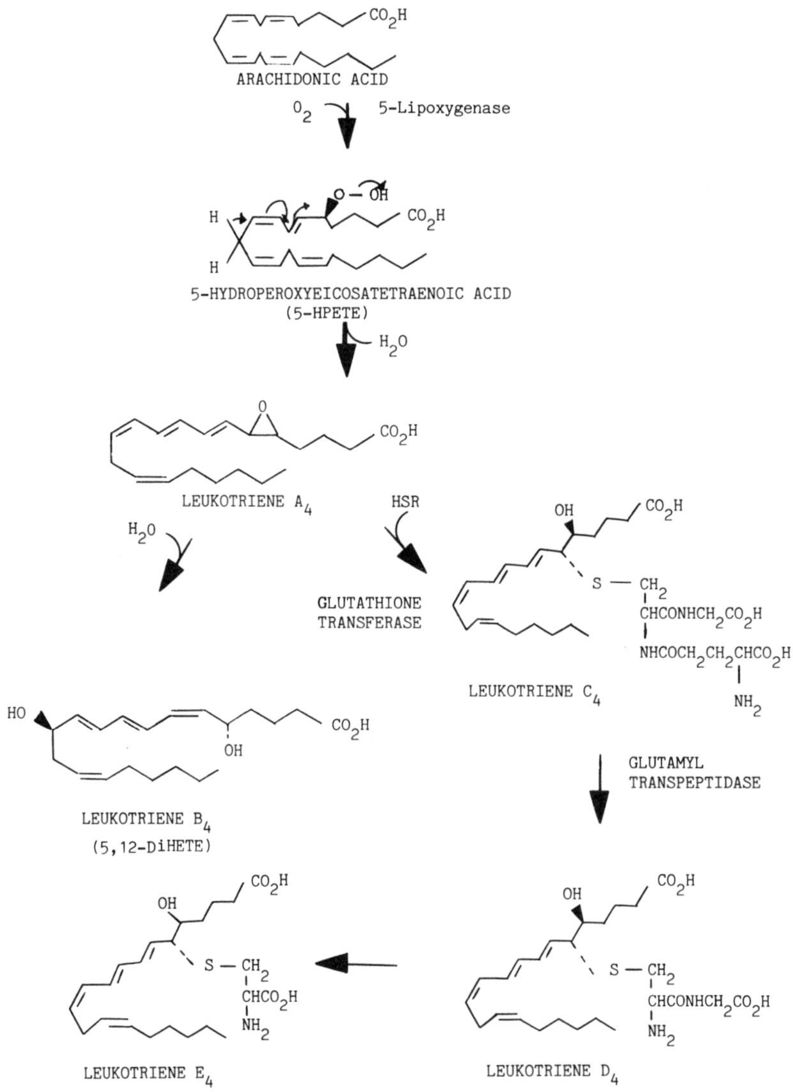

FIGURE 6. Biosynthesis of leukotrienes (4-series).

Table 1
SOME PHARMACOLOGICAL ACTIONS OF NATURALLY OCCURRING PROSTANOIDS

Prostanoid group	Pharmacological effects
PGA	Vasodilator; inhibitor of GAS[a], increases RBF[b]
PGB	Generally relatively inactive
PGC	Vasodilator
PGD	Inhibits BPA[c] (PGD_2)
PGE	Vasodilator; bronchodilator; uterine stimulant; inhibits GAS; inhibits BPA (PGE_1)
PGF	Vasodilator/vasoconstrictor (species dependent); bronchoconstrictor; uterine stimulant; luteolytic
PGG, PGH	Vasodilator/vasoconstrictor (species dependent); induces BPA; bronchoconstrictor
TXA_2	Vasoconstrictor, bronchoconstrictor, induces BPA
TXB_2	Generally relatively inactive
PGI_2	Vasodilator; inhibits BPA and GAS; increases RBF

[a] Gastric acid secretion.
[b] Renal blood flow.
[c] Blood platelet aggregation.

cofactor. PGG_2 and PGH_2 are unstable compounds with half-lives of about 5 min at 37°C, pH 7.5. PGH_2 is then further metabolized in several different ways. Three pathways give rise, respectively, to the prostaglandins PGE_2, PGD_2, and $PGF_{2\alpha}$.* PGA_2, PGC_2, and PGB_2 are probably artifacts derived by dehydration and isomerization of PGE_2 during isolation procedures (Figure 5). Thromboxane A_2, which is unstable ($t_{1/2}$, 30 sec at 37°C, pH 7.5), is formed from PGH_2 by thromboxane synthetase and is further transformed nonenzymically into the stable TXB_2. Prostacyclin (PGI_2), derived from PGH_2 by prostacyclin synthetase, is also unstable ($t_{1/2}$, 3 min at 37°C, pH 7.5) being further transformed, nonenzymically to 6-keto-$PGF_{1\alpha}$. In yet another pathway, the cyclopentane ring of PGH_2 and PGG_2 is cleaved, affording the C_{17} straight-chain acid, known as HHT, together with malondialdehyde.

The direction of the metabolism varies with different tissues; thus, in the lung all the pathways take place, whereas in blood platelets and blood vessel walls the main products are, respectively, TXA_2 and PGI_2. Further details of the biosynthesis of prostanoids is available in various books and reviews which list the numerous original references.[1,2,4]

The biosynthesis of leukotrienes from arachidonic acid is shown in Figure 6.[27,28] The initial stage, catalyzed by a lipoxygenase, gives rise to 5-hydroperoxyeicosatetraenoic acid (5-HPETE) which, on elimination of a molecule of water, affords the epoxide leukotriene A_4.

This can either undergo transformation into the dihydroxy acid LTB_4 or combination with glutathione, catalized by glutathione transferase, to give LTC_4. The latter is converted by loss of glutamine into LTD_4 and then by further degradation into LTE_4.

Biological Activity and Therapeutic Applications

Prostanoids are characterized by a wide variety of pharmacological actions and a high level of biological potency. The literature on this subject is vast and the reader is referred to reviews and books for more detailed accounts.[1-11, 17-19]

Some of the principal pharmacological actions exhibited by the main individual groups of prostanoids are shown in Table 1. Those listed have been selected mainly for their more immediate relevance to drug design in the appropriate areas; other important properties, such

* PGE_2, PGD_2, and $PGF_{2\alpha}$ can be formed nonenzymically. However, isomerases for the synthesis of PGs E_2 and D_2 have been identified, but the existence of a reductase for the formation of $PGF_{2\alpha}$ is doubtful.

Table 2
SOME PHYSIOLOGICAL FUNCTIONS ASSOCIATED WITH PROSTANOIDS

Vascular tone	Platelet aggregation
Hemostasis	Bronchial tone
GI function	Reproduction
Modulation of sympathetic nervous system	Endocrine function — pancreas, ovary, pituitary
Vascular permeability to blood proteins	Control of body temperature
	Modulation of immune response

Table 3
SOME PATHOLOGICAL STATES ASSOCIATED WITH AN IMBALANCE OF PROSTANOIDS

Hypertension	Fever
Atherosclerosis	Inflammation
Angina	Psoriasis
Asthma	Cancer
Cardiac arrythmias	Diabetes
Gastric ulcer	Thrombosis
Dysmenorrhea	Peripheral vascular disease

as effects on the autonomic and central nervous systems (CNS) and metabolic effects, e.g., inhibition of lipolysis, are also exhibited by some of the groups. It must also be emphasized that this summary is a broad generalization and only those properties which stand out as being important or relatively well known in each group are included. Some properties, e.g., contraction/relaxation of smooth muscle, are shown to a greater or lesser degree by virtually all prostanoids. Certain groups, e.g., the E series, are characterized by a broad range of biological effects, whereas others, e.g., the PGAs, have a more limited range. Table 1 also reflects the fact that some groups have been investigated more fully than others. Sometimes, opposite biological effects are seen in different species; e.g., $PGF_{2\alpha}$ is a vasodepressor in the cat and rabbit but is a pressor in the dog and rat.

The activities tend to be similar throughout the subgroups, but there are important differences; e.g., PGE_1 is an inhibitor of platelet aggregation, whereas PGE_2 is much less active or even proaggregatory.

The leukotrienes are also highly potent in their contractile effects on smooth muscle, including the lungs. LTB_4 is noteworthy for its chemotactic effects.[16,29,30] Further research in this relatively new area may well reveal a wider range of activities.

The opposing physiological effects of endogenous prostanoids play an important role in maintaining control of a variety of physiological functions, some of which are listed in Table 2. As a consequence, many disease states (e.g., those shown in Table 3) may be associated with an imbalance of the normal prostanoid levels. The employment of prostanoids in therapy is thus aimed at correcting this imbalance, either by administration of a prostanoid agonist to treat conditions where a deficiency of the appropriate natural prostanoid is involved, or a prostanoid antagonist to correct an excess of the natural substances. Excess prostanoid activity may also be blocked by the inhibition of prostanoid biosynthesis; aspirin and other nonsteroid anti-inflammatory agents are known to act by this mechanism.

Similar considerations apply with the leukotrienes. Thus, an excess of SRS-A, which is probably a mixture of leukotrienes, is a major cause of bronchospasm in asthma, and current research is aimed at the discovery of leukotriene antagonists or synthesis inhibitors suitable for administration in asthmatic attacks.

MECHANISMS OF ACTION

The wide variety of biological actions of the prostanoids and the markedly different spectra of biological activity shown by the various prostanoid groups have posed considerable problems in the elucidation of their mechanism of action. The majority of prostanoids are biosynthesized at the site where they exert their physiological effects; they are not stored in the tissue and are rapidly metabolized. Although prostanoids in some respects mimic the effects of hormones, most of them are not circulating hormones since they do not survive even a single passage through the lungs, where they undergo metabolic inactivation. However, PGA_2 and prostacyclin (PGI_2) are not inactivated in this manner. Prostacyclin blood levels are in fact increased on passage through the pulmonary circulation due to release of PGI_2 from the lung tissues.[31] This may be important for maintenance of the normal healthy state, which depends upon the correct balance between the antithrombotic PGI_2 and the thrombotic TXA_2. However, recent work has indicated that PGI_2 is in fact not a circulating hormone.[32]

Much evidence accumulated over the past 20 years supports the theory that the biological activity of prostanoids is mediated through the cyclic nucleotides, cyclic-3′,5′-adenosine monophosphate (cAMP) and cyclic-3′,5′-guanosine monophosphate (cGMP), which in turn play key second messenger roles in regulating intracellular function. The first suggestion of this mechanism came from the observation that PGE_1 blocked the action of lipolytic agents, e.g., noradrenalin, which acts by raising cAMP levels in fat cells.[33] It was also observed that noradrenalin brings about a release of material with prostaglandin activity, which suggested that prostaglandins may be formed as a consequence of cAMP action and function as negative feedback regulators.[34]

However, by contrast, more recent data showed that in several tissues prostaglandins stimulated cAMP formation.[35,36] This property was noted to be associated particularly with the E prostaglandins, which led to the conclusion that the actions of this prostanoid are largely mediated via an increase in cAMP levels by activation of the enzyme adenyl cyclase, which catalyzes its formation from adenosine triphosphate (ATP):

$$\text{ATP} \xrightarrow[\uparrow \text{ PG}]{\text{Adenyl cyclase}} \text{cAMP}$$

This is now regarded as the principal mode of action, the negative feedback mechanism, which involves a lowering of cAMP, probably being confined to only a few tissues. In fact, stimulation of prostaglandin biosynthesis by cAMP has been demonstrated in several tissues and a positive feedback between the prostaglandins and the cAMP system has therefore been suggested, which may be a more general phenomenon.

In contrast to the E series, the F-type prostaglandins have only a very weak action on cAMP, which suggests that these prostanoids act on a different receptor by a different mechanism. Although backed by less evidence than is the case with the cAMP theory and the E series, it has been proposed that the actions of the PGFs may be mediated through the nucleotide cGMP. This fits in well with the observation that cAMP and cGMP have opposing actions, which would explain the fact that the PGEs and the PGFs often exhibit opposite biological effects (see Table 1).

This hypothesis is reinforced by the discovery of separate receptors for the E and F prostaglandins in certain tissues. Thus, receptors with a specific affinity for E prostaglandins have been found for example in lipocytes,[38] affinity for which showed a good correlation with the ability of the prostaglandin to generate cAMP. Receptors for $PGF_{2\alpha}$ have been identified in several tissues, e.g., in corpora lutea.[39]

The general observations that E prostaglandins act by increasing cAMP levels to some extent links in with the effects of β-agonists and phosophodiesterase inhibitors, which also increase levels of this hormone and, like PGEs, relax smooth muscle, e.g., in the lung. Detailed studies of the relationship of prostaglandins to cAMP have been carried out in platelets. Here PGE_1, PGD_2, and PGI_2, which inhibit platelet aggregation (Table 1), appear to do so by increasing the concentration of cAMP.[40] The level of anti-aggregatory activity parallels the increase in cAMP levels, in both respects PGI_2 being much more potent than PGE_1 or PGD_2. Interestingly, here PGD_2 acts on a different receptor than does PGE_1 and PGI_2. Conversely, PGG_2, PGH_2, and TXA_2, which are proaggregatory, reduce the activity of adenyl cyclase. PGE_2 has a weakly biphasic action, which correlates with the finding that this prostanoid increases cAMP concentrations at inhibitory concentrations and produces a lowering of cAMP when present in amounts which promote aggregation. Details of the involvement of prostanoids and cyclic nucleotides in a number of different tissues have been reviewed.[1c,10]

The stimulation of smooth muscle by prostaglandins may be associated with the depolarization of cellular membranes and release of bound calcium.[41] It is possible, therefore, that they act by affecting intracellular calcium levels.

STRUCTURE-ACTIVITY RELATIONSHIPS

Numerous analogs of prostanoids have been prepared since the completion of the first total syntheses of prostaglandins in the mid-1960s. Many of these have been subjected to structure-activity studies which have afforded much information on the structural requirements necessary for biological activity, as well as revealing certain specific chemical features which bring about enhanced activity. However, published results on the biological testing have so far been very uneven, with comprehensive data being available on some analogs and very little on others, many of which have only been disclosed in the patent literature. Also, a large variety of different test methods have been used and considerable caution has to be exercised in comparing results from different investigators. For these reasons, and because of the complexity of the subject, it is often only possible to give broad generalizations, highlighting the main features of interest with appropriate examples.

Structural Features Necessary for Activity

In general, the complete molecular structure of each natural prostanoid (Figure 1) is required for the full range of biological activities. However, a substantial level of activity is retained on many biological systems so long as the following structural features are present:

1. The cyclopentane ring bearing at least one of the oxygen functions.
2. The terminal carboxylic acid or an equivalent group.
3. The ω-chain hydroxy group.
4. The correct spatial arrangements of the oxygen functions and the *trans* stereochemistry at C_8, C_{12} of the side chain attachment to the ring.
5. The side chains approximating to their normal length, although some variation is permissible, particularly in the terminal portion of the ω-chain beyond C_{15}.

Recent work with leukotriene C and D analogs (Figure 2),[42,43] using assays on guinea pig pulmonary parenchymal strip and ileum, has shown that the presence of the following features is important for the retention of a substantial level of contractile activity:

1. The thioether-linked peptide attached to the C_6 position.
2. A hydrophobic ω-portion of the molecule of which the length is more important than the precise stereochemistry.

3. The C_5 hydroxy group.
4. The free terminal amino group of the peptide.
5. The free glycine carboxyl group.
6. The conformational relationship between the carbon chain and the peptide group.

Results of leukotriene structure-activity on other systems will no doubt emerge in due course, enabling more general conclusions to be drawn in this area.

Structure-Activity of Analogs

The introduction of prostanoids into clinical practice has been seriously limited by the presence of undesirable side effects. It is unfortunate that the wide range of biological activities which has been responsible for exciting so much interest in their potential medical use has at the same time been a major hindrance to their commercial exploitation, since obviously a useful property in one clinical area is a potential side effect in another.

The principal objective has therefore been the discovery of analogs with a more favorable spectrum of activity. The approach to this work has been largely empirical, and modification has been carried out extensively on all parts of the prostanoid molecule. However, some analogs have been designed with the objective of inhibiting the metabolic inactivation of the natural prostaglandins by inserting blocking groups at or near the centers of the molecule where metabolism occurs, and others have been synthesized in order to increase the chemical stability of those natural prostanoids, e.g., the PGAs and PGIs which tend to undergo decomposition on storage. Biological evaluation has usually been carried out in primary screens designed to assess potential clinically useful properties and then in tests which are indicative of potential side effects — promising compounds being submitted to more detailed pharmacological evaluation.

From the available data the following general conclusions may be drawn:

1. Relatively few individual chemical modifications have emerged as being important in conferring a major enhancement of biological potency and/or selectivity of action.
2. The most successful of these has involved modification of the ω-chain in the C_{15}–C_{17} region, e.g., with the following groups: 15-[44,45] and 16-methyl-,[45-47] 16-methylene-,[48] 16-fluoro-,[49,50] 17-phenyl-,[51] and 16-phenoxy-[52] (Figure 7).
3. Variation of the carboxylic (α) chain has been much less successful and α-chain analogs show reduced activity in most tests (Figure 8). However, 3,7-inter-*m*-phenylene-3-oxa-PGE_1 (I), which has only weak smooth muscle spasmogenic activity, was found to be ten times more potent than PGE_1 as an inhibitor of blood platelet aggregation.[53] Also, some selectivity was found with 2-cyano and 2-fluoro substituents (II). Thus, 2-cyano-PGE_1 was half as active as PGE_1 in lowering blood pressure, but had only very weak activity on the isolated gerbil colon. 2-Fluoro-PGE_1 was very active in increasing coronary blood flow in isolated perfused rat hearts, but very weakly active on smooth muscle in vitro.[54]
4. Cyclopentane ring analogs are also relatively inactive. However, there are some notable exceptions which are illustrated in Figure 8. The 11-deoxy-11α-hydroxymethyl-PGE_2 (III) exhibits a level of rat uterine stimulant activity equivalent to that of $PGF_{2\alpha}$, but with significantly reduced side effects, thus indicating that the distance between C_{11} and C_{15} hydroxy groups can be expanded to accommodate certain receptors selectively.[55] Although replacement of the cyclopentane by heterocyclic ring systems has generally afforded relatively inactive compounds, the imidazolid-2,4-dione (hydantoin) analog (IV) has potent antithrombotic properties.[56] With PGH_2, the endoperoxide bridge may be replaced with ether[57,58] (V) and diaza[59] bridges (VI) with retention of the potent thrombotic properties characteristic of this prostanoid.

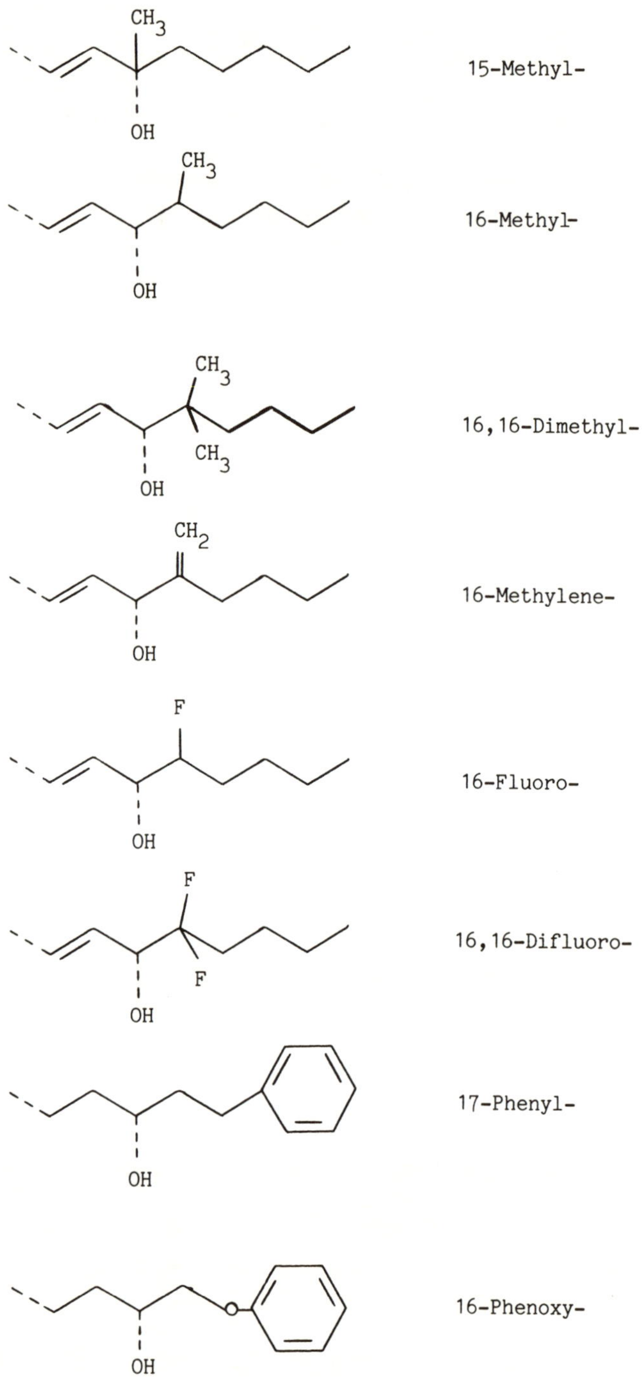

FIGURE 7. Some ω-chain modifications of prostanoids which have conferred enhanced pharmacological activity and/or selectivity.

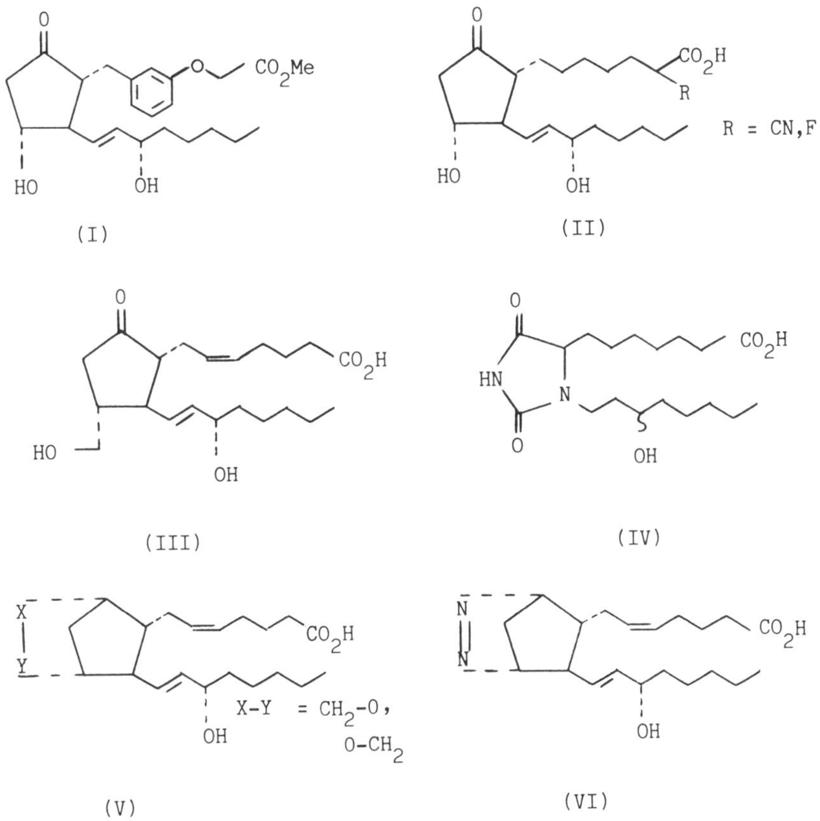

FIGURE 8. α-Chain and ring-modified prostanoids with enhanced activity and/or selectivity.

5. Substantial variations in pharmacological profile have been brought about by the combination of two or more structural changes in the molecule. The following examples of these hybrid analogs (Figure 9) illustrate this finding:

   a. 15-Deoxy-16-hydroxy-16-methyl-PGE_1 (VII) shows 30 to 100 times greater antisecretory activity than PGE_1 with only 1/8 of the potential diarrheal side effects.[60]
   b. A *cis* Δ^4-16-phenoxy-PGE_2 analog (VIII), one of a class of compounds with gastric-antisecretory properties, has a substantially lower level of the undesirable uterine stimulant and luteolytic properties than the corresponding compound having the α-chain double bond in the natural Δ^5 position.[61]
   c. A PGA_2 analog (IX) in which 16,16-dimethyl- and 18-oxo-groupings are present, was more potent as a vasodepressor than compounds having only one of these groups.[62]
   d. The *trans* Δ^2 grouping in the abortifacient prostaglandin "Gemeprost" (X) confers additional selective properties over those due to the 16-alkyl groups alone.[63]
   e. 16-Fluoro-13-dehydro-$PGF_{2\alpha}$ (XI) has similar antifertility activity to the corresponding nonfluoro compound, but only 1/5 the uterine stimulant activity.[64]
   f. The 1-methyl sulfonyl-16-phenoxy-PGE_2 (XII) has 10 to 30 times the potency of PGE_2 in tests which are indicative of antifertility properties, but with markedly lower side effects.[65]
   g. The 9-methylene-16,16-dimethyl-PGE_2 (XIII) shows a greater specificity for uterine-stimulating activity over inhibition of gastric acid output and enteropooling then does 16,16-dimethyl-PGE_2.[66]

6. The carbocyclic (carbacyclin) (XIV)[67-70] and nitrogen-ring (XV) analogs of prostacyclin

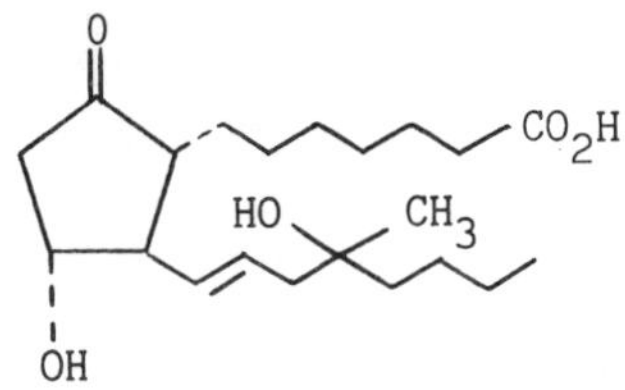

(VII)

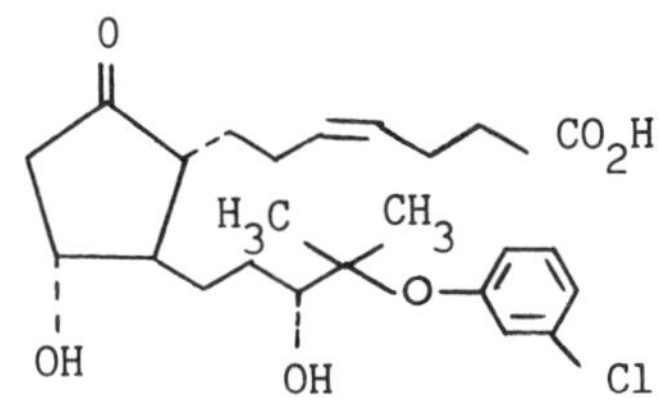

(VIII)

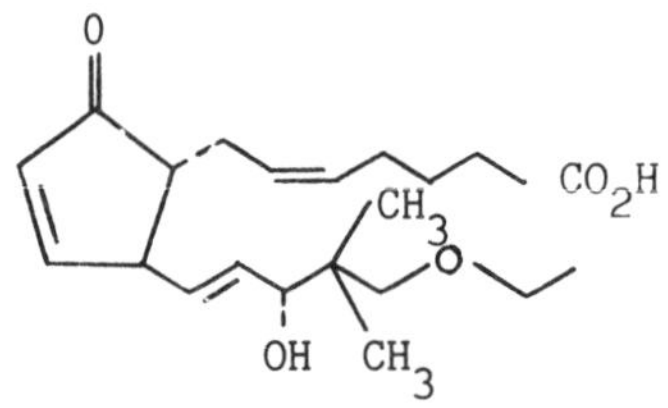

(IX)

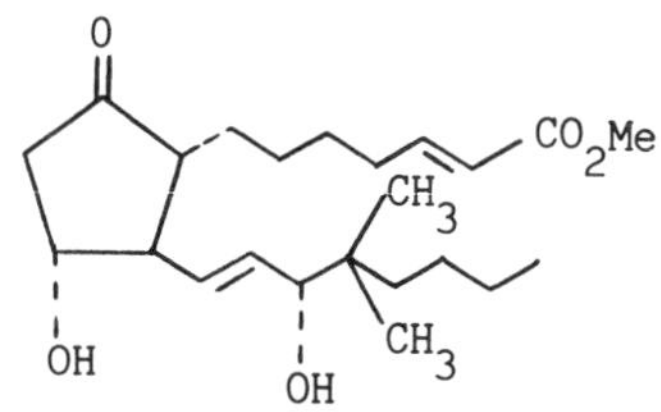

(X)

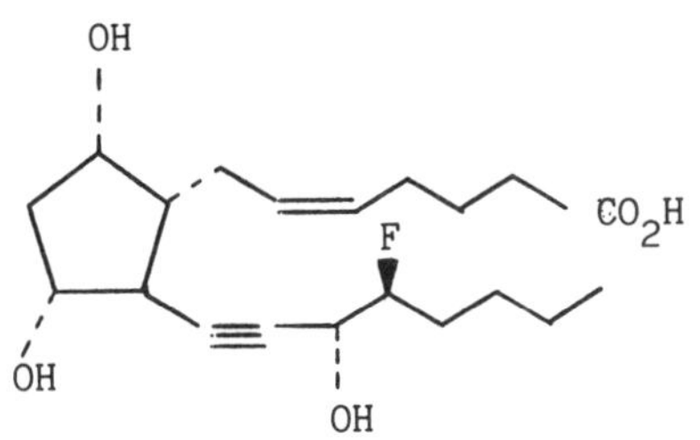

(XI)

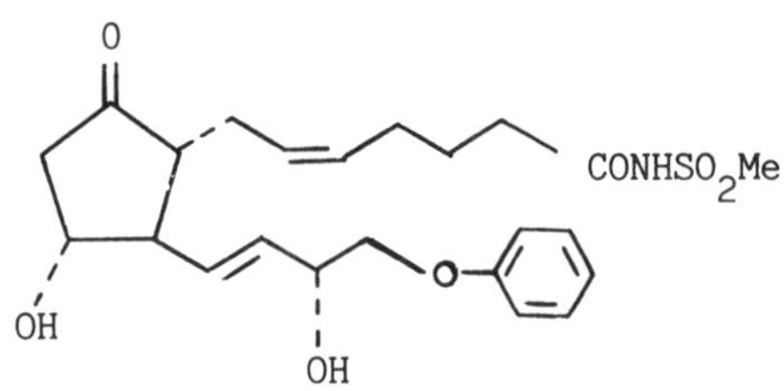

(XII)

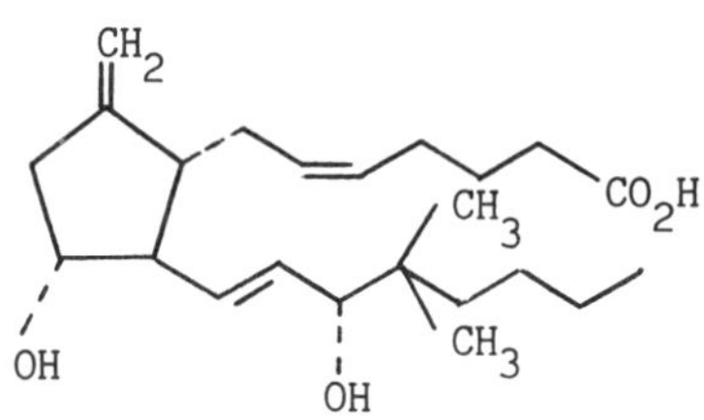

(XIII)

FIGURE 9. Some prostanoid analogs in which potentially useful properties have resulted from the combined effect of two or more structural modifications.

(Figure 10) retain the antithrombotic and antihypertensive properties of the parent substance but at a lower and somewhat modified level. These compounds are chemically stable since they do not possess the labile enol ether grouping, which is the cause of instability in PGI_2.

7. Several stable thromboxane A_2 analogs, e.g., (XVI) (Figure 10), in which one or both of the labile ring ketal oxygens have been replaced by carbon or oxymethylene bridges, retain some thromboxane-like activity.[72-75] Certain of these compounds are partial agonists, i.e., they have a degree of TXA_2-like activity, but are also able to block the action of the natural substance.
8. The 7- and 13-oxa[76,77] groupings tend to confer antagonist rather than agonist properties. Prostanoids having these features which have shown antagonist action usually have a simple structure, e.g., they lack one or more of the OH groupings (Figure 11).

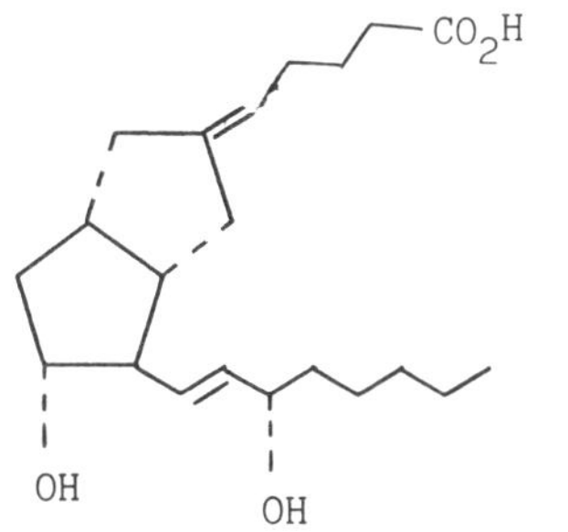

CARBACYCLIN

(XIV)

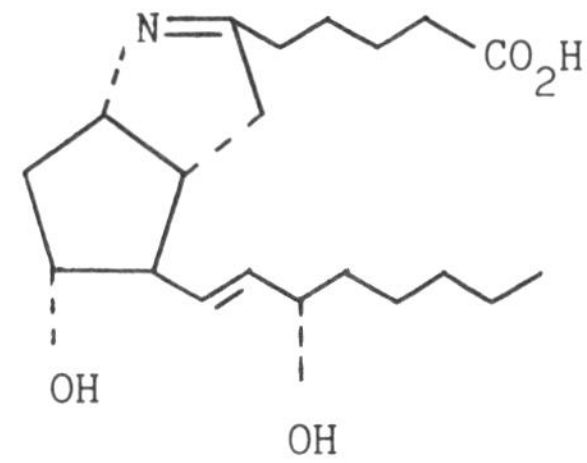

(XV)

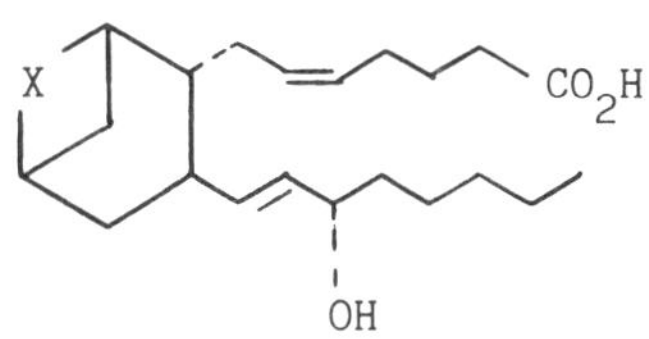

(XVI)

X = CH_2, CMe_2

FIGURE 10. Some biologically active, chemically stable analogs of prostacyclin (PGI_2) and thromboxane A_2.

Analogs modified in the C_{15}–C_{17} region of the ω-chain have been the most interesting to date and some of these have shown a sufficiently favorable pharmacological profile to merit extensive clinical evaluation. Some detailed structure-activity data illustrating the effect of structural changes in this portion of the molecule are presented in Tables 4 to 7. Table 4 shows the effect of 15-, 16-methyl, and 17-phenyl groupings on the uterotonic potency of $PGF_{2\alpha}$ and of 16-methyl and 17-phenyl groupings on that of PGE_2.[78] This study, which was carried out by a group at the Karolinska Hospital, Stockholm, used both single i.v. and i.v. infusion techniques. All of the substituents produced an enhanced activity by both routes of administration — that of 16,16-dimethyl-PGE_2 being particularly noteworthy. Table 5 illustrates the very high activity conferred by the 16-phenoxy grouping which was discovered by I.C.I. scientists.[52] This arose from a study of the luteolytic (antifertility) activity of $PGF_{2\alpha}$ analogs in the hamster. It was first noted that activity was enhanced marginally by lengthening the ω-chain by two methylene units and that a further rise in potency occurred on insertion of an ether linkage in place of the C_{17}-methylene. Further modification led to the replacement of the terminal five-carbon aliphatic chain by a phenyl group, producing a still higher level of activity. Potency of 16-phenoxy-$PGF_{2\alpha}$ in this test was increased yet again in analogs where the benzene ring contained the electron-withdrawing substituents — F, Cl, or CF_3. Interestingly, however, these substituents produced a lowering of potency in 11-deoxy-16-phenoxy-$PGF_{1\alpha}$.[79] Comparative data for 15-methyl and 16,16-dimethyl $PGF_{2\alpha}$ are also shown in Table 5. It is of interest that the 16-phenoxy $PGF_{1\alpha}$ analogs also exhibit increased uterine stimulant properties, but that the profile of activity in this test differs from that on luteolysis, thus affording a degree of selectivity.[52]

The 15- and 16-alkyl and 16-phenoxy groupings also give rise to a substantial increase in anti-ulcer and antigastric secretory activity in the E prostaglandins. Some data in which these properties are determined in the rat and compared with anticonstipatory activity in the

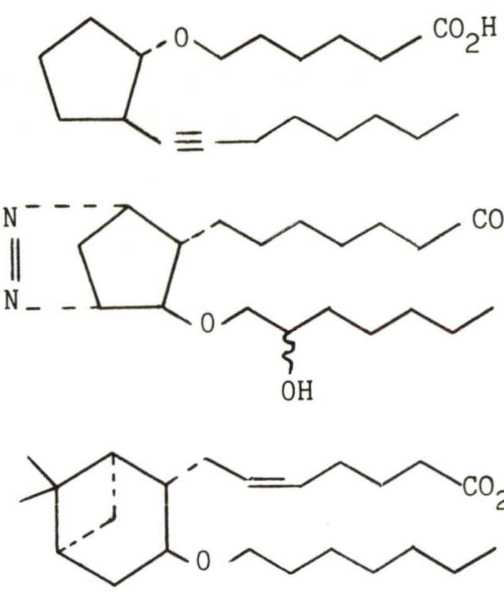

FIGURE 11. Some 7- and 13-oxa prostanoids with prostanoid antagonist properties.

Table 4
RELATIVE UTEROTONIC POTENCIES OF 15-, 16-, AND 17- SUBSTITUTED ANALOGS OF $PGF_{2\alpha}$ AND PGE_2 IN HUMAN VOLUNTEERS[78]

Compound		Single i.v. injection	i.v. Infusion
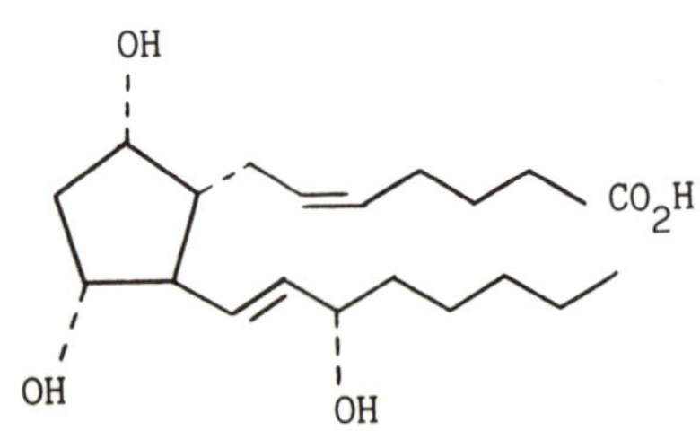	$PGF_{2\alpha}$	1	1
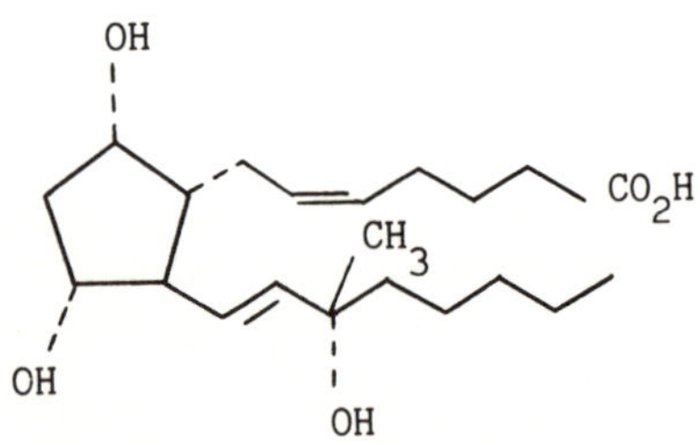		10 (5)[a]	20 (5)[a]
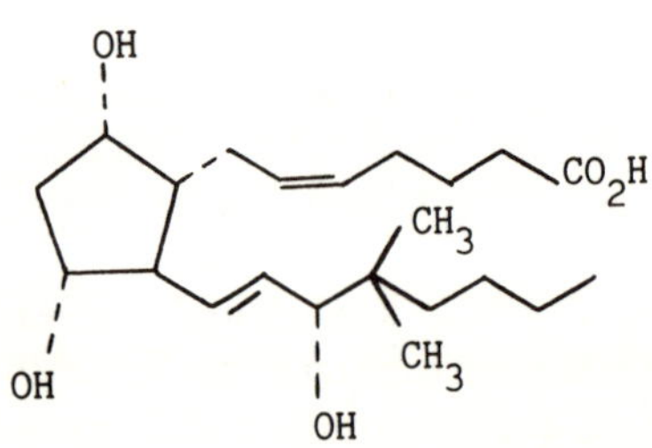		6—7 (7)	8 (4)

Table 4 (continued)
RELATIVE UTEROTONIC POTENCIES OF 15-, 16-, AND 17- SUBSTITUTED ANALOGS OF $PGF_{2\alpha}$ AND PGE_2 IN HUMAN VOLUNTEERS[78]

Compound	Single i.v. injection	i.v. Infusion
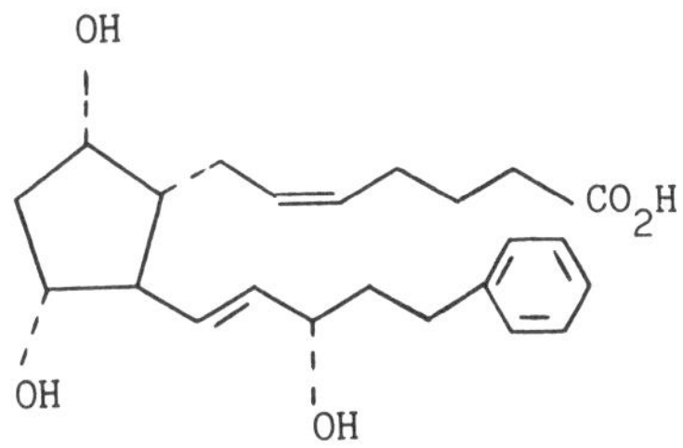	7—8 (7)	10 (3)
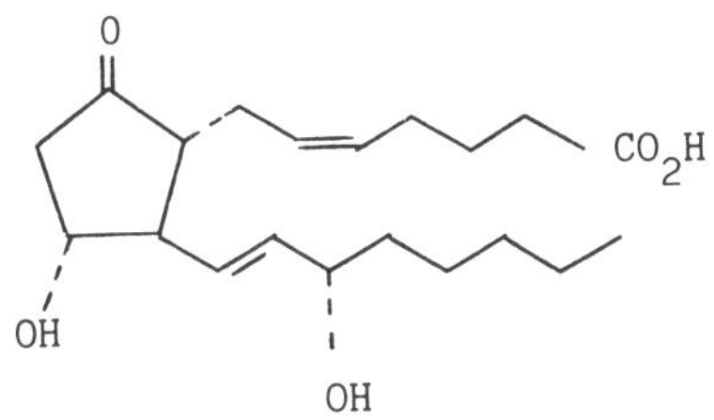 PGE_2	8	5
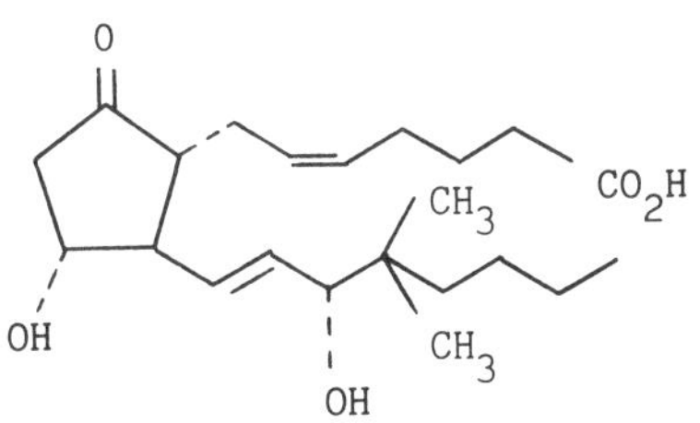	100 (8)	400 (6)
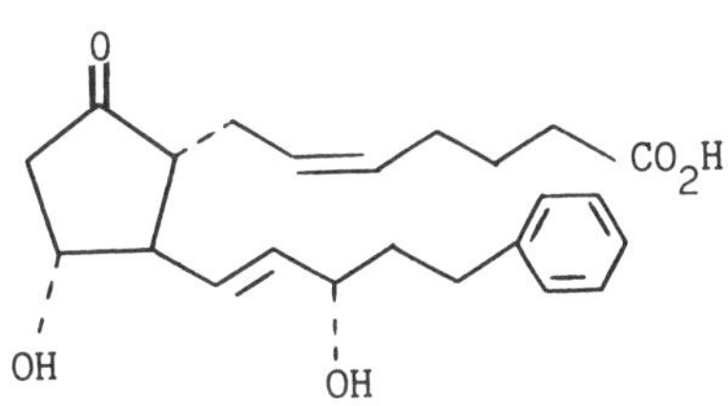	4 (9)	4 (1)

[a] Number of cases in parentheses.

mouse (which affords a measure of potential gastrointestinal side effects) are given in Table 6.[80,81] This study by May and Baker scientists (Essex, England), and work from other centers, has shown that there is a complex relationship between these groupings and their pharmacological profile. The C_{15}- and C_{16}-methyl analogs are more potent than the parent compound, but at best afford a marginal improvement in selectivity. A much higher level of both potency and selectivity is found in the 16-phenoxy compound, which is at least 300 times more potent than the parent, (±)-11-deoxy-PGE_1. Although the 16-phenoxy group also confers an enhancement in anticonstipatory potency, this increase was relatively modest, with the result that the selectivity index of the phenoxy compound is very high. Table 6 also shows the very varied results obtained when the phenyl group was substituted with *o*- *m*- *p*-chloro groups. I.C.I. workers have shown that the 16-phenoxy group also conferred an increase in gastric antisecretory potency over that of PGE_2 and have published data on a series of related

Table 5
LUTEOLYTIC ACTIVITY OF SOME (±)-$PGF_{2\alpha}$ ANALOGS IN THE HAMSTER[52]

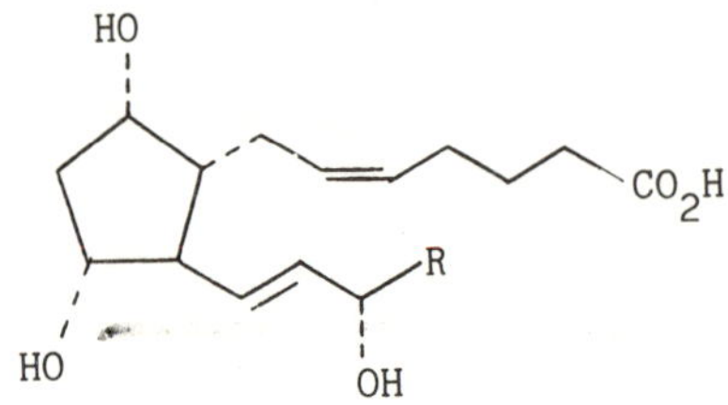

R	Lowest daily dose s.c. given on days 4 to 6 needed to terminate pregnancy (μg per hamster)
	25
	10
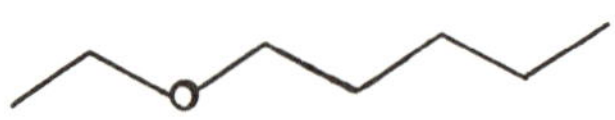	5
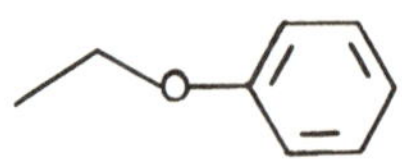	1
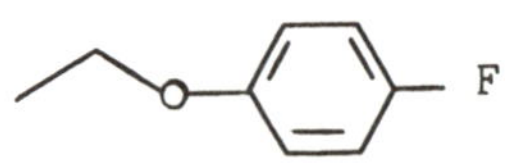	0.125
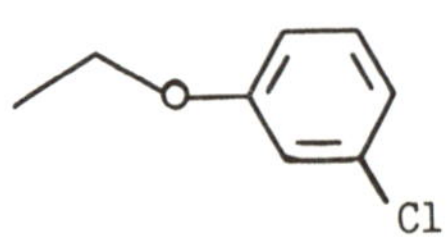	0.125
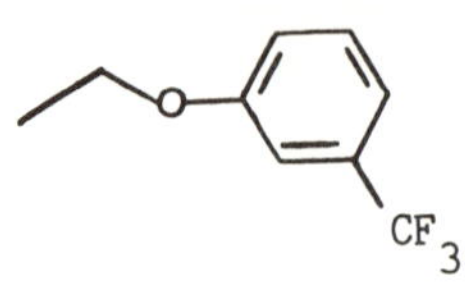	0.25
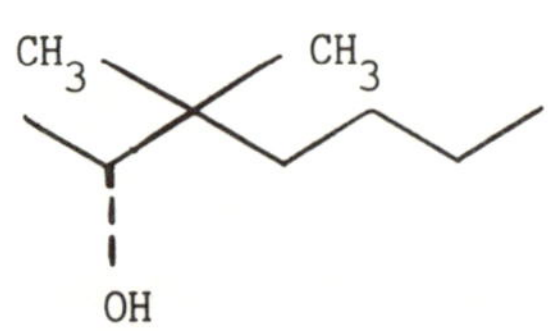	>5
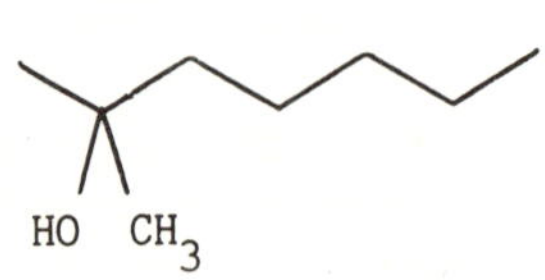	1

compounds.[61] Table 7 shows comparative data for a short series of ω-chain phenyl prostaglandin analogs as uterine stimulants, as bronchodilators, and on blood pressure.[82] The high level of antihypertensive potency with good tissue selectivity exhibited by the 13,14-dihydro-16-phenyl-PGE_2 analog is noteworthy.

It is evident from the above account that the overall pattern of the relationship between chemical structure and biological activity within the prostanoid area is very complex. Few

Table 6
ANTIULCER AND ANTICONSTIPATORY PROPERTIES OF SOME (±)-11-DEOXY 15- AND 16-METHYL AND 16-PHENOXY PROSTAGLANDINS[80,81]

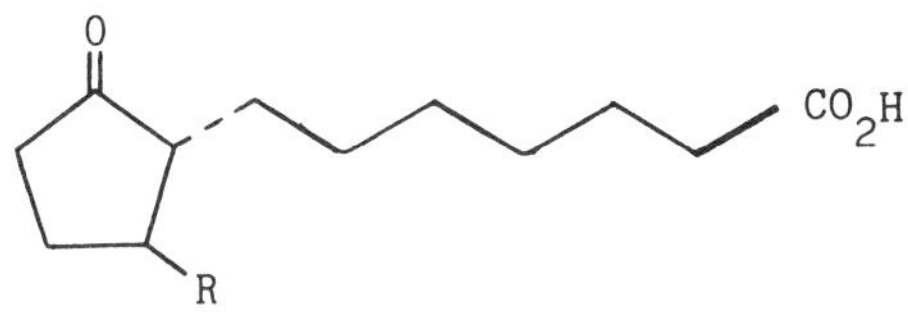

R	Rat indomethacin-induced ulcer[a] (μg/kg p.o.) A	Mouse anti-morphine constipation[a] (μg/kg p.o.) B	Selectivity index B/A
PGE_2	20	800	40
16,16-Dimethyl-PGE_2	0.3	16	53
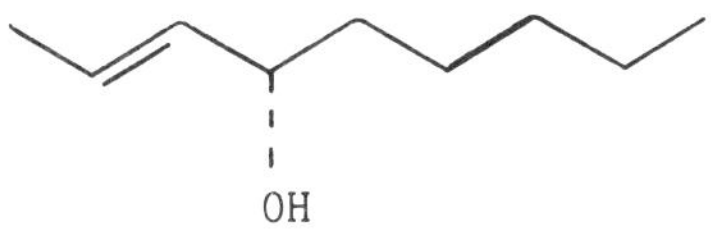 OH	<1000 (61, 70%)	>10,000 (10%)	>10
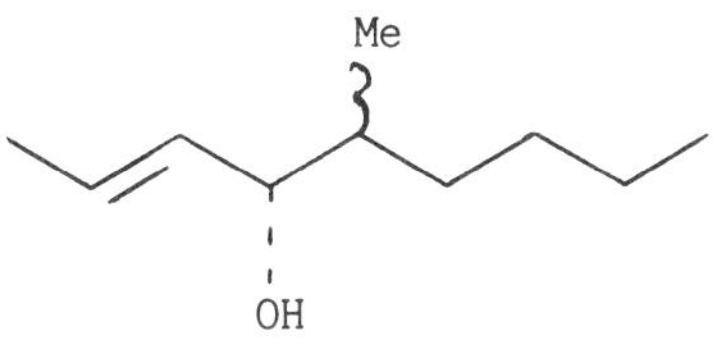 Me, OH	<100 (60%)	300	>3
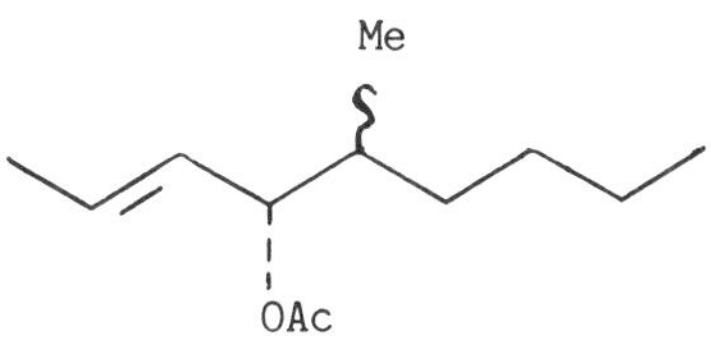 Me, OAc	10—100	1300	13—130
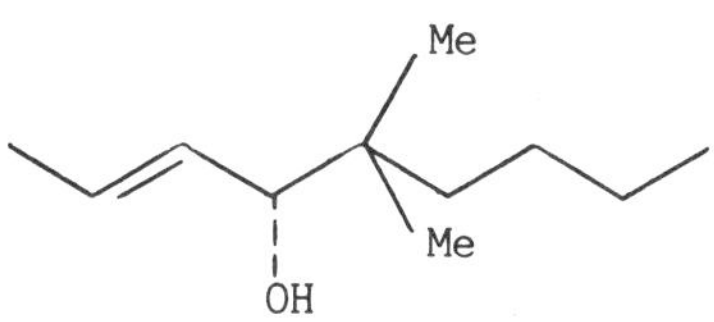 Me, Me, OH	<1000 (64%)	>10,000	>10
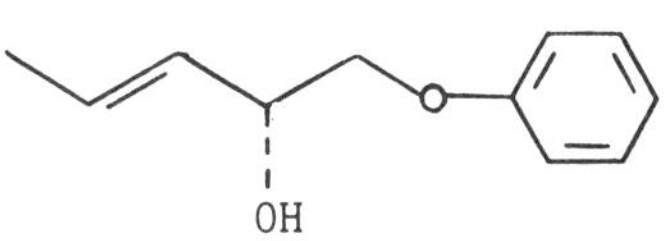 OH	3	100—400	330—470
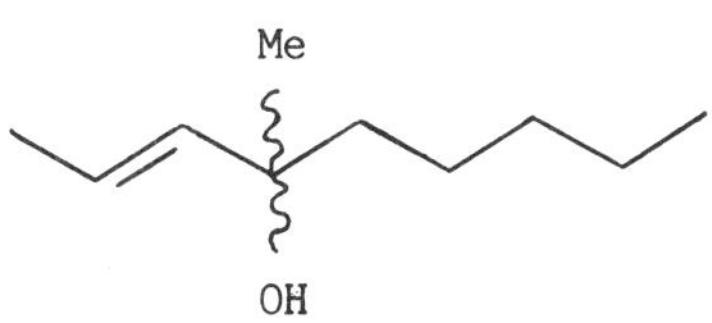 Me, OH	200	2700	14
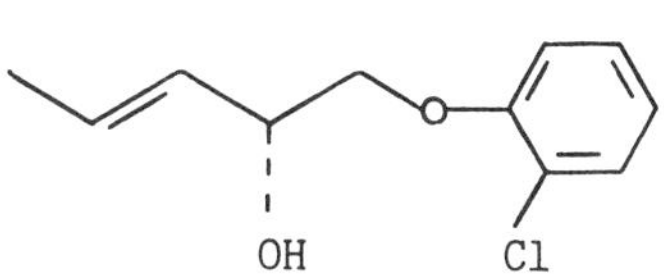 OH, Cl	100	2000	20

Table 6 (continued)
ANTIULCER AND ANTICONSTIPATORY PROPERTIES OF SOME (±)-11-DEOXY 15- AND 16-METHYL AND 16-PHENOXY PROSTAGLANDINS[80,81]

R	Rat indomethacin-induced ulcer[a] (μg/kg p.o.) A	Mouse anti-morphine constipation[a] (μg/kg p.o.) B	Selectivity index B/A
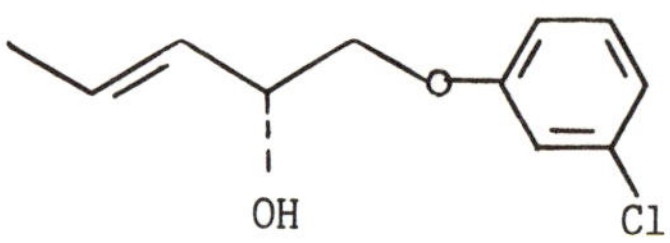	>1000	>10,000	—
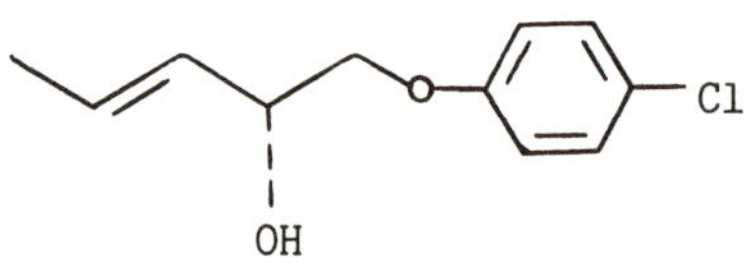	10	1300	130

[a] ED_{50} values unless otherwise shown.

Table 7
BIOLOGICAL ACTIVITIES OF ω-PHENYL-SUBSTITUTED PROSTAGLANDIN ANALOGS[82]

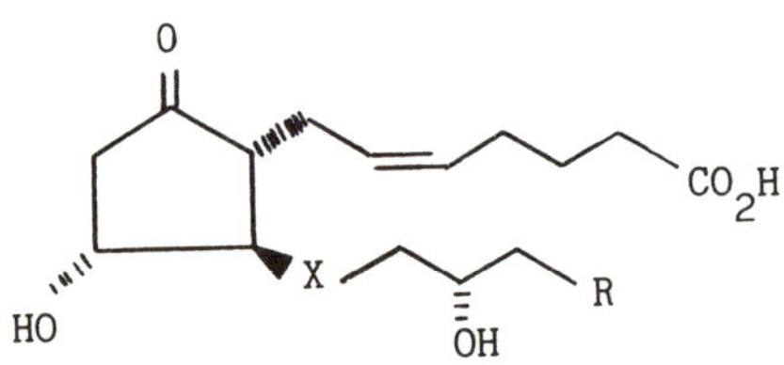

Compound	X	R	Effects: Uterine[a]	Bronchodilator[b]	Blood pressure[c]
PGE_2	CH=CH	n–C_4H_9	100	100	100
1	CH=CH	C_6H_5	3	77	500
2	CH_2CH_2	C_6H_5	1	10	2000
3	CH=CH	$CH_2C_6H_5$	100	13	25
4	CH=CH	$(CH_2)_2C_6H_5$	30	10	20
5	CH=CH	$CH_2C_6H_{11}$	5	10	100

[a] Spasmogenic effects (PGE_2 = 100) on isolated guinea pig uterus.
[b] Inhibition (PGE_2, 2.85 × 10^4 aerosol solution = 100) of histamine-induced bronchoconstriction in conscious guinea pigs.
[c] Threshold dose (PGE_2, 0.1 μg/kg, i.v. = 100) for depressor effect on blood pressure in anesthesized dogs.

unifying principles are at present available to rationalize structure-activity relationships, and the application of quantitative structure-activity techniques has been largely unsuccessful. The position is complicated further by the many stereoisomeric forms. A comparison of the activity of the various stereoisomers of PGs E_1, A_1, and $F_{1\alpha}$ on different systems have been carried out,[83] but detailed information on the variation of structure with stereochemistry is not yet available for most prostanoids. The biological activity also depends on the conformation of the molecule and it is likely that in many cases the differences in activity of analogs is in part attributable to their influence on the configuration of the molecule. It has been suggested that for some types of activity the prostanoid must adopt a configuration with the two side chains close in space and aligned in a manner which resembles a hairpin structure.[84,85] Further details of prostanoid structure-activity are available in various reviews and books.[11-14,17,86]

Leukotriene structure-activity studies are still in their early stages, the objective here being to design analogs which antagonize the bronchoconstrictor and other harmful effects of the endogenous substances. Synthesis of a leukotriene analog with antagonist properties, in which the 5,6 double bond is replaced by a methano grouping, has recently been published.[87]

METABOLISM

Prostaglandins are metabolized via the following four principal steps:

1. 15-Dehydrogenation.
2. 13,14-Hydrogenation.
3. β-Oxidation of the α-chain.
4. ω-Hydroxylation and oxidation.

These sequences are illustrated for PGE_2 in Figure 12. Stage 1 is catalyzed by the enzyme 15-hydroxyprostanoate dehydrogenase (PGDH) and stage 2 by Δ^{13} reductase. These two enzymes have a widespread occurrence in the tissues; as a consequence, prostanoids have an extremely short biological half-life, that for the primary prostaglandins being probably less than 30 sec. These first two steps give rise to 15-keto-13,14-dihydro compounds, which are the major metabolites in plasma. These have a longer half-life (10 min for the primary prostaglandins) than the parent substances and consequently occur in much larger amounts in the circulation. Further metabolism occurs via: (3) β-oxidation of the α-chain in two stages, affording dinor (C_{18}) and tetranor (C_{16}) compounds and (4) ω-hydroxylation and oxidation, to give water-soluble dicarboxylic acids which are excreted in the urine. Enzymes responsible for states 3 and 4 are found in the liver, lungs, kidney, and intestine, the major site of β-oxidation probably being in the liver.

$PGF_{2\alpha}$ is metabolized in the same way as PGE_2 to the corresponding C_{16} urinary metabolite (Figure 13), and other prostanoids follow a broadly similar pattern. In many cases, studies in man have not been carried out, and conclusions are based on animal data. In some instances, additional transformations can take place; thus, the 15-keto-13,14-dihydro metabolite may undergo reduction at C_{15} to the hydroxy compound, hydroxylation can occur at C_{19} followed by oxidation to the ketone, and the dicarboxylic acid metabolites may undergo further degradation by means of ω-chain β-oxidation. Also, PGE compounds can be reduced to PGFs by a 9-keto reductase and, conversely, oxidation of PGFs to PGEs may also take place.

Recent studies in the monkey showed that PGD_2 underwent a considerable degree of 11-keto reduction, affording $PGF_{2\alpha}$ compounds; the major metabolite found in the urine was dinor-$PGF_{2\alpha}$. The major metabolite of thromboxane B_2 in man was found to be dinor-TXB_2 (Figure 13). This departure from the normal pattern is probably a consequence of the fact that TXB_2 is a poor substrate for PGDH, this having been demonstrated in in vitro studies, so that the ω-chain is unchanged in the metabolite. Recent studies have indicated that the metabolic fate of TXA_2 is complex, although TXA_2 is hydrolyzed very rapidly to TXB_2 in aqueous medium in vitro (Figure 4). Conversion to TXB_2 in vivo appears in some cases to be only a minor pathway. It appears that this transformation may be preceded by the direct action of 15-PGDH and then Δ^{13} reductase on TXA_2; a number of pathways may be involved, in some cases including the formation of covalent derivatives with proteins.

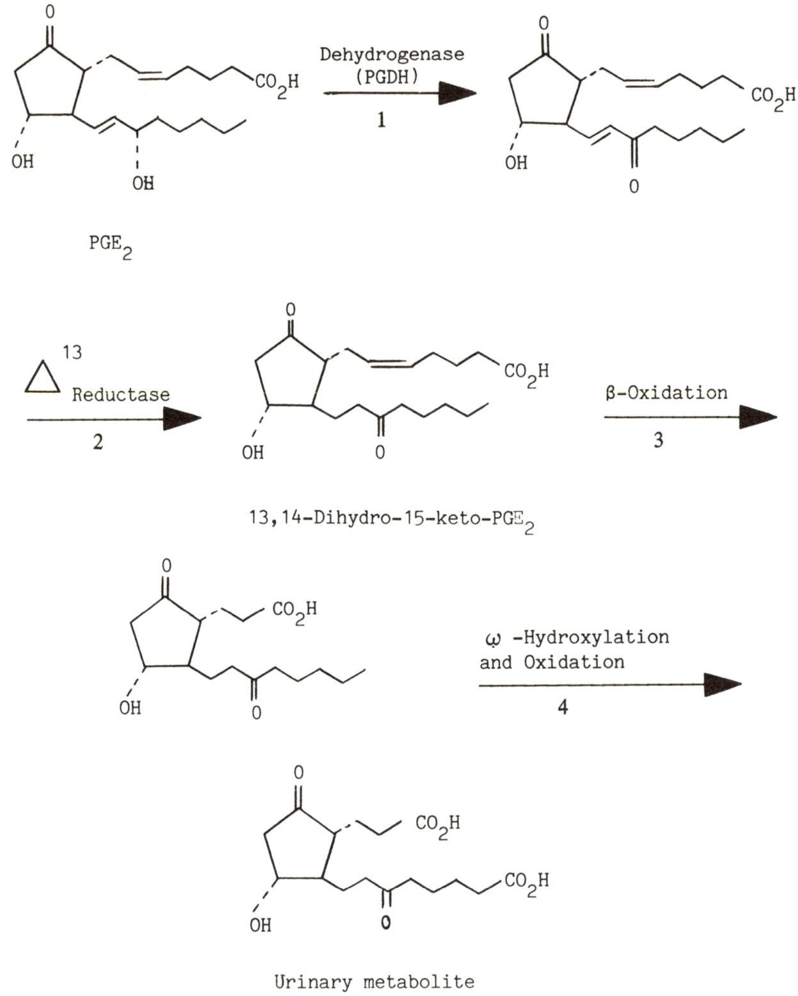

FIGURE 12. Metabolism of PGE_2 in man.

Prostacyclin metabolism has been studied in animals and recently in man. The human study indicates that PGI_2 undergoes conversion to the more stable 6-keto-$PGF_{1\alpha}$ (Figure 4), which is a poor substrate for PGDH and, therefore, is probably cleared mainly via β- and ω-oxidation. It is suggested that PGI_2 escapes enzymic degradation by PGDH initially, due to bad specificity for tissue uptake systems. It is suggested that the rate of administration may alter the direction of metabolism and products formed; further work may reveal different pathways.

The metabolism of leukotrienes has yet to be fully worked out. However, it is known that leukotrienes C, D, and E undergo biological inactivation by 15-hydroxylation, as shown for LTD_4 in Figure 14.[16] The metabolism of prostanoids is discussed in more detail in various reviews which list the original references.[5,18]

PROSTANOIDS IN CLINICAL USE

When the wide range of biological properties of the prostaglandins and their importance as physiological regulators became apparent in the 1960s, it was confidently expected that they would give rise to a new class of drugs with applications in a number of clinical areas.

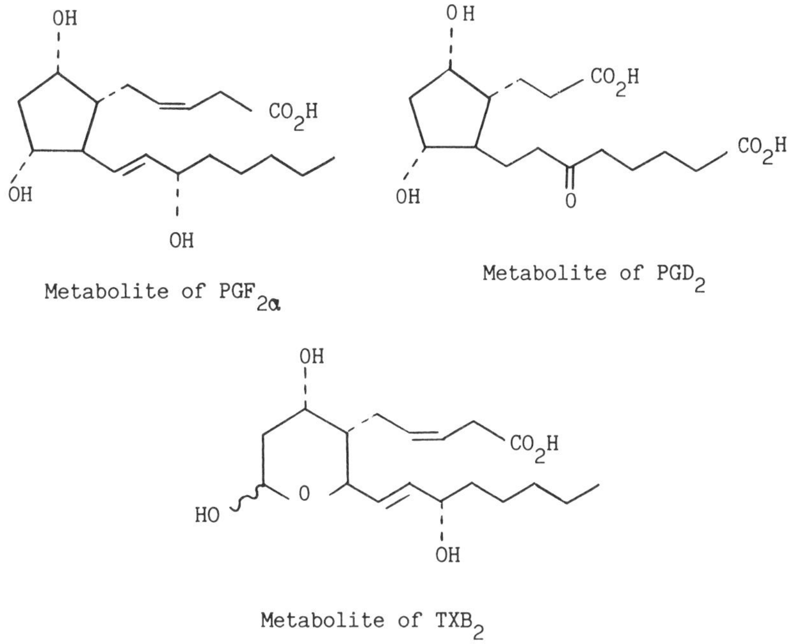

FIGURE 13. Major urinary metabolites of $PGF_{2\alpha}$, PGD_2, and TXB_2 in man.

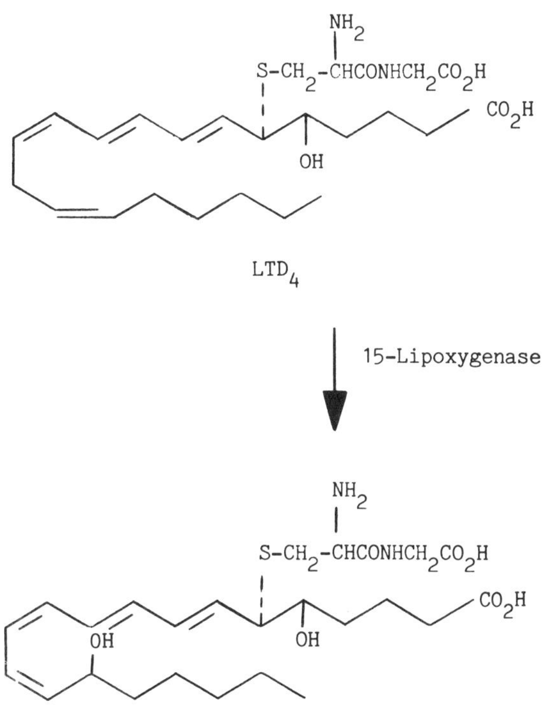

FIGURE 14. Metabolism of LTD_4.

Table 8
PROSTANOIDS — POTENTIAL CLINICAL APPLICATIONS

Biological activity	Potential clinical application
Uterine stimulation	Induction of labor, termination of pregnancy
Inhibition of gastric acid secretion, antiulcer	Treatment of peptic ulcers
Vasodilation, inhibition of platelet aggregation	Treatment of cardiovascular diseases, e.g., hypertension, thrombosis
Bronchodilation	Treatment of asthma

Considering how many successful drugs have been derived from other biologically active substances endogenous to mammalian tissues, such as steroid hormones, catecholamines, and histamine, it was reasonable to assume that prostanoids would afford an opportunity to apply this approach to drug discovery to an entirely new class of compounds. Moreover, since prostanoids are chemically very different in structure from the existing drugs, and probably act by different mechanisms, it was anticipated that they would complement rather than duplicate existing means of therapy and possibly lead to successful treatment of some disease states where no effective drugs were available.

Clinical areas in which prostanoids have aroused the most interest are in the termination of various stages of pregnancy, the induction of labor, and the treatment of peptic ulcer, asthma, and cardiovascular disease. These potential applications were suggested by the various pharmacological activities referred to in previous sections of the review, as indicated in Table 8. However, as previously indicated, the introduction of prostanoids into clinical practice has been seriously hampered by the presence of unacceptable side effects which are a consequence of their multiplicity of biological actions, as a result of which early hopes for their application to medicine have so far been realized only to a limited extent. To date, only four prostaglandins are in commercial use in the medical area. Three of these are natural prostaglandins — $PGF_{2\alpha}$, PGE_2, and PGE_1 — and the fourth is a PGE_2 analog. $PGF_{2\alpha}$ and PGE_2 have been available for gynecological use since the early 1970s, and PGE_2 has recently found limited application in cardiovascular disease. The analog, sulprostone is also used in the gynecological area.

Natural Prostaglandins — Termination of Pregnancy

The possibility of employing $PGF_{2\alpha}$ and PGE_2 for the termination of various stages of pregnancy was suggested by laboratory observations that these and other prostanoids stimulate uterine contractions in laboratory tests (Table 1) and the finding that they are endogenous regulators of the normal labor process. $PGF_{2\alpha}$ stimulates contraction of both the pregnant and nonpregnant human uterus. PGE_2 in fact relaxes the nonpregnant uterus in vitro, but contracts the pregnant uterus. During the last two trimesters of pregnancy it is considerably more potent than $PGF_{2\alpha}$.

$PGF_{2\alpha}$ was the first prostaglandin to be commercialized, as Prostin $F_{2\alpha}$ in sterile solution, by the Upjohn Company for the termination of pregnancy at the second trimester. This was followed by the introduction of PGE_2 sterile solution (Prostin E_2), also by Upjohn, for the therapeutic termination of pregnancy during the late first and second trimester, for the evacuation of the uterus in cases of missed abortion, and for evacuation of hyatidiform moles. $PGF_{2\alpha}$, which is available as a sterile solution, is more effective and better tolerated when administered by the intra-amniotic (intra-uterine) route rather than i.v. PGE_2, also available in solution, is more effective by the extra-amniotic than by the i.v. route. However, this formulation of PGE_2 has to some extent been replaced by the Prostin E_2 vaginal suppository, which has the advantage of being administered by a noninvasive route. Prostin E_2 tablets are also available for oral administration in cases of induction of labor when there are no fetal or maternal contraindications. Further details pertaining to these gynecological

Table 9
PROSTIN $F_{2\alpha}$ THAM (TROMETHAMINE) SALT (UPJOHN)

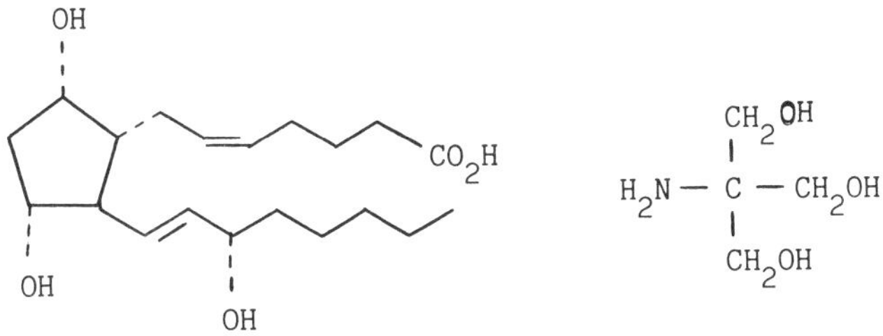

Approved name:	Dinoprost tromethamine
Brand name:	Prostin $F_2\alpha$ tromethamine
Chemical name:	Prostaglandin $F_{2\alpha}$ tromethamine (THAM) salt

Physical Properties (Tromethamine Salt)

White or slightly off-white crystalline powder, readily soluble in water at room temperature in concentrations up to at least 200 mg/mℓ. Stable for at least 24 months at 25°C. Storage should be at 4°C.

Formulation (route of administration)	Indication	Dosage	Side effects
Aqueous solution 5 mg/mℓ (intrauterine [intra-amniotic] or i.v. where intrauterine route contraindicated or less suitable)	Termination of pregnancy during the second trimester	According to individual requirement	Flushing, shivering, headache, dizziness, nausea, vomiting, diarrhea

uses of $PGF_{2\alpha}$ and PGE_2, including dosages and side effects (as presented in the manufacturers literature), are given in Tables 9 and 10.

These two prostaglandins are also marketed in Japan by the Ono Pharmaceutical Company. $PGF_{2\alpha}$ is available at Prostarmon-F Injectable for the induction of labor. It is also available for the therapeutic termination of first and second trimester pregnancy, where extra-amniotic administration is recommended since the incidence of side effects is lower than by the intra-amniotic and i.v. routes. Ono has also introduced PGE_2, as Prostarmon-E, as oral capsules and oral tablets for the induction and augmentation of labor at term. In the tablet presentation, the PGE_2, which is inherently unstable as a consequence of the β-ketol ring structure, is stabilized as a β-cyclodextrin clathrate.

$PGF_{2\alpha}$ and PGE_2 have now become well established for these obstetrical indications. In late pregnancy the response to the uterus to prostaglandins is similar to that of oxytocin. However, the prostaglandins are much more effective than oxytocin in the earlier stages of pregnancy and are regarded as agents of choice for mid-trimester abortions. For very early abortions they afford only a low success rate with an unacceptable level of side effects.

PGE_1 in Peripheral Vascular Disease and Gangrene

The finding that PGE_1 is a potent vasodilator and inhibitor of platelet aggregation (Table 1) led to its development for the treatment of peripheral vascular disease and gangrene. This compound has been approved for use and is presently available in Japan (Ono Pharmaceutical Company) for these indications under the name Prostandin. Administration is by i.v. or

Table 10
PROSTIN E_2 (UPJOHN)

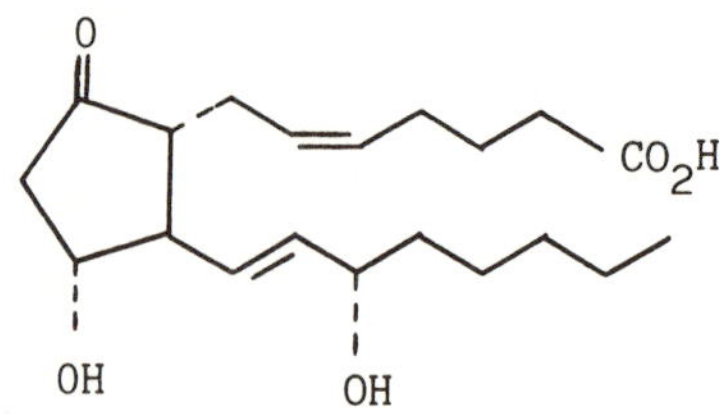

Approved name:	Dinoprostone
Brand name:	Prostin E_2
Chemical name:	Prostaglandin E_2

Physical Properties

Low-melting, white to slightly off-white crystalline solid, readily soluble in ethanol or in aqueous buffers at pH ~ 7.5. Stable in ethanol solution 12 months at 4°C.

Formulation (route of administration)	Indication	Dosage	Side effects
Tablets (oral)	Induction of labor when there are no fetal or maternal contraindications	Initial dose 0.5 mg; thereafter 0.5 mg/hr increased to 1 mg/hr if uterine motility inadequate	Nausea, vomiting, diarrhea, but seldom severe enough to necessitate discontinuation of medication
Sterile solution in normal saline 1.5 μg/mℓ solution or 0.75 mℓ ampule, 1 mg/mℓ (i.v. infusion)	Induction of labor: 1. When there are no fetal or maternal contraindications 2. In cases of fetal death *in utero* 3. When vaginal delivery is desired for safety of mother or child, e.g., Rh incompatability, maternal hypertension, preeclampsia, and premature rupture of the membranes	0.25 μg/mℓ, initially for 30 min. If satisfactory uterine contractility response, maintain this rate; otherwise, rate may be increased to 0.5 μg/min, and after 1 to 2 hr, to 1 or (rarely) 2 μg/min; fetal death *in utero* may require higher doses	Nausea, vomiting, and diarrhea commonly encountered at dose levels required to induce therapeutic termination of pregnancy i.v.; these effects markedly less frequent with doses used by the extra-amniotic route for therapeutic termination or i.v. for induction of labor. Also recorded: transient, vasovagal symptoms including flushing, shivering, headache, dizziness, local tissue irritation, erythema, temporary pyrexia, and elevated WBC
5 μg/mℓ (i.v. infusion)	1. Therapeutic termination of pregnancy 2. Missed abortion 3. Hydatatidiform mole	2.5 μg/min, initially for 30 min, then maintain this rate or increase to 5 μg/min, if necessary; after at least 4 hr the rate may be increased to 10 μg/min, if necessary, side effects permitting	
100 μg/mℓ dilute solution (extra-amniotic)	Therapeutic termination of pregnancy	1 mℓ initially; then, after 2 hr, 1 to 2 mℓ in accordance with the uterine response	

intra-arterial injection. Upjohn markets PGE_1 in the U.K. under the brand name Alprostadil (Prostin VR), for cardiac disease in neonates.

Prostaglandin Analogs

Until recently, no prostaglandin analogs had reached the medical market, although some analogs, e.g., the I.C.I. Estrumate (Table 11), are available for veterinary use, notably for synchronization of estrus in cattle.

However, two analogs, the Schering-Pfizer 16-phenoxy-PGE_2 ''sulprostone'' and the Ono *trans* Δ^2 15-alkyl-PGE_1 ''gemeprost'' (Table 11 and Figure 9), which was codeveloped by May & Baker Limited, are now marketed as medical products. Both of these compounds, which are for application in the gynecological field, have arisen from attempts, described in earlier sections of this review, to design prostaglandin analogs with greater potency and selectivity and a longer duration of action than the natural substances. Sulprostone was approved by the West German authorities in 1981 for use as an abortifacient by i.m. injection. Gemeprost was approved by the Japanese authorities in 1982 for termination of second trimester abortions, and is available as a vaginal pessary.

Analogs under Development

In addition to the above analogs, which have received approval for marketing, a number of others which have shown potentially advantageous properties over the natural compounds have undergone clinical evaluation. Some idea of those compounds which have reached an advanced stage of pharmaceutical development can be obtained from the issue of approved names as listed in the *USAN, USP Directory of Drug Names,* and a list of British Approved Names (Table 11). It is of interest that the majority of these contain the 15-, 16-alkyl or 16-phenoxy functions, which were highlighted in the structure-activity section of this review as being the most successful.

NOTE ADDED IN PROOF

A total synthesis of Thromboxane A_2 (TXA_2) has now been published (Bhagwat, S. S., Hamann, P. R., Still, W. C., Bunting, S., and Fitzpatrick, F. A., Synthesis and structure of the platelet aggregation factor thromboxane A_2, *Nature (London),* 315, 511, 1985). The following prostaglandins are now marketed in addition to those referred to previously: Epoprostol (prostacyclin), Carboprost (Table 11), and Misoprostol (antiulcer, Searle).

Table 11
PROSTAGLANDINS — APPROVED NAMES

Approved name	Chemical structure	Clinical area	Company
Dinoprost	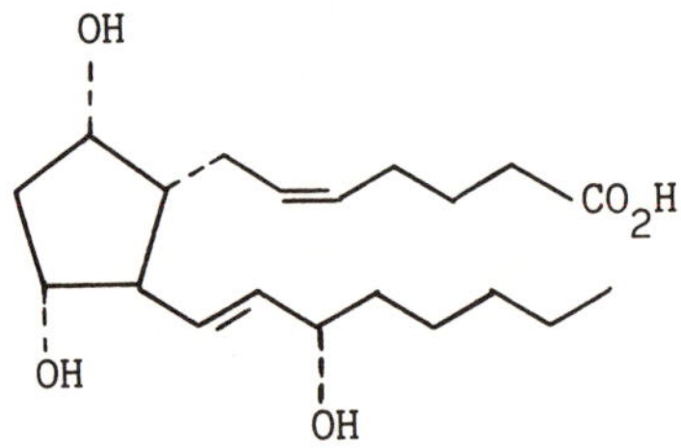 ($PGF_{2\alpha}$ Prostin $F_{2\alpha}$ Prostarmon F)	Oxytocic	Upjohn, Ono
Dinoprostone	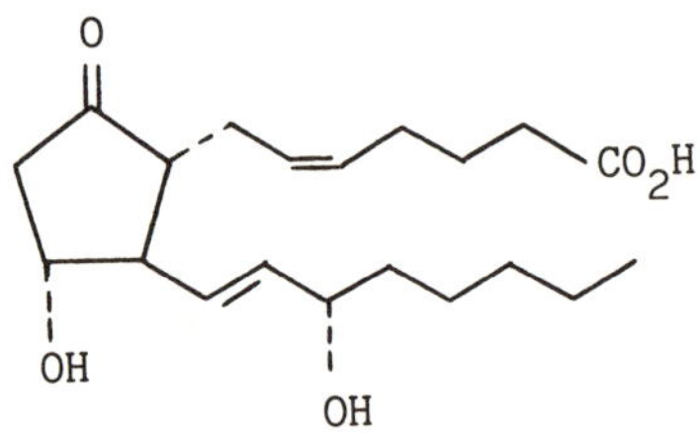 (PGE_2, Prostin E_2, Prostarmon E)	Oxytocic	Upjohn, Ono
Dinoprostone Tromethamine	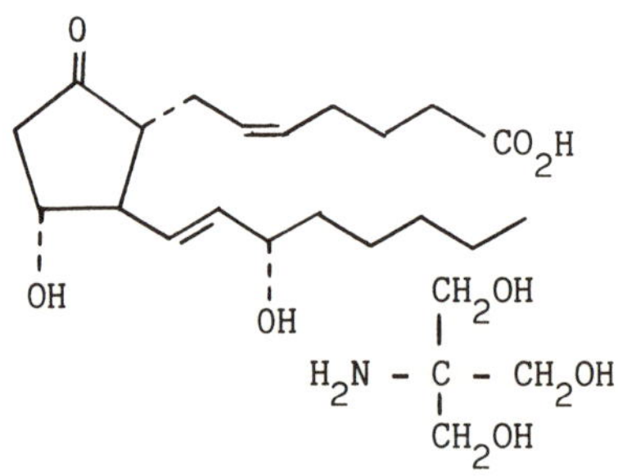 ($PGF_{2\alpha}$ tromethamine salt Prostin $F_{2\alpha}$ tromethamine)	Oxytocic	Upjohn
Sulprostone (Nalador)	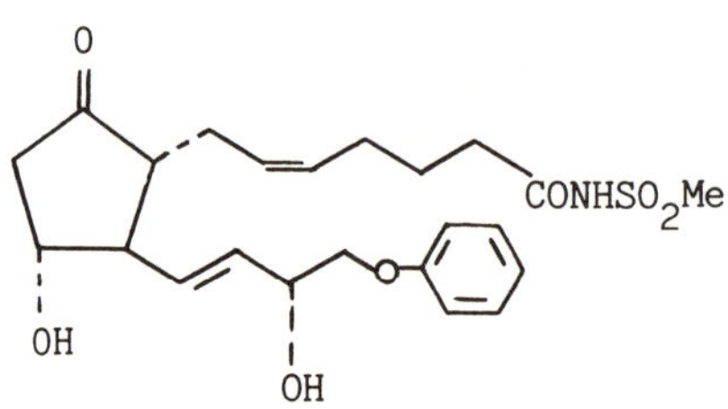	Oxytocic	Schering AG, Pfizer
Gemeprost (Preglandin)	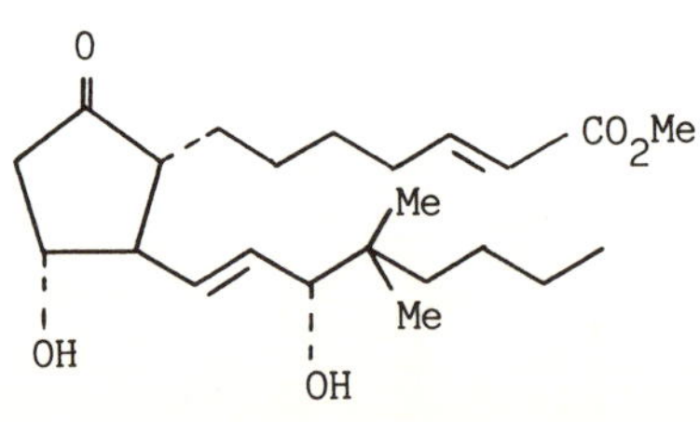	Oxytocic	Ono

Table 11 (continued)
PROSTAGLANDINS — APPROVED NAMES

Approved name	Chemical structure	Clinical area	Company
Meteneprost	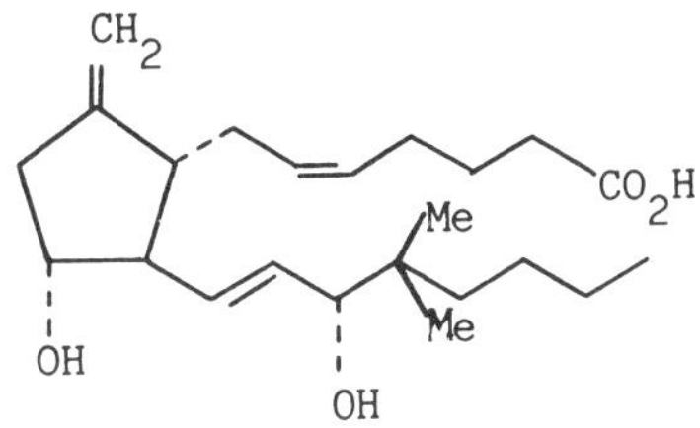	Oxytocic	Upjohn
Luprostiol (Prostianol)	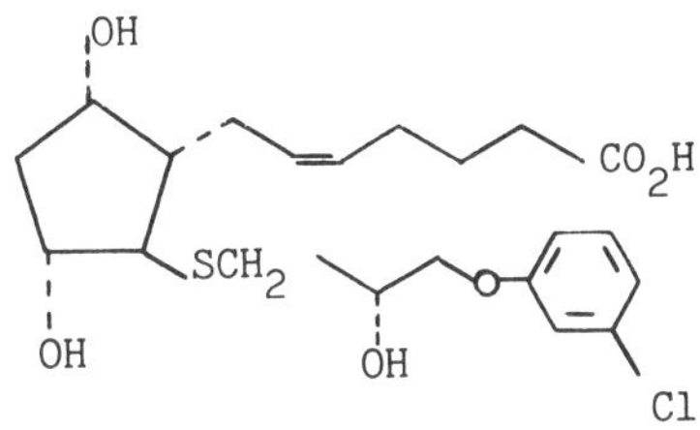	Oxytocic	E. Merck, A.G.
Carbaprost	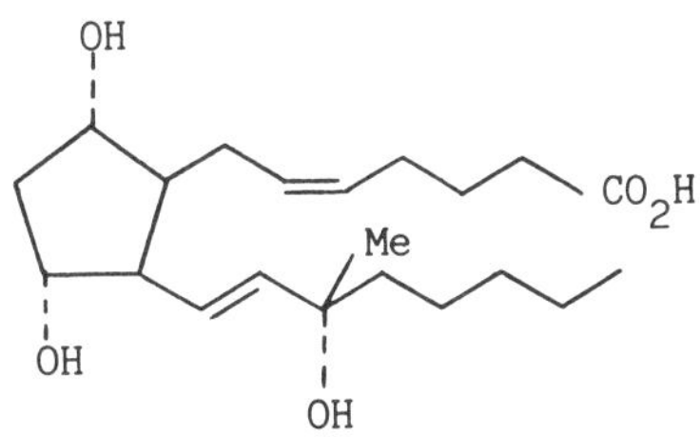	Oxytocic	Upjohn
Tiaprost	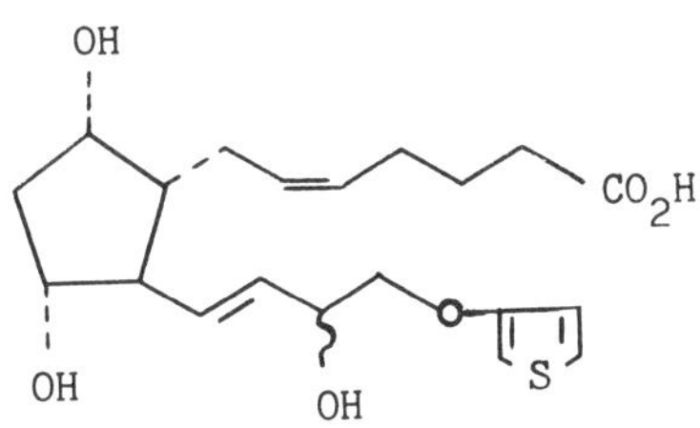		
Cloprostenol sodium (Estrumate)	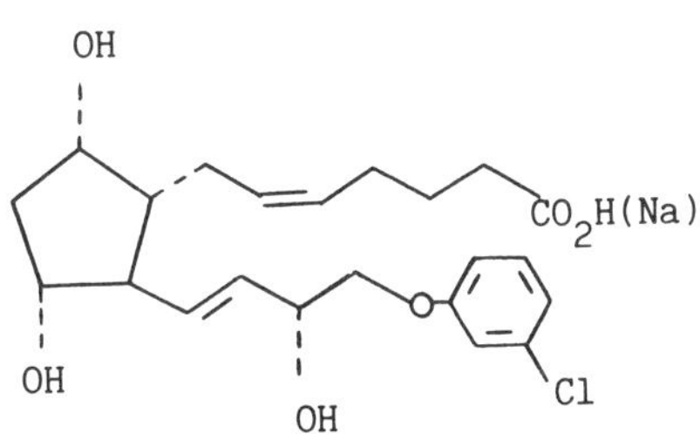	Veterinary luteolytic	I.C.I.
Delprostenate	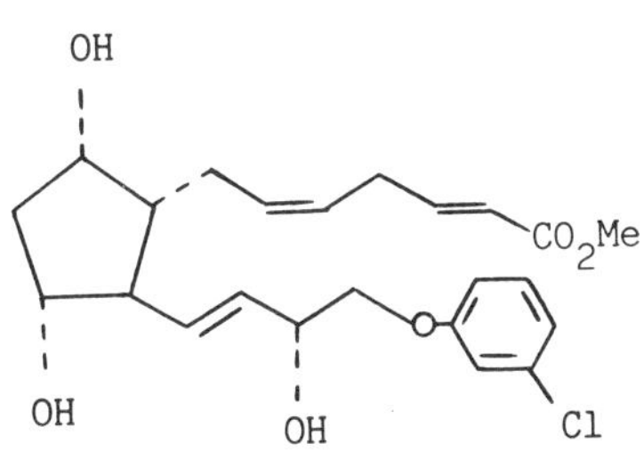	Veterinary luteolytic	Ono

Table 11 (continued)
PROSTAGLANDINS — APPROVED NAMES

Approvèd name	Chemical structure	Clinical area	Company
Fluprostenol sodium	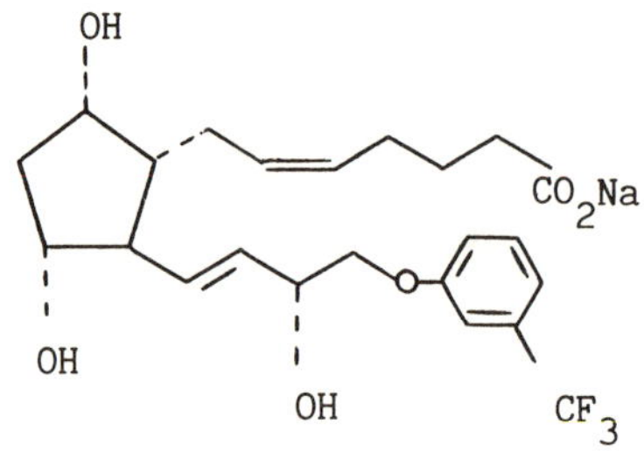	Veterinary luteolytic	I.C.I.
Prostalene	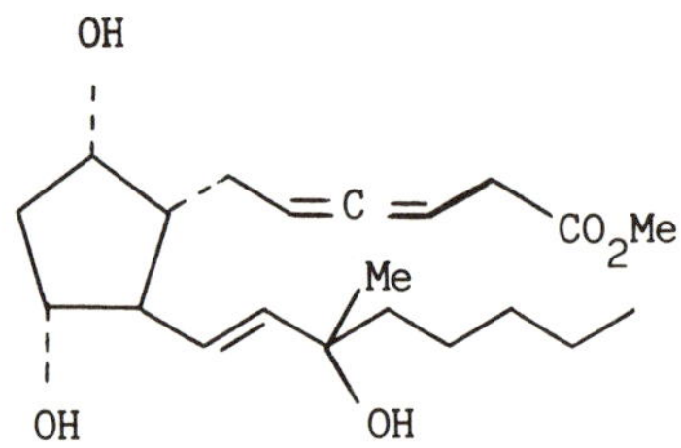	Veterinary luteolytic	Syntex
Fenprostalene	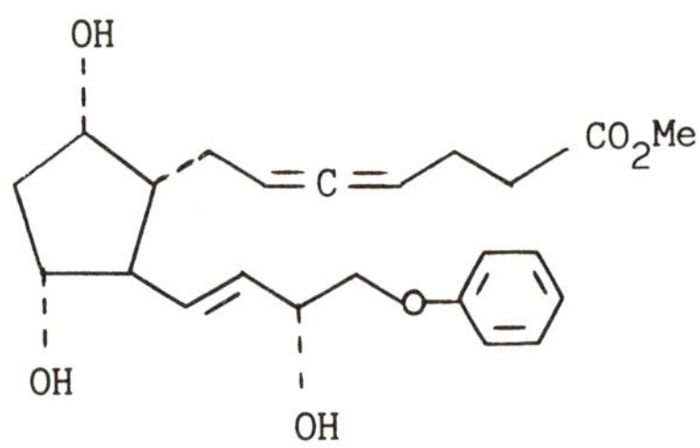	Veterinary luteolytic	Syntex
Arbaprostil	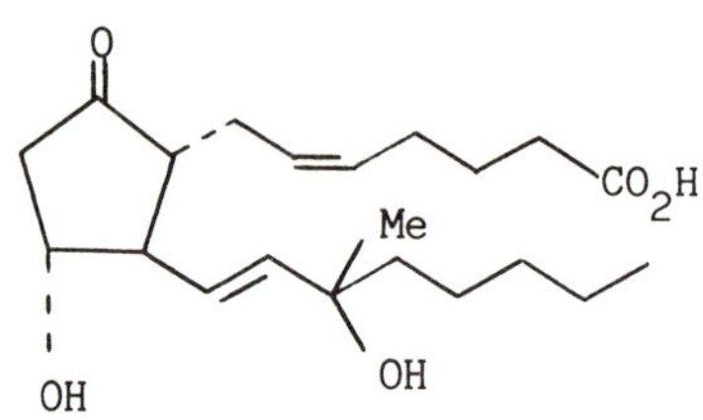	Gastric, antisecretory	Upjohn
Deprostil	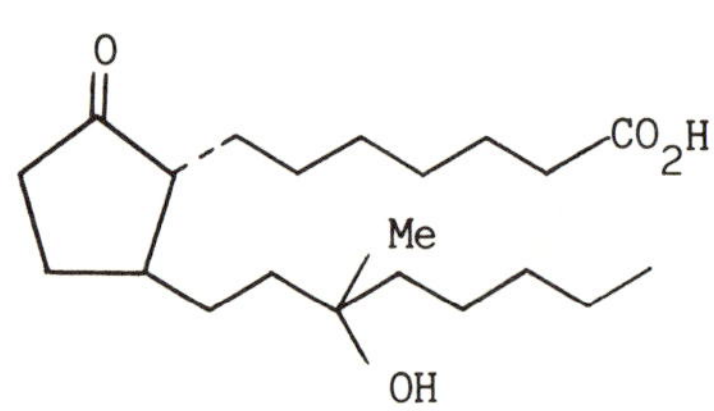	Gastric, antisecretory	Ayerst
Oxoprostol	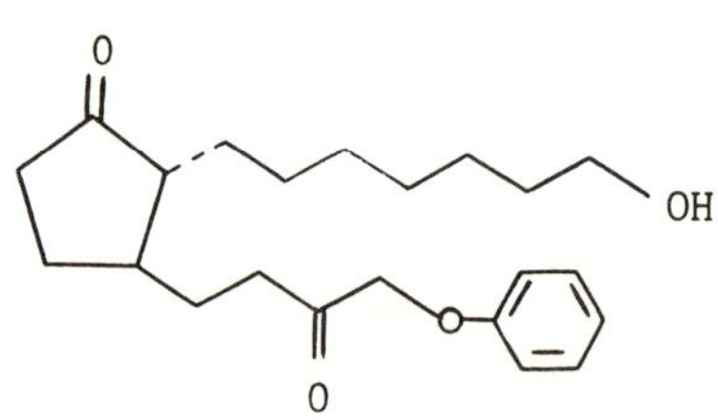	Gastric, antisecretory	May & Baker

Table 11 (continued)
PROSTAGLANDINS — APPROVED NAMES

Approved name	Chemical structure	Clinical area	Company
Epoprostenol	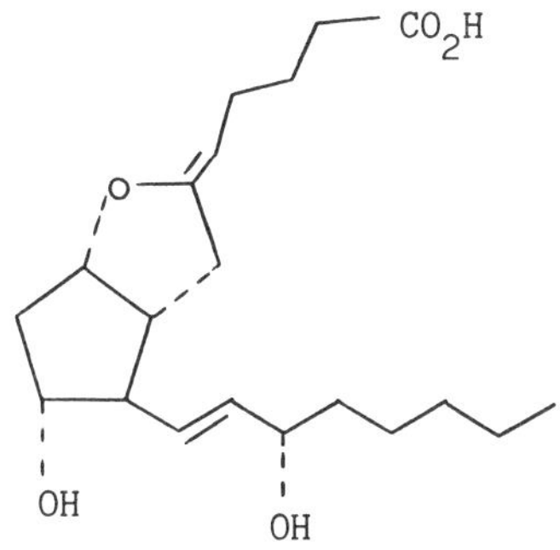 (PGI_2)	Vasodilator, platelet, aggregation inhibitor	Wellcome
Penprostene	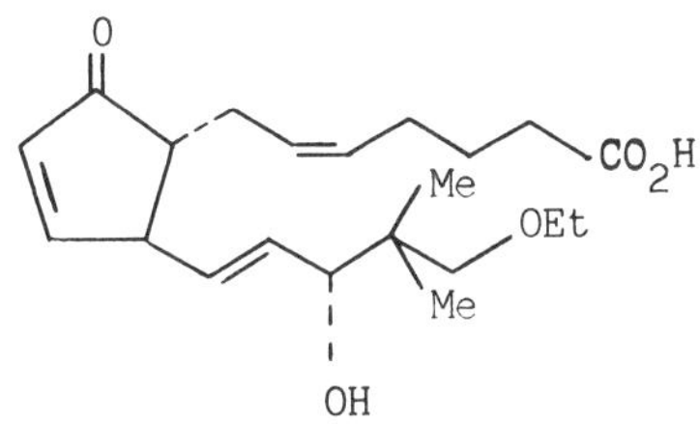	Vasodilator	Hoechst
Alprostadil	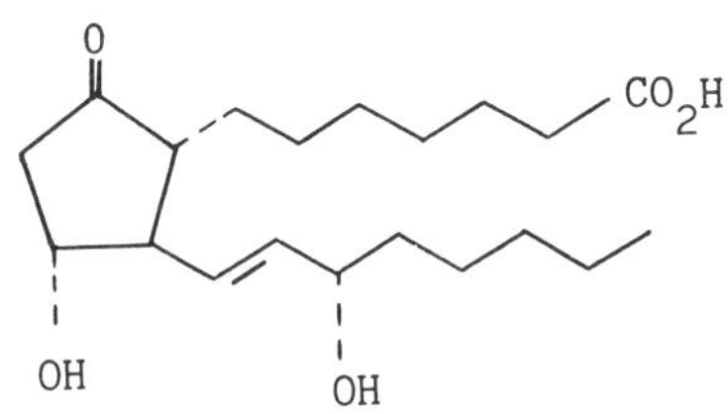 (PGE_1, Cervagem)	Vasodilator	Upjohn

REFERENCES

1. **Karim, S. M. M., Ed.,** *Advances in Prostaglandin Research,* M.T.P. Press, Lancaster, England, 1975, 1976, 1979.
2. **Samuelsson, B. and Paoletti, R.,** Series Eds., *Advances in Prostaglandin and Thromboxane Research,* Vols. 1 — 12, Raven Press, New York, 1976, 1978, 1980, 1982, 1983.
3. **Ramwell, P. W.,** *The Prostaglandins,* Vols. 1 — 3, Plenum Press, New York, 1973, 1974, 1977.
4. **Kharasch, N., Fried, J., Baumgartner, A., and Watkins, G. L.,** *Biochemical Aspects of Prostaglandins and Thromboxanes,* Academic Press, New York, 1977.
5. **Lewis, P. J. and O'Grady, J.,** *Clinical Pharmacology of Prostacyclin,* Raven Press, New York, 1981.
6. **Rainsford, K. D. and Ford-Hutchinson, A.,** *Prostaglandins and Inflammation,* Birkhauser, Stuttgart, 1979.
7. **Vane, J. R. and Bergstrom, S.,** *Prostacyclin,* Raven Press, New York, 1979.
8. **Hegyeli, R. J.,** *Prostaglandins and Cardiovascular Disease,* (Atherosclerosis Reviews), Vol. 8, Raven Press, New York, 1981.
9. **Silver, M. L., Smith, J. B., and Koesis, J. J.,** *Prostaglandins in Haematology,* Spectrum Publ., New York, 1976.

10. **Kuhr, H. and Lands, E. M.,** *Prostaglandins and Cyclic AMP,* Academic Press, New York, 1972.
11. **Roberts, S. M. M. and Scheinmann, F.,** *Chemistry, Biochemistry and Pharmacological Activity of Prostanoids,* Pergamon Press, Oxford, 1979.
12. **Bindra, J. S. and Bindra, R.,** *Prostaglandin Synthesis,* Academic Press, New York, 1977.
13. **Crabbé, P.,** *Prostaglandin Research: A Series of Monographs,* Vol. 36, Academic Press, New York, 1977.
14. **Mitra, A.,** *The Synthesis of Prostaglandin Derivatives,* John Wiley & Sons, New York, 1978.
15. **Roberts, S. M. M. and Scheinmann, F.,** *New Synthetic Routes to Prostaglandins and Thromboxanes,* Academic Press, New York, 1982.
16. **Piper, P. J.,** *S.R.S.A. and Leukotriene,* John Wiley & Sons, New York, 1981.
16a. **Piper, P. J.,** *Leukotrienes and Other Lipoxygenase Products,* John Wiley & Sons, New York, 1983.
17. **Caton, M. P. L. and Crowshaw, K.,** Prostaglandins and thromboxanes, in *Progress in Medicinal Chemistry,* Ellis, G. P. and West, G. B., Eds., North-Holland, Amsterdam, 1978, 357.
18. **Berti, F. and Velo, G. P.,** *The Prostaglandin System: Endoperoxides, Prostacyclin and Thromboxanes, NATO Advanced Study Institutes Series,* Plenum Press, New York, 1981.
18a. **Poyser, N. L.,** *Prostaglandins in Reproduction, Research Studies,* John Wiley & Sons, New York, 1981.
19. **Ramwell, P. W., Ed.,** *Prostaglandins,* Plenum Press, New York.
20. **Horrobin, D. F., Karmazyn, M., Manku, M. S., Sirois, P., and Borgeat, P., Eds.,** *Prostaglandins, Leukotrienes and Medicine,* Churchill Livingstone, Edinburgh, 1982.
21. **Nelson, N. A.,** Prostaglandin nomenclature, *J. Med. Chem.,* 17, 911, 1974.
22. **Samuelsson, B., Hamberg, M., Roberts, L. J. II, Oats, J. A. and Nelson, N. A.,** Nomenclature for thromboxanes, *Prostaglandins,* 16, 857, 1978.
23. **Samuelsson, B. and Hammarstrom, S.,** Nomenclature for leukotrienes, *Prostaglandins,* 19, 645, 1980.
24. **Weinheimer, A. J. and Spraggins, R. L.,** The occurrence of two new prostaglandin derivatives (15-epi-PGA_2 and its acetate methyl ester) in the gorgonian, *Plexaura Homomalla, Tetrahedron Lett.,* 5185, 1969.
25. **Caton, M. P. L.,** A survey of novel and useful reactions discovered through research on prostaglandins, Tetrahedron Report No. 78, *Tetrahedron,* 35, 2507, 1979.
26. **Ackroyd, J. and Scheinmann, F.,** The synthesis of leukotrienes. A new class of biologically active compounds, *Chem. Soc. Rev.,* 11, 321, 1982.
26a. **Scheinmann, F. and Ackroyd, J.,** *Leukotriene Syntheses: A New Class of Biologically Active Compounds Including SRS-A,* Raven Press, New York, 1984.
26b. **Green, R. H. and Lambeth, P. F.,** Leukotrienes, Tetrahedron Report Number 148, *Tetrahedron,* 39, 1657, 1983.
27. **Borgeat, P. and Sirois, P.,** Leukotrienes: A major step in the understanding of immediate hypersensitivity reactions, *J. Med. Chem.,* 24, 121, 1981.
28. **Bailey, D. M. and Chakrin, L. W.,** Arachidonate lipoxygenase, *Ann. Reps. Med. Chem.,* 16, 213, 1981.
29. **Sirois, P., Roy, S., Tetrault, J. P., Borgeat, P., Picard, S., and Corey, E. J.,** Pharmacological activity of leukotrienes A_4, B_4, C_4 and D_4 on selected guinea-pig, rat, rabbit and human smooth muscles, *Prostaglandins Med.,* 7, 327, 1981.
30. **Ford-Hutchinson, A. W., Bray, M. A., Doig, M. V., Shipley, M. E., and Smith, M. J. H.,** Leukotriene B, a potent chemokinetic and aggregating substance released from polymorphonuclear leukocytes, *Nature (London),* 286, 264, 1980.
31. **Moncada, S., Korbut, R., Bunting, S., and Vane, J. R.,** Prostacyclin is a circulating hormone, *Nature (London),* 273, 767, 1978.
32. **Christ, H. E. and Nugteren, D. H.,** Prostacyclin is not a circulating hormone, *Prostaglandins,* 22, 739, 1981.
33. **Steinberg, D., Vaughan, M., Nestel, P. J., Strand, O., and Bergstrom, S.,** Effects of the prostaglandins and hormone-induced mobilization of free fatty acids, *J. Clin. Invest.,* 43, 1533, 1964.
34. **Ramwell, P. W. and Shaw, J. E.,** Prostaglandin release from tissues by drug, nerve and hormone stimulation, in *Prostaglandins,* Bergstrom, S. and Samuelssen, B., Eds., Almqvist and Wiksell, Stockholm, 1967, 283.
35. **Butcher, P. W. and Baird, C. E.,** Effects of prostaglandins on adenosine 3,5-monophosphate levels in fat and other tissues, *J. Biol. Chem.,* 243, 1713, 1968.
36. **Kuel, F. A., Humes, J. L., Cirillo, V. J., and Ham, E. A.,** Cyclic AMP and prostaglandins in hormone action, in *Advances in Cyclic Nucleotide Research,* Vol. 1, Greengard, P., Robertson, G. A., and Paoletti, R., Eds., Raven Press, New York, 1972, 493.
37. **Kuel, F. A., Jr., Cirillo, V. J., Ham, E. A., and Humes, J. L.,** The regulatory role of the prostaglandin on the cyclic 3,5-AMP system, in *Advances in the Biosciences,* Vol. 9, Bergstrom, S. and Bernhards, S., Eds., Pergamon Press, Oxford, 1973, 155.
38. **Kuel, F. J., Jr. and Humes, L. L.,** Direct evidence for a prostaglandin receptor and its application to prostaglandin measurement, *Proc. Natl. Acad. Sci. U.S.A.,* 69, 480, 1972.
39. **Powell, W. S., Hammerstrom, S., and Samuelsson, B.,** Prostaglandin $F_{2\alpha}$ receptor in ovine corpora lutea, *Eur. J. Biochem.,* 41, 103, 1974.

40. **Gorman, R. R., Fitzpatrick, F. A., and Miller, O. V.,** Reciprocal regulation of human platelet c-AMP level by thromboxane A_2 and prostacyclin, *Adv. Cyclic Nucl. Res.*, 9, 597, 1978.
41. **Carsten, M. E.,** Prostaglandins part in regulating uterine contractions by transport of calcium, in *The Prostaglandins, Clinical Applications in Human Reproduction,* Southern, E. M., Ed., Futura, Mt. Kisco, N.Y., 1972, 59.
42. **Drazen, J. M., Lewis, R. A., Austen, K. F., Toda, M., Frion, F., Marfat, A., and Corey, E. J.,** Contractile activities of structural analogous of leukotrienes C and D: necessity of a hydrophobic region, *Proc. Natl. Acad. Sci. U.S.A.*, 78, 3195, 1981.
43. **Lewis, R. A., Drazen, J. M., Austen, K. F., Toda, M., Brion, F., Marfat, A., and Corey, E. J.,** Contractile activities of structural analogous of leukotrienes C and D: role of the polar substituents, *Proc. Natl. Acad. Sci. U.S.A.*, 78, 4579, 1981.
44. **Weeks, J. R., DuCharme, D. W., Magee, W. E., and Miller, W. L.,** The biological activity of the (15S)-15-methyl analogue of prostaglandins E_2 and $F_{2\alpha}$, *J. Pharm. Exp. Ther.*, 186, 67, 1973.
45. **Robert, A., Schultz, J. R., Nezamis, J. E., and Lancaster, C.,** Gastric antisecretory and antiulcer properties of PGE_2 15-methyl-PGE_2 and 16,16-dimethyl-PGE_2, intravenous, oral and intrajejunal administration, *Gastroenterology,* 70, 359, 1976.
46. **Magerlein, B. J., DuCharme, D. W., Magee, W. E., Miller, W. L., Robert, A., and Weeks, J. R.,** Synthesis and biological properties of 16-alkyl prostaglandins, *Prostaglandins,* 4, 143, 1973.
47. **Robert, A., Nylander, B., and Anderssen, S.,** Marked inhibition of gastric secretion by two prostaglandin analogs given orally to man, *Life Sci.*, 14, 533, 1974.
48. **Muryobayashi, T., Okada, K., and Shingae, J.,** Effects of new synthetic analogues of prostaglandin E_2 on gastric acid secretion and gastric ulceration, in *Advances in Prostaglandin and Thromboxane Research,* Vol. 2, Samuelsson, B. and Paoletti, R., Series Eds., Raven Press, New York, 1976, 541.
49. **Magerlein, B. J. and Miller, W. L.,** 16-Fluoroprostaglandins, *Prostaglandins,* 9, 527, 1975.
50. **Yankee, E. W., Ayer, D. E., Bundy, G. L., Lincoln, F. H., Miller, W. L., Jr., and Youngdale, G. A.,** Synthesis and biological activity of new prostanoids, in *Advances in Prostaglandin and Thromboxane Research,* Vol. 1, Samuelsson, B. and Paoletti, R., Series Eds., Raven Press, New York, 1976, 195.
51. **Miller, W. L., Weeks, J. R., Lauderdale, J. W., and Kirton, K. T.,** Biological activities of 17-phenyl-18,19,20-trinorprostaglandins, *Prostaglandins,* 9, 9, 1975.
52. **Crossley, N. S.,** The synthesis and biological activity of potent, selective, luteolytic prostaglandins, *Prostaglandins,* 10, 5, 1975.
53. **Nelson, N. A., Jackson, R. W., Au, A. T., Wynalda, D. J., and Nishizawa, E. E.,** Synthesis of *dl*-4,5,6-trinor-3,7-inter-*m*-phenylene-3-oxaprostaglandins including one which inhibits platelet aggregation, *Prostaglandins,* 10, 795, 1975.
54. **Van Dorp, D. A. and Christ, E. J.,** Specificity in the enzymic conversion of *cis*-8, *cis*-11, *cis*-14-eicosatrienoic acids into prostaglandins, *Rec. Trav. Chim. Pays-Bas.*, 94, 247, 1975.
55. **Oda, O., Kojima, K., and Sakai, K.,** Synthesis of optically active (−)11-deoxy-11_α-hydroxymethyl prostaglandin E_1, *Tetrahedron Lett.*, p. 3709, 1975.
56. **Caldwell, A. G., Harris, C. J., Stepney, R., and Whittaker, N.,** Heterocyclic prostaglandin analogues. II. Hydantoins and other imidazole analogues, *J. Chem. Soc. Perkin I,* p. 495, 1980.
57. **Bundy, G. L.,** The synthesis of prostaglandin endoperoxide analogs, *Tetrahedron Lett.*, p. 1957, 1975.
58. **Malmsten, C.,** Some biological effects of prostaglandin endoperoxide analogs, *Life Sci.*, 18, 169, 1976.
59. **Corey, E. J., Nicolaou, K. C., Machida, Y., Malmsten, C. L., and Samuelsson, B.,** Synthesis and biological properties of a 9,11-azo-prostanoid: highly active biochemical mimic of prostaglandin endoperoxides, *Proc. Natl. Acad. Sci. U.S.A.*, 72, 3355, 1975.
60. **Dajani, E. Z., Driskill, D. R., Bianchi, R. C., Collins, P. W., and Pappo, R.,** Studies on the biological properties of a prostaglandin analog, S C 29,333, a potent inhibitor of gastric secretion, in *Advances in Prostaglandin and Thromboxane Research,* Vol. 2, Samuelsson, B. and Paoletti, R., Eds., Raven Press, New York, 1976, 944.
61. **Walker, E. R. H.,** The gastric antisecretory properties of some prostaglandin analogues, in *Chemistry, Biochemistry, and Pharmacological Activity of Prostanoids,* Pergamon Press, Oxford, 1979, 326.
62. **Beck, C., Bartmann, W., Lerch, U., Teufel, H., and Scholkens, B.,** Antihypertensive activity of 16,16-dimethyl-oxa-alkyl prostaglandins of the PGA_2, PGE_2 and *trans* Δ-2-11-deoxy-PGE_1 series: structure activity relationships, *Prostaglandins,* 20, 153, 1980.
63. **Reiertsen, O., Jerve, F., and Fylling, P.,** Termination of early first trimester pregnancy with 16,16-dimethyl *trans* Δ-2-prostaglandin E_1 methyl ester, *Prostaglandins Leukotrienes Med.*, 8, 31, 1982.
64. **Fried, J., Lee, M-S., Gaede, B., Sih, J. C., Yoshikawa, Y., and McCracken, J. A.,** Chemical and biological studies on 13-dehydro-prostaglandins, in *Advances in Prostaglandin and Thromboxane Research,* Vol. 1, Samuelsson, B. and Paoletti, R., Eds., Raven Press, New York, 1976, 183.
65. **Schaaf, T. K., Bindra, J. S., Eggler, J. F., Plattner, J. J., Nelson, A. J., Johnson, M. R., Constantine, J. W., Hess, N-H., and Elger, W.,** *N*-(Methanesulfonyl)-16-phenoxyprostaglandin carboxamides: tissue selective uterine stimulants, *J. Med. Chem.*, 24, 1353, 1981.

66. **Kimball, F. A., Bundy, G. L., Robert, A., and Weeks, J. R.,** Synthesis and biological properties of 9-deoxo-16,16-dimethyl-9-methylene-PGE_2, *Prostaglandins,* 17, 657, 1979.
67. **Whittle, B. J. R., Moncada, S., Whiting, F., and Vane, J. R.,** Carbacyclin — a potent stable prostacyclin analogue for the inhibition of platelet aggregation, *Prostaglandins,* 19, 605, 1980.
68. **Aiken, J. W. and Shebuski, R. J.,** Comparison in anaesthetized dogs of the anti-aggregatory and haemodynamic effects of prostacyclin and a chemically stable prostacyclin analog, 6α-carba-PGI_2 (carbacyclin), *Prostaglandins,* 19, 629, 1980.
69. **Schror, K., Ohlendorf, R., and Darius, H.,** Beneficial effects of a new carbacyclin derivative, ZK 36 374, in acute myocardial ischemia, *J. Pharmacol. Exp. Ther.,* 219, 243, 1981.
70. **Morton, D. R., Bundy, G. L., and Nishizawa, E. E.,** Five membered ring-modified prostacyclin analogs, in *Prostacyclin,* Raven Press, New York, 1979, 31.
71. **Bundy, G. L. and Baldwin, J. M.,** The synthesis of nitrogen-containing prostacyclin analogs, *Tetrahedron Lett.,* p. 1371, 1978.
72. **Nicolaou, K. C., Magolda, A. L., Smith, J. B., Aharony, D., Smith, E. F., and Lefer, A. M.,** Synthesis and biological properties of pinane TXA_2, a selective inhibitor of coronary artery constriction, platelet aggregation, and thromboxane formation, *Proc. Natl. Acad. Sci. U.S.A.,* 76, 2566, 1979.
73. **Lefer, A. M., Smith, E. F., Araki, M., Smith, J. B., Aharony, D., Claremon, D. A., Magolda, R. L., and Nicolaou, K. C.,** Dissociation of vasoconstrictor and platelet aggregatory activities of thromboxane by carbocyclin TXA_2, a stable analog of TXA_2, *Proc. Natl. Acad. Sci. U.S.A.,* 77, 1706, 1980.
74. **Gorman, R. R., Maxey, K. M., and Bundy, G. L.,** Inhibition of human platelet thromboxane synthesis by 11α-carbathromboxane A_2 analogs, *Biochem. Biophys. Res. Commun.,* 184, 100, 1981.
75. **Schaaf, T. K., Bussolotti, D. L., Parry, M. J., and Corey, E. J.,** Synthesis of 11α, 9α-epoxymethanothromboxane A_2: a stable, optically active TXA_2 agonist, *J. Am. Chem. Soc.,* 103, 6502, 1981.
76. **Fried, J., Santhanakrishnan, T. S., Himizu, J., Lin, C. H., Ford, S. H., Robin, B., and Grigas, E. O.,** Prostaglandin antagonists: synthesis and smooth muscle activity, *Nature (London),* 223, 208, 1969.
77. **Kam, S-T., Portoghese, P. S., Gerrard, J. M., and Dunham, E. W.,** Synthesis and biological evaluation of 9,11-azo-13-oxa-15-hydroxyprostanoic acid, a potent inhibitor of platelet aggregation, *J. Med. Chem.,* 22, 1402, 1979.
78. **Wiqvist, N., Martin, J. M., Bygdeman, M., and Green, K.,** Prostaglandin analogues and uterotonic potency: a comparative study of seven compounds, *Prostaglandins,* 9, 255, 1975.
79. **Broughton, B. J., Caton, M. P. L., Coffee, E. C. J., Hambling, D. J., Palfreyman, M. N., Withnall, M. T., and Wooldridge, K. R. H.,** Synthesis and antifertility activity of ω-chain phenyl and 16-phenoxy analogues of (±)-11-deoxyprostaglandin $F_{1\alpha}$, *Prostaglandins,* 19, 559, 1980.
80. **Banerjee, A. K., Broughton, B. J., Burton, T. S., Caton, M. P. L., Christmas, A. J., Coffee, E. C. J., Crowshaw, K., Heazell, M. A., Stuttle, K. A. J., and Watkins, G. L.,** Synthesis and gastrointestinal pharmacology of some 15- and 16-modified (±)-11-deoxyprostaglandins, *Prostaglandins,* 16, 541, 1978.
81. **Banerjee, A. K., Broughton, B. J., Burton, T. S., Caton, M. P. L., Christmas, A. J., Coffee, E. C. J., Crowshaw, K., Hardy, C. J., Heazell, M. A., Palfreyman, M. N., Parker, T., Saunders, L. C., and Stuttle, K. A. J.,** Synthesis and anti-ulcer activity of 16-phenoxy analogues of (±)-11-deoxyprostaglandin E_1, *Prostaglandins,* 22, 167, 1981.
82. **Johnson, M. R., Schaaf, T. K., Constantine, J. W., and Hess, H-J.,** Structure activity studies leading to a tissue-selective hypotensive prostaglandin analog, 13,14-dihydro-16-phenyl-tetranor PGE_2, *Prostaglandins,* 20, 515, 1980.
83. **Ramwell, P. W., Shaw, J. E., Corey, E. J., and Andersen, N.,** Biological activity of synthetic prostaglandins, *Nature (London),* 221, 1251, 1969.
84. **Andersen, N. H. and Ramwell, P. W.,** Biological aspects of prostaglandins, *Arch. Intern. Med.,* 30, 133, 1974.
85. **Andersen, N. H., Imamoto, S., Subramanian, N., Picker, D. H., Ladner, D. W., De, B., Tynan, S. S., Eggerman, T. L., Harker, L. A., Robertson, R. P., Oien, H. G., and Rao, Ch. V.,** Molecular basis for prostaglandin potency. III. Tests of the significance of the "hairpin conformation" in biorecognition phenomena, *Prostaglandins,* 22, 841, 1981.
86. **Schaaf, T. K.,** Prostaglandins structure-activity relationships, *Annu. Rep. Med. Chem.,* 11, 80, 1976.
87. **Nicolaou, K. C., Petasis, N. A., and Seitz, S. P.,** 5,6-Methano leukotriene A4. A stable biologically active analogue of leukotriene A4, *J. Chem. Soc. Chem. Commun.,* p. 1195, 1981.
88. **Monada, S. and Vane, J. R.,** The pharmacology and clinical potential of prostacyclin, in *Clinical Pharmacology and Therapeutics World Conference Proceedings 1980,* Turner, P., Ed., 1980, 15.

INDEX

A

B

C

D

E

F

G

H

N

O

P

Q

R

S

T

U

V

W

X

Y

Z